The Origin and Evolution of the Genetic Code: 100th Anniversary Year of the Birth of Francis Crick

Special Issue Editor
Koji Tamura

MDPI • Basel • Beijing • Wuhan • Barcelona • Belgrade

MDPI

Special Issue Editor
Koji Tamura
Tokyo University of Science
Japan

Editorial Office
MDPI AG
St. Alban-Anlage 66
Basel, Switzerland

This edition is a reprint of the Special Issue published online in the open access journal *Life* (ISSN 2075-1729) from 2016–2017 (available at: http://www.mdpi.com/journal/life/special_issues/Francis_Crick).

For citation purposes, cite each article independently as indicated on the article page online and as indicated below:

Lastname, F.M.; Lastname, F.M. Article title. *Journal Name*. **Year**. *Article number*, page range.

First Edition 2018

ISBN 978-3-03842-769-8 (Pbk)
ISBN 978-3-03842-770-4 (PDF)

Cover photo: Sir Francis Crick, La Jolla, 1982. Cover photo courtesy of Norman Seeff Productions

Table of Contents

About the Special Issue Editor

Koji Tamura is a Japanese molecular biologist and biophysicist. He is a professor at the Department of Biological Science and Technology, Tokyo University of Science, Japan. He obtained a Bachelor of Science degree in physics from the University of Tokyo in 1989, and a Ph.D. in physics from the University of Tokyo in 1994. After being a Special Postdoctoral Researcher and a Research Scientist at RIKEN Institute, Japan, he became a Visiting Scholar (1999–2001), Research Associate (2001–2003), and Senior Research Associate (2003–2006) at The Scripps Research Institute, USA. From 2006 to 2012, he was an associate professor at the Department of Biological Science and Technology, Tokyo University of Science, Japan. Since 2012, he has been a full professor, and from 2016 to the present, he has been the head of the same department. His achievements include discovery of "chiral-selective aminoacylation of an RNA minihelix," which could be related to the origin of homochirality of biological systems. He also translated Matt Ridley's masterpiece "Francis Crick: Discoverer of the Genetic Code" into Japanese.

Preface to "The Origin and Evolution of the Genetic Code: 100th Anniversary Year of the Birth of Francis Crick"

This Special Issue is dedicated to the origin and evolution of the genetic code and to the memory of Francis Crick, the discoverer of the genetic code, in commemoration of the 100th anniversary of his birth in 2016. The genetic code is one of the greatest discoveries of the 20th century as it is central to life itself. It is the algorithm that connects 64 RNA triplets to 20 amino acids, thus functioning as the Rosetta Stone of molecular biology.

Following the discovery of the structure of DNA by James Watson and Francis Crick in 1953, George Gamow organized the 20-member "RNA Tie Club" to discuss the transmission of information by DNA. Crick, Sydney Brenner, Leslie Barnett, and Richard Watts-Tobin first demonstrated the three bases of DNA code for one amino acid. The decoding of the genetic code was begun by Marshall Nirenberg and Heinrich Matthaei and was completed by Har Gobind Khorana. Then, finally, Brenner, Barnett, Eugene Katz, and Crick placed the last piece of the jigsaw puzzle of life by proving that UGA was a third stop codon.

In the mid-1960s, Carl Woese proposed the "stereochemical hypothesis", which speculated that the genetic code derives from a type of codon–amino acid-pairing interaction. On the other hand, Crick proposed the "frozen accident hypothesis" and conjectured that the genetic code evolved from the last common universal ancestor and was frozen once established. However, he explicitly left room for stereochemical interactions between amino acids and their coding nucleotides, stating that "It is therefore essential to pursue the stereochemical theory … vague models of such interactions are of little use. What is wanted is direct experimental proof that these interactions take place…and some idea of their specificity."

The origin and evolution of the genetic code remains a mystery despite numerous theories and attempts to understand these. In this Special Issue, experts in the field present their thoughts and views on this topic. "Double helix of DNA" and "genetic code table" are the greatest gifts that Francis Crick left behind. He devoted himself to science until his death. His passion for science will continue to inspire scientists now and forever.

"Nature isn't conspiring against us to make important problems difficult, so given a finite life span, aim high—go after fundamental problems." — Francis Crick

Koji Tamura
Special Issue Editor

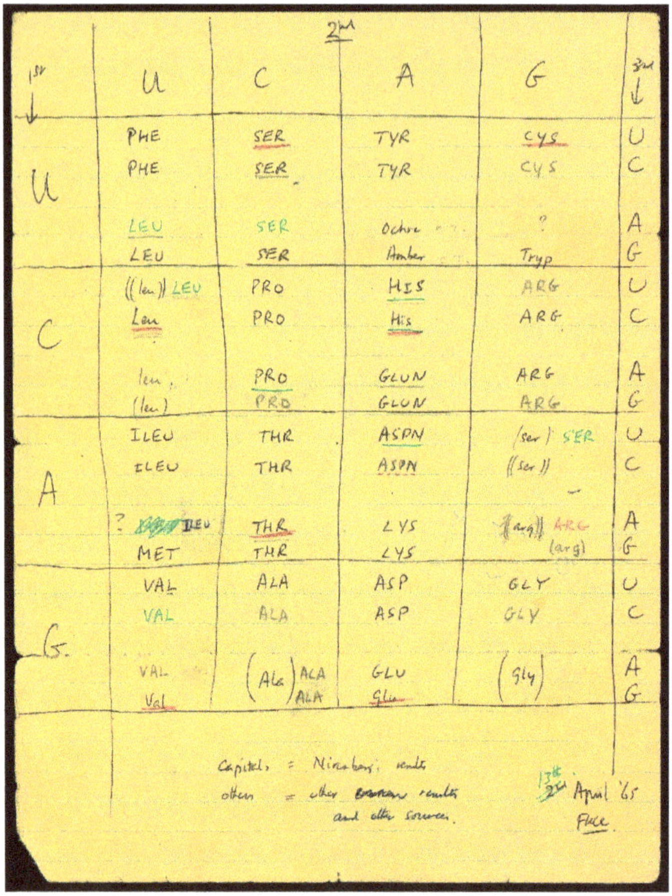

Figure 1. Crick's sketch of genetic code. Credit: Wellcome Collection.

life

MDPI

Editorial

The Genetic Code: Francis Crick's Legacy and Beyond

Koji Tamura [1,2]

[1] Department of Biological Science and Technology, Tokyo University of Science, 6-3-1 Niijuku,
    Katsushika-ku, Tokyo 125-8585, Japan; koji@rs.tus.ac.jp; Tel.: +81-3-5876-1472
[2] Research Institute for Science and Technology, Tokyo University of Science, 2641 Yamazaki, Noda,
    Chiba 278-8510, Japan

Academic Editor: David Deamer
Received: 22 August 2016; Accepted: 23 August 2016; Published: 25 August 2016

Francis Crick (Figure 1) was born on 8 June 1916, in Northampton, England, and passed away on 28 July 2004, in La Jolla, California, USA. This year, 2016, marks the 100th anniversary of his birth. A drastic change in the life sciences was brought about by the discovery of the double helical structure of DNA by James Watson and Francis Crick in 1953 [1], eventually leading to the deciphering of the genetic code [2]. The elucidation of the genetic code was one of the greatest discoveries of the 20th century. The genetic code is an algorithm that connects 64 RNA triplets to 20 amino acids, and functions as the Rosetta stone of molecular biology.

Figure 1. Sir Francis Crick, La Jolla 1982, Photograph by Norman Seeff. Credit: Norman Seeff Productions.

At the age of 60, Crick moved to La Jolla from Cambridge, England, and shifted his focus to the brain and human consciousness. He tackled this subject for the last 28 years of his life. His life-long interest was the distinction between the living and the non-living, which motivated his research career. Crick was arguably one of the 20th century's most influential scientists, and he devoted himself to science until his death.

Francis Crick continued to exercise his intellectual abilities throughout his life. His research style was characterized by collaborations with outstanding partners, James Watson in discovering the structure of DNA, Sydney Brenner in cracking the genetic code, Leslie Orgel in probing the origins of life, and Christof Koch in understanding human consciousness. Francis Crick was never modest in his choice of scientific problems [3] and was like "the conductor of the scientific orchestra" [4]. He always discussed his ideas, which helped in the progress he made in science. Interestingly, his son, Michael, then 12 years old, was the first person to read the earliest written description of the genetic code. Crick wrote the following in a letter to Michael,

> " … *Now we believe that the D.N.A. is a code. That is, the order of the bases (the letters) makes one gene different from another gene (just as one page of print is different from another). You can now see how Nature makes copies of the genes. Because if the two chains unwind into two separate chains, and if each chain then makes another chain come together on it, then because A always goes with T, and G with C, we shall get two copies where* … " (Figure 2).

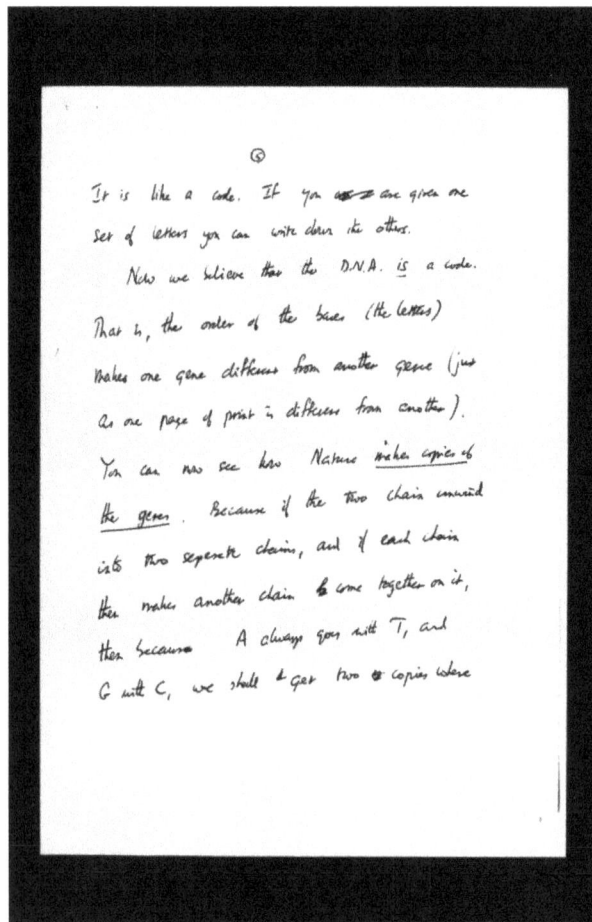

Figure 2. Letter from Francis Crick to his son, Michael, explaining his and Watson's discovery of the structure of DNA. The letter is the earliest written description of the genetic mechanism on 19 March 1953. Credit: Wellcome Library, London.

This is the fundamental principle of biology. The big questions that arose after the discovery of the structure of DNA were "how is the code used?" and "what is it a code for?" Francis Crick turned his attention to find answers to these questions for the next 13 years. George Gamow, who is famous for the Big Bang theory, founded the 20-member "RNA Tie Club" with Watson, to discuss the transmission of information by DNA. RNA-illustrated neckties were provided to all members, and a golden tiepin with the abbreviation for one of the 20 amino acids was given to each member. Crick was "TYR" (tyrosine). Crick's famous "adaptor hypothesis" was prepared for circulation in the RNA Tie Club [5], but when Paul Zamecnik and collaborators discovered transfer RNA (tRNA) [6], Crick did not believe that it was indeed the adaptor, because of its unexpectedly large size. Crick insisted that there would be 20 different adaptors for the amino acids, and that they would bring the amino acids to join the sequence of a nascent protein. A manuscript entitled "Ideas on protein synthesis (October, 1956)" remains extant (Figure 3). Crick spoke about "The Central Dogma" at a Society for Experimental Biology symposium on "The Biological Replication of Macromolecules", held at the University College London in September, 1957. The Central Dogma holds true even today, and is another example of Crick's genius.

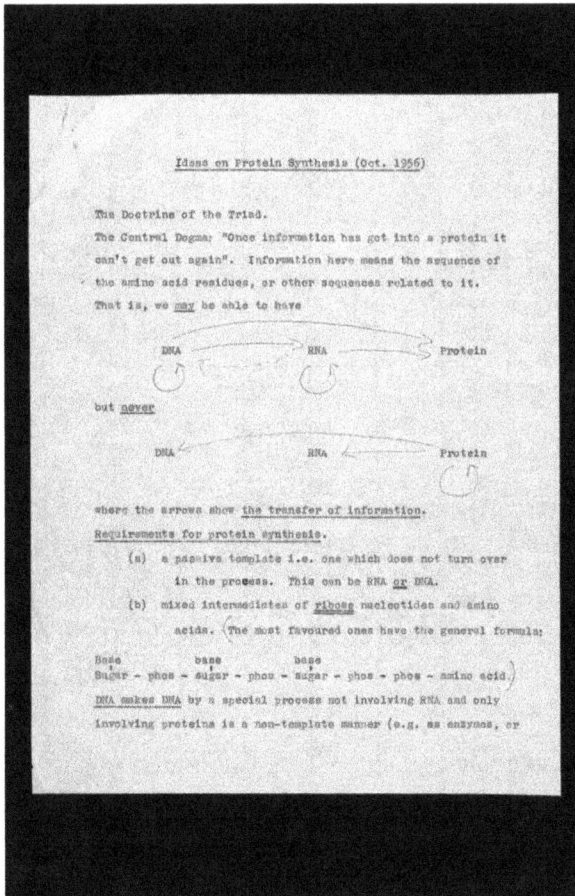

Figure 3. The earliest written description of "The Central Dogma" in a manuscript entitled "Ideas on protein synthesis (October 1956)". Credit: Wellcome Library, London.

In 1961, Francis Crick, Sydney Brenner, Leslie Barnett, and Richard Watts-Tobin first demonstrated the three bases of DNA code for one amino acid [7]. That was the moment that scientists cracked the code of life. However, ironically, the first decoding of the "word" of the genetic code was reported in the same year by a non-member of the RNA Tie Club, Marshall Nirenberg, who spoke at the International Biochemical Congress in Moscow. Matthew Meselson heard Nirenberg's 15-minute talk in a small room and told Crick about it. Crick arranged for Nirenberg to give the talk again at the end of the meeting. Starting with Nirenberg and Heinrich Matthaei's work [8], followed by that of Nirenberg and Philip Leder [9], the decoding was completed by Har Gobind Khorana [10]. Finally, Brenner, Barnett, Eugene Katz, and Crick placed the last piece of the jigsaw puzzle of life by proving that UGA was a third stop codon [11].

Thus, the genetic code was cracked, and it is the greatest legacy left behind by Francis Crick, along with the discovery of the double helical nature of DNA. As hallmarks of the foundation of molecular biology, they will continue to shine forever. However, the origin and evolution of the genetic code remain a mystery, despite numerous theories and attempts to understand them. In the mid-1960s, Carl Woese proposed the "stereochemical hypothesis", which suggested that the genetic code is derived from a type of codon–amino acid pairing interaction [12]. On the other hand, Crick proposed the "frozen accident hypothesis" and conjectured that the genetic code evolved from the last universal common ancestor and was frozen once established. However, he explicitly left room for stereochemical interactions between amino acids and their coding nucleotides, stating that "It is therefore essential to pursue the stereochemical theory ... vague models of such interactions are of little use. What is wanted is direct experimental proof that these interactions take place ... and some idea of their specificity" [13].

What is the real origin of the genetic code? tRNAs and aminoacyl-tRNA synthetases play fundamental roles in translating the genetic code in the present biological system [14], but what could have been the primitive forms of these molecules? Although Crick thought that tRNA seemed to be nature's attempt to make RNA do the job of a protein [2], the primordial genetic code prior to the establishment of the universal genetic code might have resided in a primitive form of tRNA. Such an example of "operational RNA code" [15] may be seen as a remnant in the acceptor stem of tRNA, which still functions as a critical recognition site by an aminoacyl-tRNA synthetase [16–18]. In addition, why are 20 amino acids involved in the genetic code? Discrimination of an amino acid with the high fidelity attained by modern aminoacyl-tRNA synthetases (error rate as low as $1/40,000$ [19]) would be impossible using a simple thermodynamic process alone, because the hydrophobic binding energy of a methylene group is, at the most, ~1 kcal/mol. Therefore, several sets of amino acids with similar side chains might have been coded non-selectively in the primitive stage [20]. Furthermore, the genetic code is the relationship between left-handed amino acids and right-handed nucleic acids. As non-enzymatic tRNA aminoacylation has been shown to occur chiral-selectively [21], the establishment of the genetic code might be closely associated with the evolutionary transition from the putative "RNA world" to the "RNA/protein world" in terms of homochirality [22]. All these are critical issues that should be investigated in the future.

The life force of Francis Crick was once described as similar to the "*incandescence* of an intellectual nuclear reactor" [23]. His passion for science is an inspiration for future scientific explorers. The Guest Editor of this Special Issue dedicates all articles included herein to the memory of Francis Crick.

Acknowledgments: The author thanks Kindra Crick for her valuable comments and suggestions.

References

1. Watson, J.D.; Crick, F.H.C. Molecular structure of nucleic acids: A structure for deoxyribose nucleic acid. *Nature* **1953**, *171*, 737–738. [CrossRef] [PubMed]
2. Crick, F.H.C. The genetic code–yesterday, today, and tomorrow. *Cold Spring Harb. Quant. Biol.* **1966**, *31*, 1–9. [CrossRef]

3. Sejnowski, T.J. In memoriam: Francis H.C. Crick. *Cell* **2004**, *43*, 619–621. [CrossRef] [PubMed]
4. Ridley, M. *Francis Crick: Discoverer of the Genetic Code*; HarperCollins Publishers: New York, NY, USA, 2006.
5. Crick, F.H.C. On degenerate templates and the adapter hypothesis: A note for the RNA Tie Club. 1955.
6. Hoagland, M.B.; Stephenson, M.L.; Scott, J.F.; Hecht, L.I.; Zamecnik, P.C. A soluble ribonucleic acid intermediate in protein synthesis. *J. Biol. Chem.* **1958**, *231*, 241–257. [PubMed]
7. Crick, F.H.; Barnett, L.; Brenner, S.; Watts-Tobin, R.J. General nature of the genetic code for proteins. *Nature* **1961**, *192*, 1227–1232. [CrossRef] [PubMed]
8. Nirenberg, M.W.; Matthaei, J.H. The dependence of cell-free protein synthesis in *E. coli* upon naturally occurring or synthetic polyribonucleotides. *Proc. Natl. Acad. Sci. USA* **1961**, *47*, 1588–1602. [CrossRef] [PubMed]
9. Nirenberg, M.; Leder, P. RNA codewords and protein synthesis. *Science* **1964**, *145*, 1399–1407. [CrossRef] [PubMed]
10. Khorana, H.G.; Büuchi, H.; Ghosh, H.; Gupta, N.; Jacob, T.M.; Kössel, H.; Morgan, R.; Narang, S.A.; Ohtsuka, E.; Wells, R.D. Polynucleotide synthesis and the genetic code. *Cold Spring Harb. Symp. Quant. Biol.* **1966**, *31*, 39–49. [CrossRef] [PubMed]
11. Brenner, S.; Barnett, L.; Katz, E.R.; Crick, F.H.C. UGA: A third nonsense triplet in the genetic code. *Nature* **1967**, *213*, 449–450. [CrossRef] [PubMed]
12. Woese, C.R.; Dugre, D.H.; Saxinger, W.C.; Dugre, S.A. The molecular basis for the genetic code. *Proc. Natl. Acad. Sci. USA* **1966**, *55*, 966–974. [CrossRef] [PubMed]
13. Crick, F.H.C. The origin of the genetic code. *J. Mol. Biol.* **1968**, *38*, 367–379. [CrossRef]
14. Schimmel, P. Aminoacyl tRNA synthetases: General scheme of structure-function relationships in the polypeptides and recognition of transfer RNAs. *Annu. Rev. Biochem.* **1987**, *56*, 125–158. [CrossRef] [PubMed]
15. Schimmel, P.; Giegé, R.; Moras, D.; Yokoyama, S. An operational RNA code for amino acids and possible relationship to genetic code. *Proc. Natl. Acad. Sci. USA* **1993**, *90*, 8763–8768. [CrossRef] [PubMed]
16. Hou, Y.M.; Schimmel, P. A simple structural feature is a major determinant of the identity of a transfer RNA. *Nature* **1988**, *333*, 140–145. [CrossRef] [PubMed]
17. McClain, W.H.; Foss, K. Changing the identity of a tRNA by introducing a G-U wobble pair near the 3′ acceptor end. *Science* **1988**, *240*, 793–796. [CrossRef] [PubMed]
18. De Duve, C. Transfer RNAs: The second genetic code. *Nature* **1988**, *333*, 117–118. [CrossRef] [PubMed]
19. Freist, W.; Pardowitz, I.; Cramer, F. Isoleucyl-tRNA synthetase from bakers' yeast: Multistep proofreading in discrimination between isoleucine and valine with modulated accuracy, a scheme for molecular recognition by energy dissipation. *Biochemistry* **1985**, *24*, 7014–7023. [CrossRef] [PubMed]
20. Tamura, K. Origins and early evolution of the tRNA molecule. *Life* **2015**, *5*, 1687–1699. [CrossRef]
21. Tamura, K.; Schimmel, P. Chiral-selective aminoacylation of an RNA minihelix. *Science* **2004**, *305*, 1253. [CrossRef] [PubMed]
22. Tamura, K. Toward the 'new century' of handedness in biology: In commemoration of the 100th anniversary of the birth of Francis Crick. *J. Biosci.* **2016**, *41*, 169–170. [CrossRef] [PubMed]
23. Sacks, O. Remembering Francis Crick. *The New York Review of Books.* **2005**, *52*, 24 March. Available online: http://www.nybooks.com/articles/2005/03/24/remembering-francis-crick/ (accessed on 23 August 2016).

life

MDPI

Opinion

Piecemeal Buildup of the Genetic Code, Ribosomes, and Genomes from Primordial tRNA Building Blocks

Derek Caetano-Anollés [1] and Gustavo Caetano-Anollés [2],*

[1] Department of Evolutionary Genetics, Max-Planck-Institut für Evolutionsbiologie, 24306 Plön, Germany; caetano@evolbio.mpg.de

[2] Evolutionary Bioinformatics Laboratory, Department of Crop Sciences, University of Illinois at Urbana-Champaign, Urbana, IL 61801, USA

* Correspondence: gca@illinois.edu; Tel.: +1-217-344-2739

Academic Editor: Koji Tamura
Received: 31 October 2016; Accepted: 29 November 2016; Published: 2 December 2016

Abstract: The origin of biomolecular machinery likely centered around an ancient and central molecule capable of interacting with emergent macromolecular complexity. tRNA is the oldest and most central nucleic acid molecule of the cell. Its co-evolutionary interactions with aminoacyl-tRNA synthetase protein enzymes define the specificities of the genetic code and those with the ribosome their accurate biosynthetic interpretation. Phylogenetic approaches that focus on molecular structure allow reconstruction of evolutionary timelines that describe the history of RNA and protein structural domains. Here we review phylogenomic analyses that reconstruct the early history of the synthetase enzymes and the ribosome, their interactions with RNA, and the inception of amino acid charging and codon specificities in tRNA that are responsible for the genetic code. We also trace the age of domains and tRNA onto ancient tRNA homologies that were recently identified in rRNA. Our findings reveal a timeline of recruitment of tRNA building blocks for the formation of a functional ribosome, which holds both the biocatalytic functions of protein biosynthesis and the ability to store genetic memory in primordial RNA genomic templates.

Keywords: genome evolution; origin of proteins; ribosome evolution; origin of the genetic code

---

1. Introduction

Uncovering patterns and processes responsible for the origin of life in extant macromolecules is a most challenging proposition. The biological world is largely governed by the functions of protein and nucleic acid molecules. Proteins and RNA make up the molecular machinery of the cell while DNA generally holds its historical repository, its "genetic" memory. The diversity of molecular structures and functions that have been surveyed in proteins and nucleic acids is unprecedented. As of 26 October 2016, 1221 vetted 3-dimensional fold designs defined by one protein classification [1] encompass the structure of 244,326 protein structural domains that hold individually or in combination ~5 million experimental and non-experimental annotations of molecular functions defined by ~9,000 terminal Gene Ontology definitions [2]. Only a relatively small subset of these fold structures are present in each and every organism that has been prospected [3]. Similarly, only 2,474 RNA families have been defined [4], of which only 5 are universal [5]. For decades, molecular biologists have pondered over this diversity as they attempted to explain how life originated in this planet. The genomic revolution has not been forthcoming either. No clear link has been found that explains how the 123,870 models of molecular structure deposited in the entries of the PROTEIN DATA BANK (PDB) [6] and their associated functions are encoded in the DNA of the 10,045 genomes and metagenomes that have been completely sequenced (GOLD DATABASE [7]) and that have given rise to 0.55 million UNIPROTKB/SWISSPROT and ~68 million UNIPROTKB/TREMBL protein sequence entries and information on thousands of functional RNA molecules important for probing the workings of the cell. We know there is a code in the memory of life, the genetic code. We do not know how that code maps to the memory of structure and function of proteins, the structural and functional code. Here we argue that this crucial liaison involves transfer RNA (tRNA) and was established very early in evolution once nucleotide cofactors of primordial polypeptides were lengthened into primordial RNA loops. We propose that these nucleic acid loops were capable of interacting stereochemically with evolving protein structure and responding to their molecular makeup. Increases in these interactions canalized both the appearance of genetic memory and building blocks (modules) of RNA with which to construct processive biosynthetic machinery on one hand and genomic memory storage on the other. We review phylogenetic evidence that provide support for these claims and address the properties of the emergent tRNA and rRNA molecular systems viewed fundamentally from the perspective of emerging proteins and genetic information in primordial cells. First, we examine the structures, functions and time of origin (age) of structural domains of proteins defined at the fold family (FF) and fold superfamily (FSF) levels of SCOP, the *Structural Classification of Proteins* [1]. In these studies, the ages of domains are derived from rooted phylogenomic trees built from abundance counts of domains in proteomes [8–10]; Second, we use a molecular clock of folds to convert relative age into geological time [10]; Third, the age of tRNA and ribosomal substructures calculated from an exhaustive phylogenomic analysis of thousands of molecules [3,11] is linked to the history of proteins; Finally, we assign ages of helical segments of rRNA to remote tRNA homologies recently identified in rRNA [12], establishing correlations with the ages of corresponding tRNA molecules [3]. The exercise reveals the modular role of tRNA in the early evolution of ribosomes and genomes. The results and implications are remarkable.

2. Unity and Diversity in the Evolutionary History of Biological Modules and Systems

Ever since Darwin evolution has been described using the paradigm of trees (Figure 1a), network abstractions that showcase complex historical processes of diversification (Figure 1b, bottom). The development of cladistics and advanced phylogenetic methodology has shown that biological data exhibits one universal property: vertical traces of genetic memory across time are always complemented with horizontal exchanges of that memory. Thus, the tree paradigm should be considered an oversimplification necessary for the heuristic computational search of optimal phylogenies, hypotheses of history describing the evolution of the biological entities (taxa) that are being studied. Instead, trees with reticulations (sometimes making up reticulated nets or rhizomes; Figure 1b, top) may be more appropriate, especially when studying the evolution of taxa in which processes of horizontal exchange of genetic information override vertical genetic signatures. These scenarios are common in the evolution of bacteria and archaea. Central to evolutionary tree and network thinking is the notion of a common ancestor to the group of evolving entities, a *"radix communis"* that unifies the phylogeny (Figure 1b). This usually takes the form of a "trunk", a branch leading to a root node exemplifying the hypothetical common ancestor of the entities that are evolving along the branches of the tree or network.

Phylogenetic trees or networks are built from useful biological features of evolving taxa, which are known as phylogenetic "characters". These characters are usually building blocks (parts) of more complex physical or functional systems (wholes). Molecular examples include amino acids of proteins or nucleotides of nucleic acids. Because parts and wholes are interrelated, trees describing the evolution of systems also describe the evolution of their building blocks (Figure 1c). Under this new paradigm, the evolutionary unification of building blocks results in new emerging systems (defined below), which then diversify. We exemplify this process with a mathematical abstraction (Figure 1d) in which the edges of a primordial root network join to form an ancestor trunk edge. This trunk then diversifies into a crown network of extant entities and their ancestors. Here we focus on the root network of this new abstraction, using structural domains of proteins and central nucleic acid molecules as the subjects of study. We note that this new "hourglass" network paradigm applies to each and every component part of a biological system and that each hourglass does not necessarily occur contemporaneously in evolution. For example, the rise of multidomain proteins from the combination of individual structural domains (reviewed in [3]) was likely preceded by the combination of lower level structural parts to form each protein domain. Here we discuss how this can be made explicit to help us understand processes of macromolecular emergence.

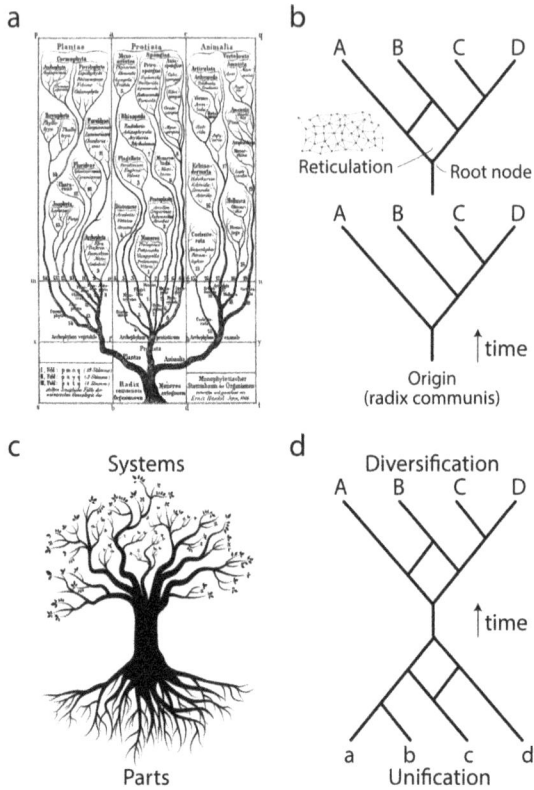

Figure 1. Paradigms governing evolution. (**a**) Tree of life drawn by German zoologist Ernst Haeckel (ca. 1866) depicting the existence of a common ancestor or "*radix communis organismorum*" (the common root of all organisms) unifying diversified cellular life embodied in the leaves of the tree or any transect along its crown; (**b**) In mathematics (graph theory), a tree abstraction can be used to describe the evolution of biological entities, which can be considered either parts of systems or entire wholes. The tree must be rooted to impart a direction and "arrow of time" to its statement of diversification and change. However, tree descriptions can be faulty because multiple evolutionary origins (convergences) are possible when the initial memory of systems is tangled by recruitment or other complicating processes of horizontal exchange. These convergences cause reticulations (see tree with reticulation in top) and in extreme cases "rhizomes" (inset). For example, taxon B has two possible ancestors (one shared with taxon A and the other with taxon C and D), which converge to form its lineage; (**c**) A new paradigm describes the rise of biological parts (modules) from more primordial components and their subsequent diversification. This is illustrated with a tree that shows its trunk separating its root and crown. When considering all biological parts, the tree-like structure describes the evolution of biological systems; (**d**) The abstraction of panel c can be defined by two networks (root and crown networks) joined by a common edge (trunk). This common edge represents the last common ancestor of systems A, B, C and D (members of the crown) as it arises from modular parts a, b c and d (members of the root).

3. Memory and the Evolutionary Drivers of Abundance, Recruitment and Accretion

An emerging system in biology must be dynamic and persistent. It must be a natural object with behavior and makeup delimited by a set of interacting component parts (subsystems). Its behavior and makeup must be characterized and individuated from other systems by its cohesion, i.e., by the dynamical stabilities of the component parts when constrained by the system as a whole [13,14]. Persistence refers to the ability of the system to display memory, i.e., to preserve a behavior and make up despite constant perturbation from environments internal and external to the system in question. Within these confines, the emerging system exploits the three fundamental properties of any engineered object, economy, flexibility and robustness [15]. Since these properties are strongly impacted by the way the system perceives both the environment and its internal state, the trade-off solutions that are achieved vary with time and context and have been modeled by a "triangle of persistence" and the system's environmental history, its "scope" [16]. We note that scope has two components, "umwelt" (the system's perception of that history) and "gap" (the system's blind spot, the scope that is not covered by its umwelt). The triangle of persistence was recently used to mathematically explain the existence of a Menzerath-Altmann law of language in the domain makeup of proteins [17]. This law, which states that larger systems hold smaller component parts, manifests by decreasing the length of structural domains when their number increases in multidomain proteins. Thus, the interplay of economy, flexibility and robustness can be made explicit at the biomolecular and biophysical level.

In biology, the memory of a hierarchical biological system (α) increases by increasing the abundance of both its nested parts and wholes. Equation (1) summarizes the process of increasing memory by increasing the number of parts and wholes (that we label a) to higher abundance levels ($a' > a$, that we label A).

$$a \xrightarrow{\alpha 1} A \tag{1}$$

Highly abundant parts and wholes have higher chances of remaining persistent and by doing so enhancing the survival and memory of the system under consideration. For that reason, it is generally unlikely that once high abundance levels A are achieved these levels will return to lower levels by loss, unless strong reductive evolutionary forces are at play that are beneficial to the system. This is particularly so when the views of hierarchical systems are global and focus on the higher hierarchical level rather than the local and lower level. Abundance can be increased in many ways but it generally involves the existence of compositional or informational bias. For example, the famous Urey-Miller spark experiments of the 1950's demonstrate the facile generation of only a limited set of amino acids from the simulated gaseous environments of early Earth [18]. These sets include alanine, glycine, aspartic and glutamic acid, valine, leucine, isoleucine and serine. These same amino acids are overrepresented in salt-induced experimental formation of small dipeptides and polypeptides under prebiotic conditions [19]. Similarly, peptides enriched in alanine, glycine, aspartic acid and valine hold hydrolytic functions and can be produced experimentally by repeated dry-heating cycles and by solid phase peptide synthesis [20]. Finally, these same amino acids are overrepresented in the dipeptide constitution of proteins when globally surveyed in proteomes [21]. Thus, a memory implanted by compositional biases in plausible chemical reactions manifests at different and increasing levels of the hierarchy of life. In our example, they even express in proteins that are encoded by modern genomes. It is particularly noteworthy that this memory has been made mathematically explicit by computer simulations that describe how compositional biases relate to information storage [22].

Memory can also be enhanced by recruitment (also known as cooption or exaptation), the ability to use existent parts in new different contextual environments. Equation (2) summarizes how the process of memory of recruited parts a increases when these are recruited by parts b.

$$a + b \xrightarrow{\alpha 2} Ab \tag{2}$$

Since recruitment is usually associated with increases of abundance, the abundance of older parts that are coopted by newer ones follow the trends of Equation (1). Examples of recruitment of these kinds are many. In metabolism, a simple analysis of the distribution of protein fold structures in metabolic pathways suggests metabolic networks grow piecemeal and form evolutionary patchworks [23]. More recently, the use of an algorithmic implementation that derives the most plausible ancestry of an enzyme from structural and evolutionary annotations revealed that the recruitment of ancient domain structures by modern enzymes is widespread in metabolic networks [24]. Note that parts need not be physically associated to fulfill Equations (1) and (2) and enable increases of abundance by recruitment. In the case of metabolic networks, the enzymatic parts that are recruited are often loosely associated in the cell but their metabolic functions are established cohesively. Memory can also be enhanced by a related process, accretion. This occurs when parts are recruited pervasively into a system, one at a time, and remain tightly linked with each other. We note that in most cases the recruited parts become physically associated with the growing system, with b of Equation (2) representing the growing system and a the accreted part. Examples of accretion of these kinds are macromolecular complexes such as the ATP synthases and the ribosome [25,26]. The F-type and A/V-type synthases are multi-subunit complexes responsible for membrane-coupled energy conversion reactions. They produce most of the ATP needed to power cellular processes. Using phylogenomic methods we have shown that the synthase complexes developed gradually by addition of structural domains, starting with the ring structures of the rotating head, followed by the central stalk (the axle), and ending with the structures that regulate their motion (the stators) [25]. Similarly, the ribosome has been shown to grow in evolution by addition of helical segments to the evolving molecules [26], complying with the principle of continuity that sustains evolutionary thinking and explaining the formation of highly sophisticated macromolecular machinery.

Since parts and wholes in biology are highly dynamic entities, only the subset of them that expresses some long-term stability and arise as cohesive element within the expanding hierarchical system will be stable enough to display memory. These parts generally represent *modules*. A module can be defined as a set of integrated parts that cooperate to perform a task [27]. These parts interact more with each other than with other parts of the system, including parts of other modules. Since modules result from the emergent properties of the hierarchical system [15], they must hold history. While defining modules in biology can be challenging, the fact that modules must be "evolutionary units" enables the use of phylogenetic methods to appropriately test definitions of modularity. In this regard and in the absence of clear statements of topographic correspondence, definition of modules through homologies often requires dynamic homology analysis [28] or the use of hidden Markov models [29]. These methods are capable of distinguishing similarity due to common ancestry from similarities due to other causes that are not evolutionary.

4. The Usefulness of Abundance in Phylogenomic Analysis

We study the crucial role of abundance by focusing on the evolution of proteins and nucleic acids and reconstructing phylogenetic hypotheses that are grounded in data and computational optimization. Both of these macromolecules are responsible for the rise of biology and genetic memory that we here explore. Methodologically, we take an ideographic (historical and retrodictive) approach that uses information in the protein repertoires of thousands of genomes and advanced tools of phylogenetic analysis to build statements of history, phylogenies of protein parts [3]. The approach takes advantage of the benefits that both molecular structure and abundance provide to the many challenges of phylogenetic reconstruction. These benefits have been discussed in detail elsewhere [30]. Because protein domain structure is several orders of magnitude more resistant to the effects of mutation than its sequence [31], its high conservation levels make structure more suitable for deep phylogenetic exploration. In particular, the slow evolutionary pace of structural change diminishes the chances that the consequences of the "Markov chain convergence theorem" and the "data processing inequality", which define how time erases useful historical information, limit retrieval of evolutionary history [32].

Furthermore, the use of domain abundance as phylogenetic character avoids the need of alignments in phylogenetic analysis of sequences (the search for similarities of sets of sequences with unknown correspondences but restricted by the lineal order of residues in the sequence). It also offsets the limitations of character independence of phylogenetic reconstruction that plague the use of molecular sequences in phylogenetic analysis; the mere existence of folding and 3-dimensional structure in macromolecules implies character non-independence in a sequence alignment.

The methodology not only builds phylogenetic trees or networks that describe the origin and evolution of protein and nucleic acid repertoires but also tests if the molecular data that is being analyzed holds historical information. This second aspect of the analysis is crucial since it provides experimental support to the link between molecular abundance and history that enables the construction of powerful trees describing the evolutionary history of molecular parts (e.g., the entire world of protein domains or substructures that describe the history of RNA molecules). It also dispels the possibility that molecular structures emerged from lower component parts (e.g., loops, or nucleic acid motifs) by their fortuitous association, driven solely by improvements in functional versatility. Finding phylogenetic signal in the data supports the existence of historical contingency. We note the importance of understanding the meaning of trees of component parts. For example, a "tree of structural domains" holds domains (suitably defined) as taxa (the leaves of the tree) and defines historical relationships (tree topologies) based on domain abundance in the proteomes analyzed (the phylogenetic characters and data analyzed). Since the data matrix makes use of molecular abundance, the tree that is built from the data does not arise from a model of change that involves structural transformations of domains (e.g., [33]). Instead, the historical relationship of different domains is inferred directly from quantitative information in genomic makeup. This "criterion of primary homology" rests exclusively on genomic abundance of individual domains in proteomes, and its validity is permanently tested by mutual optimization of phylogenetic signal in characters and tree reconstruction (an exercise known as Hennigian illumination). Thus, the methodology operates under the Popperian pillars of content of theories and degree of corroboration (see [21] for an explicit elaboration), acknowledging the need of more modern technical definitions of "verisimilitude" for scientific inquiry. The historical signal and reliability of phylogenies of structural domains that have been published are being gradually strengthened by more data (e.g., more proteomes and better sampling of the world of cells and viruses) and better optimization (e.g., improved hidden Markov models and increased background knowledge).

Operationally, the phylogenomic methodology of building trees of parts starts by defining taxa and doing so exhaustively (e.g., all suitably defined structural domains present in proteomes, all substructures of RNA molecules). Properties of this finite taxon set are then studied, such as abundance of structural domains in proteomes or structural or thermodynamic features of RNA substructures in RNA molecules. This generates taxon-character data matrices with data encoded by transformation into alpha-numeric values suitable for optimization with phylogenetic reconstruction software. The software optimizes character changes in all possible unrooted trees (portraying phylogenetic relationships of taxa) during exhaustive or branch-and-bound tree searches or uses heuristic approaches to find optimal solutions according to the maximum parsimony criterion. The most parsimonious trees that are retained are then rooted *a posteriori* using Weston's generality criterion of derived character states being less widespread in tree branches implemented with the Lundberg method, which reorients and roots the tree by pulling down the branch that yields the minimum increase in character state change. The rooted trees that are recovered are comb-like and can be converted into chronologies of taxa by backward-counting nodes (branch points or bifurcations) from leaves to the root of the optimal trees. This counting defines a relative *node distance* (*nd*) and a scale from $nd = 0$ (most basal and old) to $nd = 1$ (most recent and young), which is used to define how evolutionarily derived is each taxon in the tree of parts. In the case of protein structural domains, we have shown that *nd* correlates strongly with actual time for folds that are linked to markers of the geological record [10]. We use this molecular clock of folds to define the ages of structural domains as a chronology in billions of years (Gy). Given protein-RNA domain interactions that are known, the chronology can be used to transfer age from proteins to interacting RNA by assuming that the age of the RNA molecule is the age of the protein-RNA interaction.

We end by noting that abundance can be coarse-grained into occurrence for phylogenomic analysis, i.e., quantitative valued characters can be reduced to a data matrix of 0 s and 1s, a binary system. While coarse-graining results in some loss of phylogenetic signal, occurrence and abundance generally produce congruent historical statements that can be separately optimized (e.g., [34]).

5. The Early Primacy of Peptides, Polypeptides and Proteins in Cellular Environments

> *"All language is a set of symbols whose use among its speakers assumes a shared past. How,*
> *then, can I translate into words the limitless Aleph, which my floundering memory can scarcely*
> *encompass?"* —El Aleph and Other Stories, Jorge Luis Borges.

In modern biology, peptides and proteins are encoded in genomes and are translated from mRNA into folded polymers that are functional. Similarly, transcribed non-coding RNA folds into functional forms. The 3-dimensional molecular structures of these macromolecules tend to become compact when they collapse into stable conformations and abandon the benefits of interacting with the aqueous environment that forces them to maintain the unfolded state. The folded structures provide a fundamental scaffold to constrain in favorable conformation the small subset of amino acid residues responsible for protein functions. These residues are generally lodged in pockets on the surface of the protein, though networks of residues throughout the molecule play also roles in allosteric regulation and protein stability. The small subset of residues is mostly associated with the unstructured regions of proteins, suggesting that the complex arrangement of secondary structures (helix, strand and turn) collapses into 3-dimensional topologies that best respond to the needs of the more dynamic and functional regions of the protein. In fact, phylogenomic analysis has shown that folding speed and flexibility are beneficial traits that are fostered in evolution [35] and that flexible loop regions were enriched in proteins by the rise of genetics [21]. Moreover, structural flexibility is even a conserved feature in the assembly of protein complexes [36]. Thus, protein flexibility appears a crucial property in protein evolution at different levels of the hierarchical molecular system, even in the absence of a primordial biology that could translate nucleic acid information into proteins.

Structural domains in proteins are considered evolutionary units. Any statement about their history that is obtained using phylogenetic approaches or hidden Markov model libraries (e.g., timelines of the evolutionary appearance of domains, trees or networks of domains, domain groupings) will relate solely to their history and not to the history of other modules and evolutionary units that exist at lower or higher level of the hierarchical molecular system. During the past decade we have generated phylogenomic reconstructions of the evolution of structural domains at the different structural abstractions of SCOP: folds, FSFs and FFs, from deeper to shallower evolutionary views of molecular structure (reviewed in [3]). The emergent picture of molecular evolution derived from domain history is largely congruent regardless of the level of abstraction [9] or the classification system that is used to define structural domains [37]. The very early and oldest domains are fully dependent on cellular membranes. Thus, the first proteins appear to have emerged enclosed in primordial containers (cells) and evolved from there to form the wide diversity of globular proteins that currently exist. A devil's advocate however could challenge any inference derived from annotations of historical timelines by claiming bioinformatic associations of functions and structures say nothing about early historical processes. Indeed, one must assume that modern definitions of functions and structures can be used to interpret those that existed in the past. In other words, we must consider that viewing past events with a modern "lens" is a valid approach. This may not always hold [32] and more philosophical, mathematical and biological elaborations of the implications of considering modern entities as relics of the past await development. The early rise of cellular containers as single or multi-layer vesicles is however supported by the existence of amphiphilic molecules in meteorites that organize themselves spontaneously into liposomes in the laboratory and the possible important role of meteoritic influx on the environments of early Earth (see discussions in [9,25]). Similarly, plausible prebiotic synthesis of membrane constituents exist that could explain their early and abundant formation, notably aided by the effects of clays and other mineral deposits. Finally, structural canalization can be invoked as important force that freezes in time the structure of molecules making these structures highly conserved at evolutionary level [38]. While the activity of this force throughout all levels of structural complexity remains to be explored, canalization appears an important and general principle of conservation in biology that shields the effects of the environment on the organism and may represent an inevitable consequence of complex processes [39].

A previous study mapped in detail the first evolutionary appearance of the oldest 54 FFs and traced a number of properties of these protein structures, including their ability to bind cofactors, interact with RNA, and display broad molecular movements and flexibility [9]. The set was selected because it laid the foundations for both the metabolic and translation machineries [8,9]. The very early timeline of FFs, which is described in Figure 2, showed that the first four FFs were the ABC transporter ATPase domain-like family (c.37.1.12), the extended and tandem AAA-ATPase domain families (c.37.1.20 and c.37.1.19) and the tyrosine-dependent oxidoreductase domain family (c.2.1.2), all of which exist in highly structured cellular environments. The functions of these FFs are linked to the start of modern metabolic networks, providing hydrolase and transferase functions needed for nucleotide interconversion, storage and phosphate transfer-mediated recycling of chemical energy, and terminal production of beneficial cofactors (e.g., [40]). The early evolution of metabolism in association with nucleotide cofactors culminated with the appearance of the first enzymes of the biosynthetic pathways of nucleotide metabolism 3.5 Gy-ago and the completion of a functional biosynthetic pathway ~3 Gy-ago, which coincides with the rise of a functional ribosome [41]. These coordinated developments suggest the coevolutionary need of a steady supply of nucleotide precursors internal to the cell for the synthesis of large RNA molecules, large enough to store genomic information and fulfill the ribosomal role of processive biosynthesis. The rise of aerobic metabolism at about that time (~2.9 Gy-ago) ultimately results in the great oxygenation event (GOE) of our planet that occurred 2.45 Gy-ago [42], which coincides with the rise of superkingdom-specific domain structures and early organismal diversification that we call the epoch of "superkingdom specification" (Epoch 2; Figure 2).

The P-loop containing nucleotide triphosphate (NTP) hydrolase fold (c.37) is the first folded structure of the timeline of FFs (Figure 2). It appears for the first time associated with a primordial bundle, the predominant structure associated with membranes. The c.37 structure holds a "Rossmann-like" $\alpha/\beta/\alpha$-layered design that "sandwich" a sheet of strands between helical segments. This layered design dominates the topologies of many subsequent basal FF structures, including 36 of the 54 oldest FFs. The primordial appearance of this fold confirms once again abundant evidence from phylogenetic reconstruction suggesting that the layered structure was responsible for the first compact protein modules (beginning with [43]). The primordial $\alpha/\beta/\alpha$-layered structure has special properties related to lower level structural organization that are very relevant. The structural and functional diversity of proteins can be described by a combinatorial interplay of "supersecondary" structures, modular-like arrangement of helix, strand and turn segments (e.g., $\alpha\alpha$-hairpins, $\beta\beta$-hairpins, $\beta\alpha\beta$-elements), that act as lower level evolutionary building blocks of protein folds and biochemical diversity [44,45]. These supersecondary motifs are generally ~25–30 amino acid residues long and in most cases form recurrent loop structures, many of which determine biochemical diversity [46] and protein flexibility [21]. In evolution, these so called "elementary functional loops" (EFLs) likely combine with each other to better bind cofactors and exert molecular functions (Figure 3a). In fact, the history of these EFLs can be traced back to a small set of loop prototypes, which represent collectives of many sequences embedded in proteins and capable of collapsing into stable loop structures. These EFLs are likely stabilized by the formation of van der Waals locks [47]. An analysis of the most abundant of these EFL prototypes revealed they were associated with a small set of folds defined at the FSF level of the SCOP hierarchy [48].

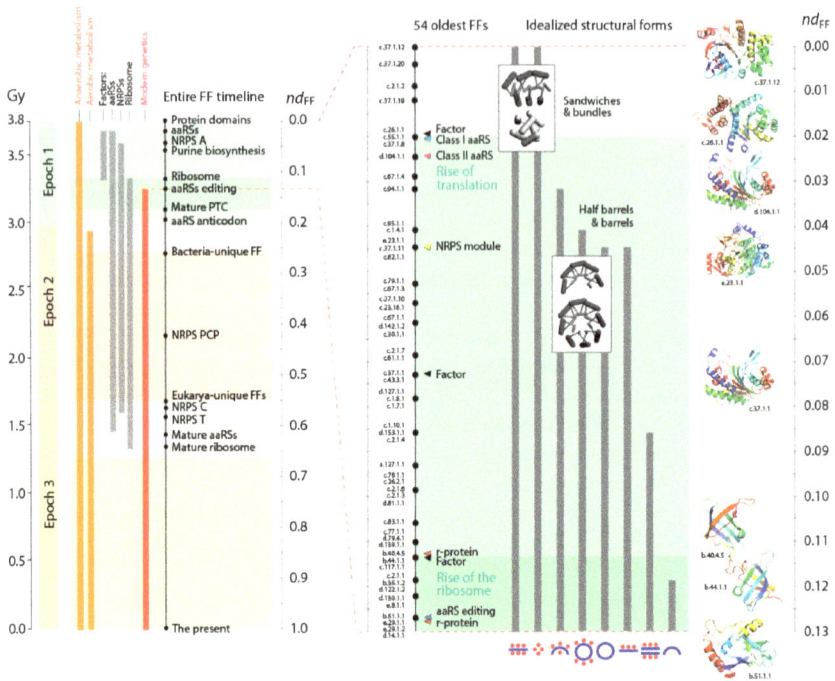

Figure 2. Timeline describing the evolution of structural domains responsible for the primordial components of metabolism and translation. The timeline was derived directly from the tree of FFs reconstructed from free-living organisms. Ages are given as node distances (nd_{FF}) and geological time in billions of years (Gy). Time flows from top to bottom. The three evolutionary epochs of the protein world, "architectural diversification" (epoch 1), "superkingdom specification" (epoch 2), and "organismal diversification" (epoch 3) (see definition in [8,9]) are indicated with different color shades. Fundamental structural and functional discoveries are identified with circles along the timeline. The inset describes in detail the evolutionary timeline of the 54 most ancient FFs, showing examples of 3-dimensional models and idealized structures with diagrams representing helices with red dots and sheets of strands with blue lines. Colored arrowheads indicate FFs associated with the listed functional discoveries. aaRS, aminoacyl-tRNA synthetase; NRPS, non-ribosomal peptide synthetase; PTC, peptidyl transferase center; ;PCP, peptidyl carrier protein.

Figure 3b shows a small subnetwork of the most abundant EFL prototypes and their FSFs, with the ages of FSF mapped onto the network and transferred to EFLs [49]. The bipartite network showed that the P-loop hydrolase FSF (c.37.1) was the most connected hub, benefiting from the assembly of numerous EFLs. In particular, the EFL 536 hub links c.37.1 to the NAD(P)-binding Rossmann fold FSF (c.2.1) that holds the ancient tyrosine-dependent oxidoreductase domain FF. Remarkably, the small subnetwork contains the oldest FSFs of the EFL-FSF network, supporting the fundamental evolutionary link of abundance and time of origin of modules and the special properties of the $\alpha/\beta/\alpha$-layered design for building domains. The fact that the history of domains matches inferences from the bipartite network suggests domains structures assembled from loops to form larger and more stable folded structures and that loop ligations were more prone to form stable folded structures (given biases of prebiotic amino acid constituents) in areas of sequence space that materialized into the c.37 fold.

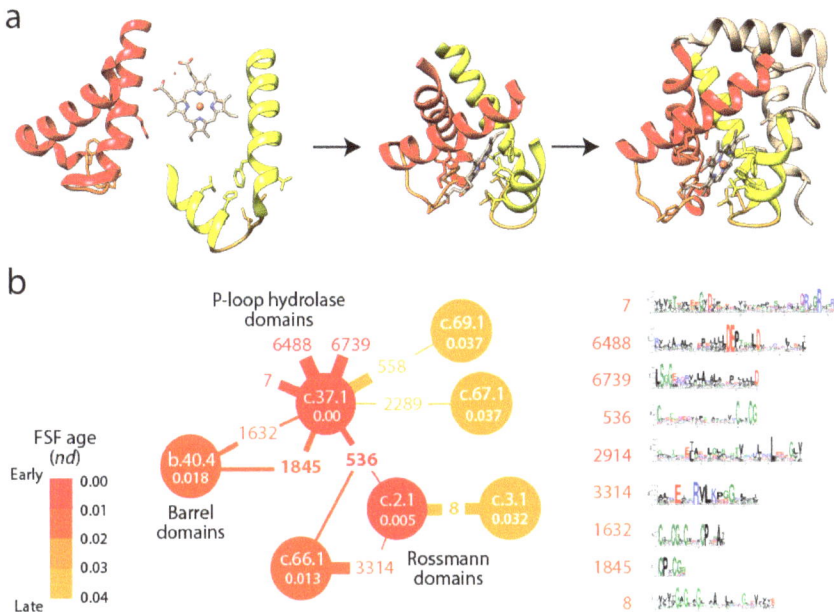

Figure 3. The evolutionary rise of structural domains from supersecondary structural motifs. (a) An illustration of the process with the haemoglobin molecule (PDB entry 1THB). Two main elementary polypeptide loops (colored red and yellow) with $\alpha\alpha$-hairpin structures have sites capable of binding to a protoporphyrin IX-iron complex (heme) for oxygen transport when they come together in space. The evolutionary joining of these loop structures into a single molecule and further growth by addition of extra α-helices produces the modern structural domain of the α-chain haemoglobin; (b) Linking evolution of the oldest and most abundant FSF domains (circles) and EFL motifs (numbers) using a bipartite graph. The graph describes the most connected subnetwork of a bipartite network that describes how FSF share EFLs in proteins [48]. This subnetwork is also the oldest. The age of FSFs (*nd*) is indicated inside circles. The P-loop hydrolase fold (c.37.1, *nd* = 0) is the most connected FSF and EFL 536 and EFL 1845 the most connected loops in the subgraph and in the entire network that is not shown. Prototype logos in the right show amino acid residue frequencies in sequence sites of the most ancient EFLs and a clear pattern of EFL length decrease with age that extends to the rest of the graph. Edges represent EFL matches to domains; their width is proportional to the number of matches. FSFs are labeled with SCOP *concise classification strings* and EFLs with prototype numbers.

This primordial $\alpha/\beta/\alpha$-layered design of the most ancient FFs has also special properties related to amino acid usage. The c.2.1.2 structure for example uses almost exclusively amino acids encoded by the GC-rich half of the codon table and its genes have multiple open reading frames [50]. This appears to indicate that these enzymes acquired their fold structures earlier than a diversified genetic code. Similarly, the dipeptide make-up of protein domains appearing early in the timeline is enriched in hydrophobic amino acids and underrepresented in dipeptides participating in flexible loop regions, suggesting protein flexibility was an important driver for the rise of genetics [9,21]. Enrichment patterns suggest hydrophobicity of dipeptide make-up of the first FFs and a primordial association with membranes. Their rigid protein structures lacking flexible arms and showing limited motions is compatible with standard enzymatic functions. We note however that mutation saturation of sequences has probably replaced the amino acid repertoires present in ancient domain structures with amino acids of the modern 20+ repertoire and that FF structural cores have been decorated with additional structures of much more recent origin, probably harboring all possible amino acid sites. Thus, more modern processes of change complicate inferences derived from sequence analysis.

6. The Late Appearance of Interactions with RNA

A number of FFs appeared ~3.7–3.6 Gy-ago (nd_{FF} = 0.02–0.045) after the rise of metabolism (Figure 2). These structures catalyzed crucial acylation and condensation reactions involved in aminoacylation of tRNA bound to aminoacyl-tRNA synthetases (aaRSs) or phosphopantetheinyl arms of carrier proteins that are part of non-ribosomal peptide synthetase (NRPS) complexes. These structures, which made their debut before ribosomal proteins in the timeline of FFs, are also part of the catalytic makeup of enzymes important for fatty acid biosynthesis. The first four FFs of this group involve class I aaRS catalytic domain (c.26.1.1), class II aaRS and biotin synthetases (d.104.1.1), G proteins (c.37.1.8) and actin-like ATPase domain (c.55.1.1) FFs. All of them have the α/β/α-layered Rossmannoid design and three of them define the catalytic domains of aaRSs and structures of elongation factors that are central for translation and the specificity of the genetic code. Translation therefore appears to have metabolic origins that predate the appearance of the ribosome [8]. We note the profound implications of the phylogenomic timeline, especially for proponents of the ancient "RNA world" theory that dominates current thinking in origin of life research. While we have discussed the feasibility of this theory elsewhere [25], we ask the reader to keep an open mind when considering the alternatives suggested by phylogenetic evidence that will follow.

Before discussing further implications of the timeline, we want to emphasize the putative environment that fostered all of these structural innovations. Without a genetic memory, the systems had to rely exclusively on biases of the emerging polymers, the prebiotic and biotic chemistries surrounding their functions, and the physical constraints imposed by the emerging cellular systems. The sequence and structures of proteins that we study today have been the subject of up to ~3.8 Gy of continued optimization, of course within the constraints of structural canalization. During the first few hundred million years, those same macromolecules could have not achieved the levels of compactness of modern folds nor the functional efficiency and specificity of the modern macromolecules for several reasons. Evolutionary optimization through mutational change (read compositional variation) could have not covered enough sequence space and any diffusion by random walk had to be faulty and limited by frequent loss and absence of strong selective constraint. We therefore envision that molecules spent considerable time in conformations that were unproductive but were still able to advance optimization through the compositional codes that were slowly materializing. This probably involved favoring limited sets of building blocks, smaller molecules, and smaller patches of inter-molecular interactions. It is highly likely that a multitude of reactants and chemical reactions was available for probing in billions of combinations throughout the entire planet and that only those fortuitous successes would have spread to the rest of the cellular systems through rather free cellular exchange. This necessitates "porous" membranes and smaller molecules than those of today's biology. For example, it would be non-productive to combine emerging domains into larger ensembles during that time. Indeed, phylogenetic analysis suggests that it took and additional ~2 Gy to fully develop the benefits of domain combinations in multidomain proteins [51].

Implicit in the evolutionary appearance of tRNA-associated FFs is the development of stereochemical interactions between molecules that could jumpstart both "translation specificity" and "genetic memory". We have proposed a model of emergence of genetics in which molecular interactions define: (i) specificities of an emerging genetic code in "identity elements" of the nucleic acid molecule; and (ii) corresponding FF enzymatic activities responsible for tRNA aminoacylation and the formation of peptide bonds [21]. A corollary that follows from this model is that stereochemical interactions were established between small polypeptide and nucleic acid molecules that were already "structured" by molecular folding. This implies that the FF structural cores had already assembled from small EFLs by statistically biased condensation reactions and were developing archaic aminoacylation and ligation activities for cellular persistence. We stress that the dual role of stereochemical interactions is needed to explain the hidden evolutionary link between the specificity of tRNA identity elements and information in the structure of proteins, and at the same time, explain the selective forces that could be at play. We posit that the hidden link is the formation of dipeptide molecules from pairs of aminoacylated tRNAs by primordial aaRS urzymes [21]. Figure 4 shows that class II aaRS and biotin synthetases (d.104.1.1) and class I aaRS catalytic (c.26.1.1) domains responsible for SerRS and TyrRS aminoacylation activities, respectively, have close structural homologues in amino acid-[acyl carrier protein]-ligases (aaACPLs) and cyclodipeptide synthases (CDPSs), respectively. Note that aaACPLs are relatives of NRPSs and CDPSs are dipeptidases that produce dipeptides from sets of two aminoacylated tRNA. These strong structural homologies are evolutionarily deep. They reveal highly conserved structural protein cores that are putative founders of archaic biosynthetic activities needed to jumpstart primordial genetic and structural codes. To test if indeed dipeptidases were involved in providing building blocks for the structuring of protein domains, we looked for biases in the dipeptide make-up of FFs appearing prior to anticodon binding domains, and found significant biases ($p < 0.05$) against flexible loop regions but favoring turns and bends in the initial FFs [21]. This suggests that genetics arose from biases in the 400+ word vocabulary of dipeptides that makes up proteins and a transition from rigid to flexible protein structural cores.

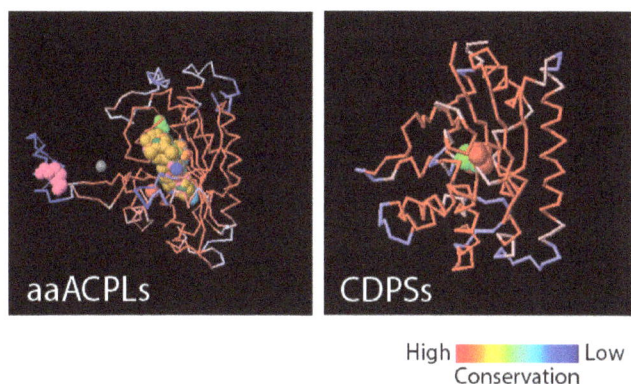

High ▰▰▰▰▰▰▰▰ Low
Conservation

Figure 4. Structural relics of ancient aaRS ligases and dipeptidases. A structural alignment of aaACPL B110957 (PDB entry 3PZC) to the d.104.1.1 FF catalytic core of class II SerRS enzyme from a metanogenic archaeon (entry 2CJ9), which is its closest structural neighbor (Z = 26.8; RMSD = 2.7 Å), is shown in the left. A structural alignment of CDPS AlbC (entry 3OQV) to its best match, the c.26.1 FF catalytic core of a class Ic TyrRS enzyme from an archaeon (Z = 10.0; RMSD = 3.2 Å) is shown in the right. A color scale shows structural alignment conservation used in the tracing of polypeptide backbones. The proximity in DALI structural neighborhoods of aaACPLs and CDPSs to aaRS catalytic cores with major groove specificities suggests a deep evolutionary link to archaic founders of aaRS biosynthetic activities [21].

7. Defining a Natural History of Protein Catalytic Mechanisms and Their Interaction with Cofactors

Biocatalytic mechanisms are chemical transformations of organic compounds facilitated by protein enzymes and other natural catalysts. In turn, cofactors are "helper" non-protein chemical compounds required for biomolecular activity. Mechanisms and cofactors must reside in special pockets of the enzymatic structure (e.g., active sites) for them to be effective. Recent studies traced the appearance of biocatalytic mechanisms and associated cofactors in structural domain evolution [9,52]. Phylogenomic trees reconstructed from a structural census at the "homology superfamily" level of the CATH classification system (analogous to the FF level of SCOP) allowed to trace the mechanistic step types of the fold structures [52]. Each mechanistic step type is one of 51 mechanistic annotations in the MACiE database that are used to describe the chemistries underlying enzymatic activities.

The basal P-loop containing NTP hydrolase fold (3.40.50.300) introduced the mechanistic steps that are most widely spread in enzymes, including "proton transfer", "bimolecular nucleophilic addition", "bimolecular nucleophilic substitution" and "unimolecular elimination by the molecular base". However, it was two of the following three CATH structures that added almost half of all 51 mechanistic annotations, the NAD(P)-binding Rossmann-like domain (3.40.50.720) and FAD/NAD(P)-binding domain (3.50.50.60) (Figure 5a). These structures preceded the inception of domains that interact with RNA, the Hups (3.40.50.620) α/β-layered domains of aaRSs, which introduce the single mechanistic step of "intramolecular elimination" needed to fulfill their aminoacylation reactions.

A similar progression can be seen by studying the use of cofactors by SCOP FFs in the timelines [9] (see Figure 2). The first appearance of domain interactions with cofactors inferred by cofactor annotations in entries of the PROCOGNATE and PDB databases revealed the primordial use of ATP and ADP by the c.37.1 structure (Figure 5b), an observation previously intimated from the distribution patterns of small molecule ligands in proteins [53]. However, the Rossmann fold of the c.2.1.2 FF that followed added almost half of all known cofactors of proteins. This burst matches the substantial rise of mechanistic steps immediately preceding the appearance of catalytic domains of aaRSs and protein interactions with tRNA. The finding supports the long held idea of RNA originating from ligation of precursors that were acting as cofactors (e.g., [54]). However, and in contrast with many RNA world-inspired proposals, these ribotide cofactors were being synthesized in pockets of the primordial $\alpha/\beta/\alpha$-layered structures. This is compatible with the observation that aaRSs are able to form a wide variety of dinucleoside oligophosphates in the presence of amino acids (e.g., [55]), a property that is also shared by NRPS domains. We therefore hypothesize that the $\alpha/\beta/\alpha$-layered structures fostered nucleotide ligations that extended suitable combinations of nucleotides to form longer polymers and that this interplay naturally materialized in rudiments of the genetic code within the confines of an increasingly more complex ribonucleoprotein molecular world. Conversely and in parallel, the $\alpha/\beta/\alpha$-layered structures could have also facilitated the ligation of dipeptide and small peptides to form larger molecules. As mentioned above, the structures of catalytic domains of aaRSs can form dipeptides with the aid of tRNA molecules (e.g., [56]), a molecular function that left relics in the dipeptide makeup of proteins [21].

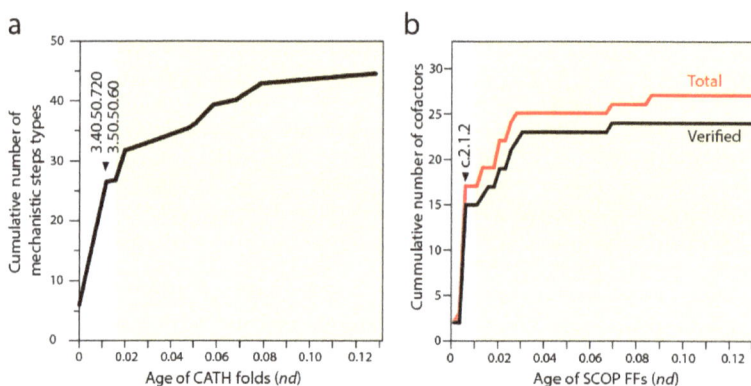

Figure 5. The very early accumulation of new mechanistic steps (**a**); and cofactors (**b**) in evolutionary timelines of structural domains. Mechanistic step types were taken from annotations in MACiE [52]. Relationships that exist between cofactors and FFs were derived from the PROCOGNATE and the PDB databases [9]. The total cofactor dataset contains both experimentally verified cofactor-structure relationships and relationships that are not. The most ancient CATH homologous superfamilies and SCOP FFs were arranged by their age (*nd* values). White shaded areas involve domain structures that do not interact with RNA.

8. The Coevolutionary History of Emerging tRNA, rRNA and Proteins and the Rise of Genetics

If indeed emerging domains interacted with initial tRNA cofactors, then it is possible to envision that aaRS enzymes coevolved with tRNA during the rise of genetic code specificities and that tRNA coevolved with the emerging ribonucleoprotein structure of the ribosomes. Coevolution is here defined as the coordinated succession of structural changes mutually induced by the increasingly interacting and growing protein and nucleic acid molecules in their quest to fold into more stable and functionally efficient structures that would provide enhanced stability to primordial cells. Using phylogenomic reconstruction we have been able to support both of these coevolutionary assertions with considerable data. Phylogenetic analysis of thousands of RNA molecules and millions of protein structural domains allowed reconstruction of phylogenies and evolutionary timelines of the history of tRNA amino acid charging and anticodon-binding specificities of tRNA [20] and the history of ribosomal accretion [11]. The relative ages of structures of aaRS domains, ribosomal proteins, tRNA and rRNA drawn directly from the phylogenetic trees were indexed with structural, functional and molecular contact information and mapped (by color) onto three-dimensional models of individual molecules and ribosomal complexes (Figure 6). Four important coevolutionary patterns were revealed.

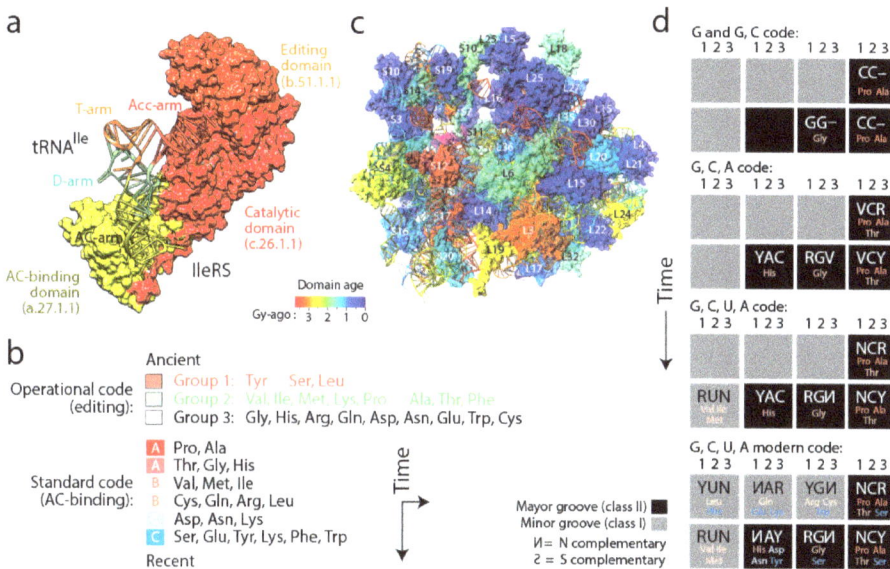

Figure 6. Coevolution of proteins and nucleic acids to form the evolutionary cores of translation machinery and genetics. (**a**) The age of the domains of aaRSs, exemplified by IleRS (PDB entry 1qu2), match the age of the interacting arms of their tRNA isoacceptors. The oldest acceptor (Acc) arm interacts with the oldest catalytic domain and the more recent anticodon (AC) arm interacts with the more recent AC-binding domain [21]; (**b**) Two codon systems evolved sequentially but acted redundantly, one delimiting amino acid charging and the other codon specificity. Phylogenomic analysis dissects their history [21]; (**c**) The ribosomal complex with ages of ribosomal proteins and rRNA helices traced on an *Escherichia coli* structural model of the ribosomal core [11]. Note the very ancient and central translocation core of helix 44 and ribosomal proteins S12 and S17; (**d**) Alphabet evolution of the "standard" genetic code [21]. Ancestries of tRNA-aaRS binding were mapped most parsimoniously onto the condensed Rodin & Rodin's *vis-a-vis* degenerate genetic code representation, taking into consideration anticodon loop identity elements. This timeline of late genetic code expansion was indexed with major and minor groove modes of tRNA recognition in the aaRS enzymes. N = G, C, U, A; V = G, C, A; R = G, A; Y = C, U; S = G, C.

(i) The history of aaRS catalytic, editing and anticodon-binding domains matched the history of tRNA charging and encoding [21] (Figure 6a). These coevolutionary patterns allowed to infer a history of progression of specificities for both the "operational" genetic code of the acceptor arm of tRNA and the "standard" genetic code of the more derived anticodon-binding stem of tRNA (Figure 6b). Since specificity determinants in tRNA result from interaction with the synthetases, the progression describes the rise of the aminoacylation specificities of tRNA isoacceptors. The first specificities involved pre-transfer and post-transfer editing and trans-editing activities responsible of sieving amino acids by size in the active sites of the catalytic domains of the synthetases. These specificities involved 11 of the 20 standard amino acids, which were split into two groups. Group 1 specificities were associated with the older type II tRNA structures holding a variable arm. Group 2 were associated with standard type I tRNA cloverleaf structures. These interactions involved the acceptor stem of the tRNA molecule, the oldest of the molecule [57]. They delimited the operational genetic code, probably in absence of a fully functional ribosome and a full cloverleaf structure. In turn, codon specificities were determined by specific anticodon binding domains in interaction with the more modern anticodon stem of tRNA and appeared much later in the timeline, ~3 Gy-ago. The development of this more modern "standard" genetic code produced its own timeline of codon specificities (Figure 6b). Thus, protein history unfolded separate timelines of amino acid charging and codon recognition, which we had already intimated in an earlier study [58], and revealed coevolution of the emerging domains and nucleic acid cofactors.

(ii) A similar analysis of the evolution of the structure of rRNA and ribosomal proteins of the small (SSU) and large (LSU) subunits of the ribosome produced an evolutionary timeline of accretion of the universally conserved ribosomal complex [11] (Figure 6c). The age of rRNA helical regions (see Figure 7a) and interacting domains of ribosomal proteins coevolved to form a fully functional ribosomal core. The oldest protein (S12, S17, S9, L3) appeared together with the oldest rRNA substructures responsible for decoding and ribosomal dynamics 3.3–3.4 Gy-ago. These structures include the ratchet and two hinges of SSU rRNA and the L1 and L7/L12 stalks of LSU rRNA important for ribosomal movement of tRNA in the complex. While protein-RNA coevolution manifested throughout the timeline, the appearance of RNA substructures at first occurred in orderly fashion until the formation of a 10-way LSU and 5-way SSU junctions, at which point a "major transition" in ribosomal evolution occurred 2.8–3.1 Gy-ago. This transition brought ribosomal subunits together through inter-subunit bridge contacts. It also stabilized loosely evolving ribosomal components and developed tRNA-interacting structures and a fully-fledged peptidyl transferase center (PTC) with exit pore capable of protein biosynthesis. Thus, ribosomal history also showed gradual coevolution between RNA and proteins.

(iii) Coupling the evolutionary timelines of tRNA and rRNA structure with annotations of their interactions with protein domains revealed that the tRNA cloverleaf structure was already fully formed when the PTC appeared in evolution [59]. This was previously intimated directly from phylogenetic analysis of ribosomal history [11]. Thus, fully formed tRNA molecules played other roles before being recruited for processive protein biosynthesis, perhaps as cofactors of peptide-producing dipeptidases and ligases. A more detailed elaboration of our data-driven hypothesis for the origin of translation and genetics can be found elsewhere [9,21].

(iv) Finally, tracing ancestries of tRNA-aaRS binding in a condensed code representation of primordial complementarity indexed with major/minor groove modes of tRNA recognition revealed gradual evolution of the genetic code (Figure 6d). Mappings showed the early use of major groove recognition and the second and first codon positions. The early codes were associated with small and hydrophobic amino acids. The coding of Pro, the founder, was based only on C and already used second and first code positions (identity elements G35 and G36). The code soon expanded into a duplex code by adding G to its alphabet. The use of a third codon position (G34) for the first time with Thr and then His (the last two initial recruitments of the c.51.1.1 FF) expanded the alphabet to a triplex code that used C, G and A. Finally, the "yin-yang" complementarity pattern of the condensed code representation was finally fulfilled with the last recruitment of the a.27.1.1 FF once the modern tetraplex code was in place.

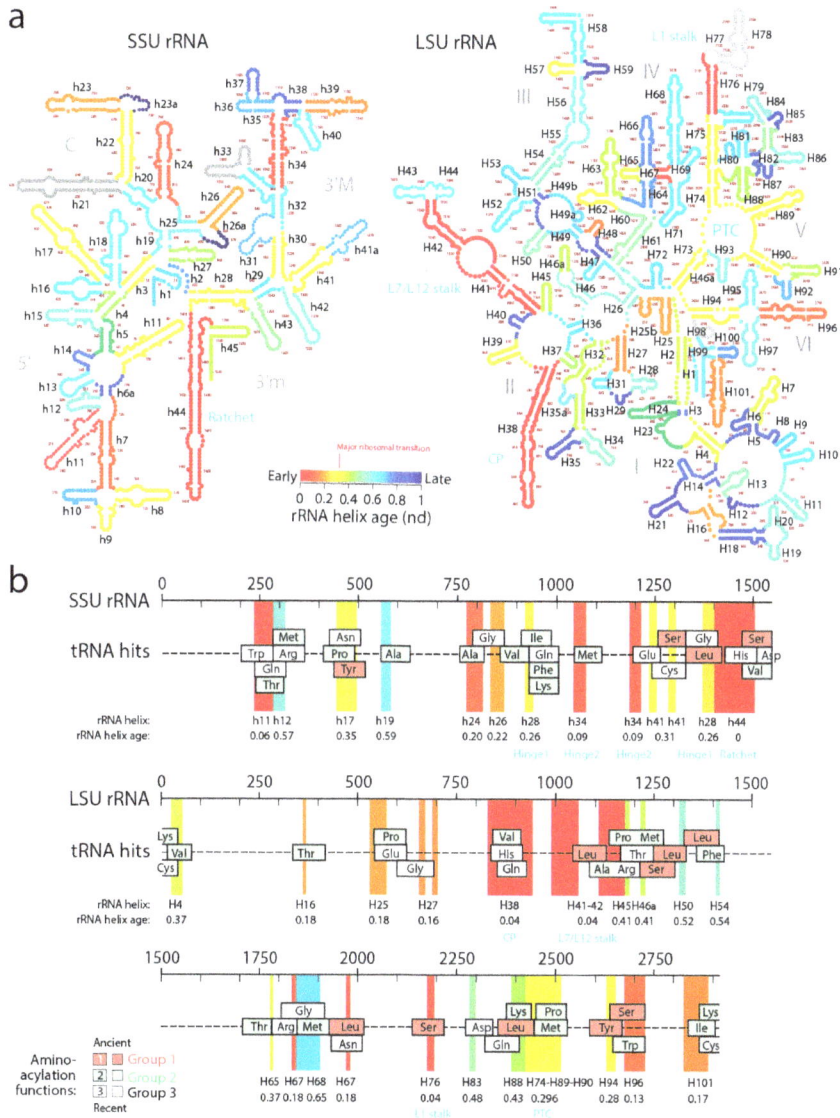

Figure 7. Revealing the gradual formation of a functional ribosome by accretion of tRNA building blocks. (**a**) Secondary structure models of the small (SSU) and large (LSU) subunits of ribosomal rRNA from *Escherichia coli* with helical segments colored according to their relative age (*nd*); (**b**) Mapping of tRNA homologies onto rRNA sequences and tracings of the projection of the oldest helical segments of rRNA that encompass the tRNA homology hits, colored according to age. tRNA homologies as indicated with squares colored according groups of aminoacylation function, with Groups 1 and 2 holding editing functions.

The existence of two codes embedded in the acceptor stem and in the anticodon stem of tRNA has recently received additional support from a study that shows that the acceptor and anticodon stem determinants code for size and polarity of amino acid residues, respectively [60]. This matches the differential encoding of information in the top and bottom half of the tRNA molecule and the role of editing and anticodon binding recognition that differentiate these two sequential and apparently redundant codes [21].

9. Accretion of tRNA Building Blocks Forms Functional Ribosomes

A recent study generated lists of non-overlapping alignments between tRNA and rRNA molecules using a pairwise global alignment method implemented with the LALIGN algorithm without end gap penalties and using default parameters [12]. The study uncovered a number of remote homology hits, often overlapping, which suggested both subunits of the ribosome were built piecemeal from primordial tRNA molecules (Figure 7). The finding is significant as it supports the hypothesis anticipated by David Bloch and his colleagues in the 80s that tRNA and rRNA shared a common history [61]. It also supports recent findings of sequential and overlapping homologies of reconstructed tRNA with the PTC core of LSU rRNA [62].

In order to explore how the tRNA accretion process gave rise to functional rRNA, we traced the age of rRNA regions associated with relics of ancient tRNA building blocks (Figure 7). The ages of rRNA substructures were taken directly from ref. [11]. The oldest structural regions present in tRNA relics, for each relic, were highlighted as projections in the sequence of SSU and LSU rRNA and colored according to rRNA helix age (nd), from red ($nd = 0$; oldest) to blue ($nd = 1$; youngest). Relics were enriched in projections of old ribosomal regions (red, orange and yellow hues of the projections) that preceded the rise of the PTC and the "major transition" in ribosomal evolution (Figure 7). Note how projections of these old regions usually unify the many overlapping tRNA homologies, suggesting tRNA building blocks may have been at the beginning smaller and then slowly materialized into larger cloverleaf-like forms. In fact, and as we previously commented, tracing ribosomal protein history and tRNA interactions with domains in the phylogenetic timelines of ribosomal accretion revealed that a full-blown tRNA molecule was already interacting with the ribosome at the time of the major ribosomal transition [59]. Note the existence of substantial tRNA homology embedded in the PTC of LSU rRNA. Many tRNA homologies also showed substantial matching to other functional regions, including the central ratchet and hinges of SSU rRNA and the L1 and L7/12 stalks and the central protuberance (CP) of LSU rRNA that are involved in ribosomal dynamics. The most numerous overlapping tRNA matches coincided with old structures and involved tRNA with aminoacylation functions corresponding to the oldest Groups 1 and 2 that hold pre-transfer and post-transfer editing and trans-editing activities. In particular, tRNA relics encoding Ser and Leu were the most abundant (6 Leu and 5 Ser tRNAs, respectively) and matched the old central functional regions, supporting the ancestrality of editing specificities for the charging of these two amino acids (Figure 6b) and the proposal that they jumpstarted the "operational" genetic code [21]. This is expected since anticodon-binding domains responsible for major specificities of the standard genetic code appeared after the ribosomal transition (Figure 2).

The ages of rRNA helices of the tRNA relics was also plotted against the age of tRNA isoacceptors derived from phylogenetic constraint analysis [58], which dissects the history of Groups 1 and 2 specificities in the timeline of tRNA accretion (Figure 8). The plot makes evident the ancestral nature of tRNA homologies and also shows the more recent tRNA recruitments. When considering homologies in the oldest rRNA segments, a coevolutionary pattern between the age of tRNA and tRNA building blocks of the ribosome appears evident (dashed line, Figure 8). The pattern suggests ribosomal construction by tRNA recruitment began very early and made use very quickly of the entire repertoire of tRNA isoacceptors derived from the editing specificities of their acceptor arms.

A close examination of Figure 7 allows postulating a succession of early cooption steps involving emerging Groups 1 and 2 isoacceptors into the growing ribosome. The primordial ribosomal ratchet of SSU and LSU moving parts ($nd = 0$–0.04) appeared to have been developed by cooptions of tRNA^{Leu}, tRNA^{Ser}, tRNA^{Val}, tRNA^{Pro}, and tRNA^{Ala} homologies. Similarly, the more derived SSU rRNA hinges (nd = 0.09–0.26) involved tRNA^{Met}, tRNA^{Ile}, tRNA^{Phe}, and tRNA^{Lys} homologies. Finally, the rise of the ribosomal PTC (nd = 0.28–0.30) involved accretion of tRNA^{Leu}, tRNA^{Ser}, tRNA^{Tyr}, tRNA^{Pro}, tRNA^{Met}, tRNA^{Lys} and tRNA^{Thr} homologies. The oldest Group 1 aminoacylation specificities appeared to have been remembered in the oldest structures of the ribosome, while the more derived Group 2 specificities are more abundant in later accretion steps.

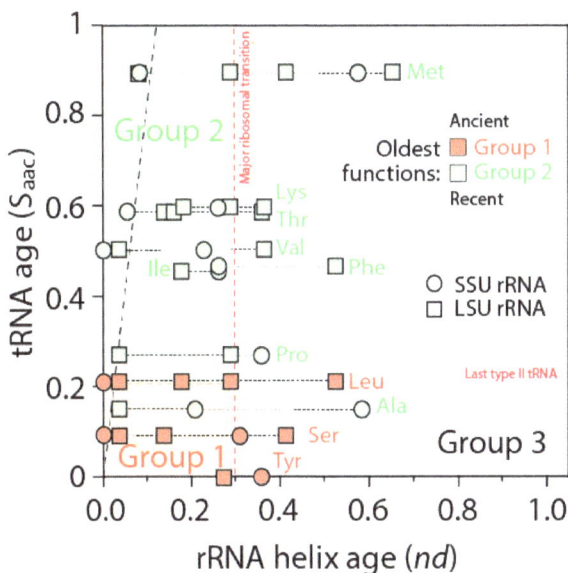

Figure 8. Coevolution of the most ancient rRNA substructures and tRNA holding the oldest aminoacylation functions (Groups 1 and 2) as these are pervasively coopted in rRNA. Ages were derived from phylogenetic constraint analysis of tRNA molecules [58] and from the timeline of ribosomal accretion inferred from the mappings of tRNA homologies to rRNA (see Figure 7). The more recent Group 3 aminoacylation functions are not plotted but are enriched in the white quadrant of the plot.

10. Genomic Accretion of tRNA Building Blocks

When ancestral tRNA was translated in silico into proteins, its sequences showed homologies to elongation factors, aaRS enzymes, enzymes of nucleotide biosynthesis pathways and RNA polymerases [12]. Similar results were found in a separate study [63]. These remarkable results suggest both tRNA and tRNA ribosomal relics hold deep phylogenetic information indicating they both stored genetic information for ancient proteins and acted as ancient genomes. Modern biology provides important clues to this very primordial role of tRNA. Dispersed repetitive elements, especially those associated with tRNA, have the potential to spatially and functionally organize the genome by providing barriers to chromatin structure, DNA replication, and contributing to fragile sites prone to genomic rearrangements [64]. Synteny blocks in genomes, believed to be the result of chromosomal rearrangements, are often flanked by tRNA genes (e.g., [65]), suggesting an active role of tRNA encodings in genomic make up. Transposable elements often exhibit homologies to tRNA and have also active roles in the evolutionary restructuring of genomes [66]. The 3′-terminal ends of mRNAs in mitochondrial DNA are often immediately continuous to tRNA genes, which likely punctuate the polycistronic transcripts by endonucleolytic cleavage [67]. On this point, there is evidence that many aaRSs not only bind to their respective tRNA in order to catalyze esterification of the appropriate cognate amino acid, but also bind to homologous sequences on their own mRNA in order to carry out autogenous regulation of synthetase production. Examples include many of the Group 2 and 3 tRNAs, including the aaRS for tRNA[Thr] [68–79], tRNA[Asp] [80–83], tRNA[His] [84], tRNA[Met] [74], and tRNA[Phe] [85]. Many other aaRSs (especially those in Groups 1 and 2) are also regulated by direct binding of the protein to genetic regulatory elements but not directly to their own mRNA [79,83]. Such autogenous control of synthetases provides additional evidence that tRNA may have played a central role not only in the origins of the ribosome, but also in the origins of the genome that encodes ribosome-related proteins. Finally, recent analysis of mimivirus transcripts shows tRNA genes are expressed as polyadenylated messengers and follow a stringent "hairpin rule", which extends to the entire genome [86]. The ancestrality of giant viruses, and viruses in general [87], now suggests this oddity is an ancient (not derived) feature of the mimivirus genome. All of these properties support the crucial functional and structural role of genomic tRNA, boosting their ancient role as genomic building blocks.

11. Ribosomal Structure Supports rRNA and Genomic Evolutionary Growth from Primordial tRNA Pieces

The structural makeup of the ribosome provides information about its possible growth by covalent joining of primordial tRNA pieces [26]. Identification of putative insertions of "branch" helices onto preexisting coaxially stacked "trunk" helices in crystallographic models of the ribosome showed that not all insertions support the outward and gradual growth of ribosomal structures [88]. Seventeen putative insertions suggest either evolutionary events of inward growth or the existence of "structural grafting" of building blocks to build larger rRNA structures. The fact that these putative insertions flank regions with numerous tRNA homologies supports the idea that those building blocks were in fact primordial tRNA molecules (D. Caetano-Anollés, ms. in preparation).

12. Conclusions

We have postulated a phylogenomic data-driven evolutionary scenario describing the rise of translation and genetics [9]. It involves the lengthening of primordial cofactors into short RNA hairpins, which slowly gained compositional specificities and evolved into longer nucleic acid polymers protected by catalytic sites of the $\alpha/\beta/\alpha$-layered structures of archaic protein domains (summarized in Figure 9). Similarly, primordial protein domains likely assembled from smaller loop subunits, the EFLs [46]. This process of accretion of loop structures produced crucial domains, exemplified in the recently proposed emergence of class II aaRSs from three hairpin structures [89]. The initial protein-nucleic acid interactions resulted in "ternary complexes" of primordial aaRSs, translation factors, and tRNA, which aminoacylated tRNAs, ligated charged amino acids into dipeptides and longer polymers, and gradually gained specificities to ensure compositional memories would be preserved in proteins and interacting RNA [9]. These complexes were then "vectorially" transferred to other molecular contexts, which would give rise to more complex NRPS-like and ribosomal-like machinery. In particular, their interaction with newly formed OB-fold barrel structures produced an ancestor of the central ribosomal ratchet of SSU rRNA and its S12 and S17 ribosomal protein partners (the oldest of the ribosome) [11]. One important corollary of this scenario is that the specificities of the genetic code developed through stereochemical interactions between nucleic acid and protein molecules that were fully structured. In this regard and in line with the "self-referential model" for the origin of the genetic code [90], pockets in the $\alpha/\beta/\alpha$-layered structures of archaic synthetases were able to accommodate pairs of interacting RNA hairpins that were aminoacylated, catalyzing peptide bond formation. We believe the molecular environment of structural pockets resembled those of modern CDPSs, which foster the formation of tRNA-mediated dipeptidyl enzyme intermediates to produce a wide variety of dipeptides [91].

In the present study, phylogenetic tracings of ancient tRNA homologies in the ribosome reveal that cooption of emerging tRNA modules appears to be a protracted phenomenon responsible for both ribosomal structure and RNA "templating" memory. It is likely that the dynamics of cooption at RNA level responsible for rRNA and genomes, also brought with it interactions with emerging proteins domains. This resulted in a growing ribonucleoprotein ribosomal complex that was built gradually and from smaller pieces through protein-nucleic acid coevolution. It is also likely that numerous regulatory interactions involving tRNA mimicry at genomic level may have been established at this very early stage as a primordial and labile epigenetic ("paragenetic" *sensu* Alexander Brink) mechanism. These interactions evolved hand-in-hand with the emerging genetic machinery and ultimately gave rise to "*field(s) of possibilities*", the genes of genomes [92].

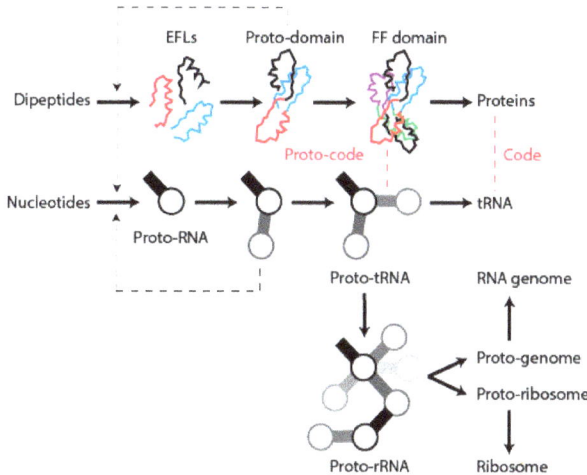

Figure 9. Model of evolutionary growth of macromolecules from component parts leading to translation machinery and genomes. Longer polypeptide molecules would have assembled from amino acids and dipeptides by statistically biased condensations. Some of these produced elementary functional loops (EFLs) capable of interacting with ligands and forming larger protein ensembles (EFLs with variant sequences and structures are illustrated with differently colored loop backbones). Similarly, proto-RNA molecules folding into small hairpins (stems are illustrated with solid bars and loops with open circles) assembled from nucleotides in EFL-delimited pockets and were later ligated to form larger RNA molecules serving as proto-genomes and proto-ribosomes. Interactions between RNA and emerging proteins establish primordial structural correspondences. This code of genetic memory is illustrated with red dashed lines. Black dashed arrows illustrate feed-forward catalytic activities.

Acknowledgments: Computational biology in the Evolutionary Bioinformatics lab is supported by grants from NSF (OISE-1132791) and USDA (ILLU-802-909). Derek Caetano-Anollés is a recipient of NSF postdoctoral fellowship award 1523549. We thank Robert Root-Bernstein for his constructive suggestions.

Author Contributions: D.C.A. mapped the age of tRNA homologies in rRNA and re-analyzed annotations of the evolutionary timeline of protein domains. G.C.A. proposed the evolutionary tracings and wrote the manuscript with the help of D.C.A.

Conflicts of Interest: The authors declare no conflict of interest.

Abbreviations

The following abbreviations are used in this manuscript:

aaACPL	amino acid-[acyl carrier protein]-ligase
aaRS	aminoacyl-tRNA synthetase
CDPS	cyclodipeptide synthase
EFL	Elementary functional loop
FF	Fold family
FSF	Fold superfamily
NRPS	non-ribosomal peptide synthetase
PTC	peptidyl transferase center
SCOP	Structural classification of proteins

References

1. Murzin, A.G.; Brenner, S.E.; Hubbard, T.; Chothia, C. SCOP: A structural classification of proteins database for the investigation of sequences and structures. *J. Mol. Biol.* **1995**, *247*, 536–540. [CrossRef]

2. The Gene Ontology Consortium. Gene ontology consortium: Going forward. *Nucleic Acids Res.* **2014**, *43*, D1049–D1056.

3. Caetano-Anollés, G.; Wang, M.; Caetano-Anollés, D.; Mittenthal, J.E. The origin, evolution and structure of the protein world. *Biochem. J.* **2009**, *417*, 621–637. [CrossRef] [PubMed]

4. Nawrocki, E.P.; Burge, S.W.; Bateman, A.; Daub, J.; Eberhardt, R.Y.; Eddy, S.R.; Floden, E.W.; Gardner, P.P.; Jones, T.A.; Tate, J.; et al. Rfam 12.0: Updates to the RNA families database. *Nucleic Acids Res.* **2014**. [CrossRef] [PubMed]

5. Hoeppner, M.P.; Gardner, P.P.; Poole, A.M. Comparative analysis of RNA families reveals distinct repertoires for each domain of life. *PLoS Comput. Biol.* **2012**, *8*, e1002752. [CrossRef] [PubMed]

6. Berman, H.M.; Westbrook, J.; Feng, Z.; Gilliland, G.; Bhat, T.N.; Weissig, H.; Shindyalov, I.N.; Bourne, P.E. The Protein Data Bank. *Nucleic Acids Res.* **2000**, *28*, 235–242. [CrossRef] [PubMed]

7. Reddy, T.B.K.; Thomas, A.; Stamatis, D.; Bertsch, J.; Isbandi, M.; Jansson, J.; Mallajosyula, J.; Pagani, I.; Lobos, E.; Kyrpides, N. The Genomes OnLine Database (GOLD) v.5: A metadata management system based on a four level (meta)genome project classification. *Nucleic Acids Res.* **2014**. [CrossRef]

8. Caetano-Anollés, D.; Kim, K.M.; Mittenthal, J.E.; Caetano-Anollés, G. Proteome evolution and metabolic origins of translation and cellular life. *J. Mol. Evol.* **2011**, *72*, 14–33. [CrossRef] [PubMed]

9. Caetano-Anollés, G.; Kim, K.M.; Caetano-Anollés, D. The phylogenomic roots of modern biochemistry: Origins of proteins, cofactors and protein biosynthesis. *J. Mol. Evol.* **2012**, *74*, 1–34. [CrossRef] [PubMed]

10. Wang, M.; Jiang, Y.-Y.; Kim, K.M.; Qu, G.; Ji, H.-F.; Zhang, H.-Y.; Caetano-Anollés, G. A molecular clock of protein folds and its power in tracing the early history of aerobic metabolism and planet oxygenation. *Mol. Biol. Evol.* **2011**, *28*, 567–582. [CrossRef] [PubMed]

11. Harish, A.; Caetano-Anollés, G. Ribosomal history reveals origins of modern protein synthesis. *PLoS ONE* **2012**, *7*, e32776. [CrossRef] [PubMed]

12. Root-Bernstein, M.; Root-Bernstein, R. The ribosome as a missing link in the evolution of life. *J. Theor. Biol.* **2015**, *367*, 130–158. [CrossRef] [PubMed]

13. Root-Bernstein, R.S.; Dillon, P.F. Molecular complementarity, I: The molecular complementarity theory of the origin and evolution of life. *J. Theor. Biol.* **1997**, *188*, 447–479. [CrossRef] [PubMed]

14. Collier, J. Hierarchical dynamical information systems with a focus on biology. *Entropy* **2003**, *5*, 100–124. [CrossRef]

15. Mittenthal, J.E.; Caetano-Anollés, D.; Caetano-Anollés, G. Biphasic patterns of diversification and the emergence of modules. *Front. Genet.* **2012**, *3*, 147. [CrossRef] [PubMed]

16. Yafremava, L.S.; Wielgos, M.; Thomas, S.; Nasir, A.; Wang, M.; Mittenthal, J.E.; Caetano-Anollés, G. A general framework of persistence strategies for biological systems helps explain domains of life. *Front. Genet.* **2013**, *4*, 16. [CrossRef] [PubMed]

17. Shahzad, K.; Mittenthal, J.E.; Caetano-Anollés, G. The organization of domains in proteins obeys the Menzerath-Altmann's law of language. *BMC Syst. Biol.* **2015**, *9*, 44. [CrossRef] [PubMed]

18. Miller, S.L. A production of amino acids under possible primitive earth conditions. *Science* **1953**, *117*, 528–529. [CrossRef] [PubMed]

19. Jakschitz, T.; Rode, B.M. Evolution from simple in- organic compounds to chiral peptides. *Chem. Soc. Rev.* **2012**, *41*, 5484–5489. [CrossRef] [PubMed]

20. Ikehara, K. Possible steps to the emergence of life: The [GADV]-protein world hypothesis. *Chem. Rec.* **2005**, *5*, 107–118. [CrossRef] [PubMed]

21. Caetano-Anollés, G.; Wang, M.; Caetano-Anollés, D. Structural phylogenomics retrodicts the origin of the genetic code and uncovers the evolutionary impact of protein flexibility. *PLoS ONE* **2013**, *8*, e72225. [CrossRef] [PubMed]

22. Segré, D.; Lancet, D. Composing life. *EMBO Rep.* **2000**, *1*, 217–222. [CrossRef] [PubMed]

23. Teichmann, S.A.; Rison, S.C.G.; Thornton, J.M.; Riley, M.; Gough, J.; Chothia, C. Small-molecule metabolism: An enzyme mosaic. *Trends Biotechnol.* **2001**, *19*, 482–486. [CrossRef]

24. Kim, H.S.; Mittenthal, J.E.; Caetano-Anollés, G. Widespread recruitment of ancient domain structures in modern enzymes during metabolic evolution. *J. Int. Bioinform.* **2013**, *10*, 214.

25. Caetano-Anollés, G.; Seufferheld, M.J. The coevolutionary roots of biochemistry and cellular organization challenge the RNA world paradigm. *J. Mol. Microbiol. Biotechnol.* **2013**, *23*, 152–177. [CrossRef] [PubMed]

26. Caetano-Anollés, G.; Caetano-Anollés, D. Computing the origin and evolution of the ribosome from its structure—Uncovering processes of macromolecular accretion benefiting synthetic biology. *Comput. Struct. Biotech. J.* **2015**, *13*, 425–447. [CrossRef] [PubMed]

27. Hartwell, L.H.; Hopfield, J.J.; Leibler, S.; Murray, A.W. From molecular to modular cell biology. *Nature* **1999**, *401*, c47–c52. [CrossRef] [PubMed]

28. Grant, T.; Kluge, A.G. Parsimony, explanatory power, and dynamic homology testing. *Syst. Biodivers.* **2009**, *7*, 357–363. [CrossRef]

29. Yang, Z. A space-time process model for the evolution of DNA sequences. *Genetics* **1995**, *139*, 993–1005. [PubMed]

30. Caetano-Anollés, G.; Nasir, A. Benefits of using molecular structure and abundance in phylogenomic analysis. *Front. Genet.* **2012**, *3*, 172. [CrossRef] [PubMed]

31. Illegård, K.; Ardell, D.H.; Elofsson, A. Structure is three to ten times more conserved than sequence—A study of structural response in protein cores. *Proteins* **2009**, *77*, 499–508. [CrossRef] [PubMed]

32. Sober, E.; Steel, M. Time and knowability in evolutionary processes. *Philos. Sci.* **2014**, *81*, 537–557. [CrossRef]

33. Efimov, A.V. Structural trees for protein superfamilies. *Proteins* **1997**, *28*, 241–260. [CrossRef]

34. Kim, K.M.; Caetano-Anollés, G. The evolutionary history of protein fold families and proteomes confirm that the archaeal ancestor is more ancient than the ancestor of other superkingdoms. *BMC Evol. Biol.* **2012**, *12*, 13. [CrossRef] [PubMed]

35. Debès, C.; Wang, M.; Caetano-Anollés, G.; Gräter, F. Evolutionary optimization of protein folding. *PLoS Comput. Biol.* **2013**, *9*, e1002861. [CrossRef] [PubMed]

36. Marsh, J.A.; Teichmann, S.A. Parallel dynamics and evolution: Protein conformational fluctuations and assembly reflect evolutionary changes in sequence and structure. *BioEssays* **2014**, *36*, 209–218. [CrossRef] [PubMed]

37. Bukhari, S.A.; Caetano-Anollés, G. Origin and evolution of protein fold designs inferred from phylogenomic analysis of CATH domain structures in proteomes. *PLoS Comput. Biol.* **2013**, *9*, e1003009. [CrossRef] [PubMed]

38. Ancel, L.W.; Fontana, W. Plasticity, evolvability, and modularity in RNA. *J. Exp. Zool. (Mol. Dev. Evol.)* **2000**, *288*, 242–283. [CrossRef]

39. Siegal, M.L.; Bergman, A. Waddington's canalization revisited: Developmental stability and evolution. *Proc. Natl. Acad. Sci. USA* **2002**, *99*, 10528–10532. [CrossRef] [PubMed]

40. Caetano-Anollés, G.; Kim, H.S.; Mittenthal, J.E. The origin of modern metabolic networks inferred from phylogenomic analysis of protein architecture. *Proc. Natl. Acad. Sci. USA* **2007**, *104*, 9358–9363. [CrossRef] [PubMed]

41. Caetano-Anollés, K.; Caetano-Anollés, G. Structural phylogenomics reveals gradual evolutionary replacement of abiotic chemistries by protein enzymes in purine metabolism. *PLoS ONE* **2013**, *8*, e59300. [CrossRef] [PubMed]

42. Kim, K.M.; Qin, T.; Jiang, Y.-Y.; Chen, L.-L.; Xiong, M.; Caetano-Anollés, D.; Zhang, H.-Y.; Caetano-Anollés, G. Protein domain structure uncovers the origin of aerobic metabolism and the rise of planetary oxygen. *Structure* **2012**, *20*, 67–76. [CrossRef] [PubMed]

43. Caetano-Anollés, G.; Caetano-Anollés, D. An evolutionarily structured universe of protein architecture. *Genome Res.* **2003**, *13*, 1563–1571. [CrossRef] [PubMed]
44. Söding, J.; Lupas, A.N. More than the sum of their parts: On the evolution of proteins from peptides. *Bioessays* **2003**, *25*, 837–846. [CrossRef] [PubMed]
45. Trifonov, E.N.; Frenkel, Z.M. Evolution of protein modularity. *Curr. Opin. Struct. Biol.* **2009**, *18*, 335–340. [CrossRef] [PubMed]
46. Goncearenco, A.; Berezovsky, I.N. Protein function from its emergence to diversity in contemporary proteins. *Phys. Biol.* **2015**, *12*, 045002. [CrossRef] [PubMed]
47. Berezovsky, I.N.; Trifonov, E.N. Van der Waals locks: Loop-n-lock structure of globular proteins. *J. Mol. Biol.* **2001**, *307*, 1419–1426. [CrossRef] [PubMed]
48. Goncearenco, A.; Berezovsky, I.N. Prototypes of elementary functional loops unravel evolutionary connections between protein functions. *Bioinformatics* **2010**, *26*, i497–i503. [CrossRef] [PubMed]
49. Aziz, M.F.; Caetano-Anollés, K.; Caetano-Anollés, G. The early history and emergence of molecular functions and modular scale-free network behavior. *Sci. Rep.* **2016**, *6*, 25058. [CrossRef] [PubMed]
50. Duax, W.L.; Huether, R.; Pletnev, V.; Langs, D.; Addlagatta, A.; Connare, S.; Habegger, L.; Gill, J. Rational genomics I. Antisense open reading frames and codon bias in short oxidoreductase enzymes and the evolution of the genetic code. *Proteins* **2005**, *61*, 900–906. [CrossRef] [PubMed]
51. Wang, M.; Caetano-Anollés, G. The evolutionary mechanics of domain organization in proteomes and the rise of modularity in the protein world. *Structure* **2009**, *17*, 66–78. [CrossRef] [PubMed]
52. Nath, N.; Mitchel, J.O.B.; Caetano-Anollés, G. The natural history of biocatalytic mechanisms. *PLoS Comput. Biol.* **2014**, *10*, e1003642. [CrossRef] [PubMed]
53. Ji, H.F.; Kong, D.X.; Shen, L.; Chen, L.L.; Ma, B.G.; Zhang, H.Y. Distribution patterns of small molecule ligands in the protein universe and implications for origins of life and drug discovery. *Genome Biol.* **2007**, *8*, R176. [CrossRef] [PubMed]
54. Yarus, M. Getting pass the RNA world: The initial Darwinian ancestor. *Cold Spring Harb. Perspect. Biol.* **2010**, *1*, a003590.
55. Goerlich, O.; Foeckler, R.; Holler, L. Mechanism of synthesis of adenosine (5′) tetraphospho (5′) adenosine (AppppA) by aminoacyl-tRNA synthetases. *Eur. J. Biochem.* **1982**, *126*, 135–142. [CrossRef] [PubMed]
56. Gondry, M.; Sauguet, L.; Belin, P.; Thai, R.; Amouroux, R.; Tellier, C.; Tuphile, K.; Jaquet, M.; Braud, S.; Courçon, M.; et al. Cyclodipeptide synthetases are a family of tRNA-dependent peptide-bond-forming enzymes. *Nat. Chem. Biol.* **2009**, *5*, 414–420. [CrossRef] [PubMed]
57. Sun, F.-J.; Caetano-Anollés, G. The origin and evolution of tRNA inferred from phylogenetic analysis of structure. *J. Mol. Evol.* **2008**, *66*, 21–35. [CrossRef] [PubMed]
58. Sun, F.-J.; Caetano-Anollés, G. Evolutionary patterns in the sequence and structure of transfer RNA: A window into early translation and the genetic code. *PLoS ONE* **2008**, *3*, e2799. [CrossRef] [PubMed]
59. Caetano-Anollés, G.; Sun, F.-J. The natural history of transfer RNA and its interactions with the ribosome. *Front. Genet.* **2014**, *5*, 127. [PubMed]
60. Carter, C.W., Jr.; Wolfenden, R. tRNA acceptor stem and anticodon bases form independent codes related to protein folding. *Proc. Natl. Acad. Sci. USA* **2015**, *112*, 7489–7494. [CrossRef] [PubMed]
61. Bloch, D.; McArthur, B.; Widdowson, R.; Spector, D.; Guimarães, R.C.; Smith, J. tRNA-rRNA sequence homologies: A model for the origin of a common ancestral molecule, and prospects for its reconstruction. *Orig. Life* **1984**, *14*, 571–578. [CrossRef] [PubMed]
62. Farias, S.T.; Rêgo, T.G.; José, M.V. Origin and evolution of the peptidyl transferase center from proto-tRNAs. *FEBS Open Bio* **2014**, *4*, 175–178. [CrossRef] [PubMed]
63. Farias, S.T.; Rêgo, T.G.; José, M.V. tRNA core hypothesis for the transition from the RNA world to the ribonucleoprotein world. *Life* **2016**, *6*, 15. [CrossRef] [PubMed]
64. McFarlane, R.J.; Whitehall, S.K. tRNA genes in eukaryotic genome organization and reorganization. *Cell Cycle* **2009**, *8*, 3102–3106. [CrossRef] [PubMed]
65. Dietrich, F.S.; Voegeli, S.; Brachat, S.; Lerch, A.; Gates, K.; Steiner, S.; Mohr, C.; Pöhlmann, R.; Luedi, P.; Choi, S.; et al. The *Ashbya gossypii* genome as a tool for mapping the ancient *Saccharomyces cerevisiae* genome. *Science* **2004**, *304*, 304–307. [CrossRef] [PubMed]
66. Hughes, A.L.; Friedman, R. Transposable element distribution in the yeast genome reflects a role in repeated genomic rearrangement events on an evolutionary time scale. *Genetica* **2004**, *121*, 181–185. [CrossRef] [PubMed]

67. Ojala, D.; Montoya, J.; Attardi, G. tRNA punctuation model of RNA processing in human mitochondria. *Nature* **1981**, *290*, 470–474. [CrossRef] [PubMed]

68. Lestienne, P.; Plumbridge, J.A.; Grunberg-Manago, M.; Blanquet, S. Autogenous repression of *Escherichia coli* threonyl-tRNA synthetase expression in vitro. *J. Biol. Chem.* **1984**, *259*, 5232–5237. [PubMed]

69. Springer, M.; Plumbridge, J.A.; Butler, J.S.; Graffe, M. Autogenous control of *Escherichia coli* threonyl-tRNA synthetase expression in vitro. *J. Mol. Biol.* **1985**, *185*, 93–104. [CrossRef]

70. Butler, J.S.; Springer, M.; Dondon, J.; Grunberg-Manago, M. Posttranscriptional autoregulation of *Escherichia coli* threonyl tRNA synthetase expression in vivo. *J. Bacteriol.* **1986**, *165*, 198–203. [CrossRef] [PubMed]

71. Springer, M.; Graffe, M.; Dondon, J.; Grunberg-Manago, M. tRNA-like structures and gene regulation at the translational level, a case of molecular mimicry in *Escherichia coli*. *EMBO J.* **1989**, *8*, 2417–2424. [PubMed]

72. Moine, H.; Ehresmann, B.; Romby, P.; Ebel, J.P.; Grunberg-Manago, M.; Springer, M.; Ehresmann, C. The translational regulation of threonyl-tRNA synthetase. Functional relationship between the enzyme, the cognate tRNA and the ribosome. *Biochim. Biophys. Acta* **1990**, *1050*, 343–350. [CrossRef]

73. Brunel, C.; Caillet, J.; Lesage, P.; Graffe, M.; Dondon, J.; Moine, H.; Romby, P.; Ehresmann, C.; Ehresmann, B.; Grunberg-Manago, M. Domains of the *Escherichia coli* threonyl-tRNA synthetase translational operator and their relation to threonine tRNA isoacceptors. *J. Mol. Biol.* **1992**, *227*, 621–634. [CrossRef]

74. Romby, P.; Brunel, C.; Caillet, J.; Springer, M.; Grunberg-Manago, M.; Westhof, E.; Ehresmann, C.; Ehresmann, B. Molecular mimicry in translational control of E. coli threonyl-tRNA synthetase gene. Competitive inhibition in tRNA aminoacylation and operator-repressor recognition switch using tRNA identity rules. *Nucleic Acids Res.* **1992**, *20*, 5633–5640. [CrossRef] [PubMed]

75. Gendron, N.; Putzer, H.; Grunberg-Manago, M. Expression of both *Bacillus subtilis* threonyl-tRNA synthetase genes is autogenously regulated. *J. Bacteriol.* **1994**, *176*, 486–494. [CrossRef] [PubMed]

76. Romby, P.; Caille, J.; Ebel, C.; Sacerdot, C. The expression of *E. coli* threonyl-tRNA synthetase is regulated at the translational level by symmetrical operator-repressor interactions. *EMBO J.* **1996**, *15*, 5976–5987. [PubMed]

77. Nogueira, T.; de Smit, M.; Graffe, M.; Springer, M. The relationship between translational control and mRNA degradation for the *Escherichia coli* threonyl-tRNA synthetase gene. *J. Mol. Biol.* **2001**, *310*, 709–722. [CrossRef] [PubMed]

78. Torres-Larios, A.; Dock-Bregeon, A.C.; Romby, P.; Rees, B.; Sankaranarayanan, R.; Caillet, J.; Springer, M.; Ehresmann, C.; Ehresmann, B.; Moras, D. Structural basis of translational control by E. coli threonyl-tRNA synthetase. *Nat. Struct. Biol.* **2002**, *9*, 343–347. [CrossRef] [PubMed]

79. Romby, P.; Springer, M. Bacterial translational control at atomic resolution. *Trends Genet.* **2003**, *19*, 155–161. [CrossRef]

80. Frugier, M.; Giegé, R. Yeast aspartyl-tRNA synthetase binds specifically its own mRNA. *J. Mol. Biol.* **2003**, *331*, 375–383. [CrossRef]

81. Frugier, M.; Ryckelynck, M.; Giegé, R. tRNA-balanced expression of a eukaryal aminoacyl-tRNA synthetase by an mRNA-mediated pathway. *EMBO Rep.* **2005**, *6*, 860–865. [CrossRef] [PubMed]

82. Ryckelynck, M.; Masquida, B.; Giegé, R.; Frugier, M. An intricate RNA structure with two tRNA-derived motifs directs complex formation between yeast aspartyl-tRNA synthetase and its mRNA. *J. Mol. Biol.* **2005**, *354*, 614–629. [CrossRef] [PubMed]

83. Ryckelynck, M.; Giegé, R.; Frugier, M. tRNAs and tRNA mimics as cornerstones of aminoacyl-tRNA synthetase regulations. *Biochimie* **2005**, *87*, 835–845. [CrossRef] [PubMed]

84. Goldberger, R.F. Autogenous regulation of gene expression. *Science* **1974**, *183*, 810–816. [CrossRef] [PubMed]

85. Plumbridge, J.A.; Springer, M. *Escherichia coli* phenylalanyl-tRNA synthetase operon, transcription studies of wild-type and mutated operons on multicopy plasmids. *J. Bacteriol.* **1982**, *152*, 661–668. [PubMed]

86. Byrne, D.; Grzela, R.; Larigue, A.; Audic, S.; Chenivesse, S.; Encinas, D.; Claverie, J.M.; Abergel, C. The polyadenylation site of Mimivirus transcripts obeys a stringent "hairpin rule". *Genome Res.* **2009**, *19*, 1233–1242. [CrossRef] [PubMed]

87. Nasir, A.; Caetano-Anollés, G. A phylogenomic data-driven exploration of viral origins and evolution. *Science Adv.* **2015**, *1*, e1500527. [CrossRef] [PubMed]

88. Caetano-Anollés, D.; Caetano-Anollés, G. Ribosomal accretion, apriorism and the phylogenetic method: A response to Petrov and Williams. *Front. Genet.* **2015**, *6*, 194. [PubMed]

89. Smith, T.F.; Hartman, H. The evolution of Class II aminoacyl-tRNA synthetases and the first code. *FEBS Lett.* **2015**, *589*, 3499–3507. [CrossRef] [PubMed]

90. Guimarães, R.C. Essentials in the life process indicated by the self-referential genetic code. *Orig. Life Evol. Biosph.* **2014**, *44*, 269–277. [CrossRef] [PubMed]
91. Moutiez, M.; Schmitt, E.; Seguin, J.; Thai, R.; Favry, E.; Belin, P.; Mechulam, Y.; Gondry, M. Unraveling the mechanism of non-ribosomal peptide synthesis by cyclodipeptide synthases. *Nat. Commun.* **2014**, *5*, 5141. [CrossRef] [PubMed]
92. Jorgensen, R.A. Epigenetics: Biology's quantum mechanics. *Front. Plant Sci.* **2011**, *2*, 10. [CrossRef] [PubMed]

life

MDPI

Review

Homocysteine Editing, Thioester Chemistry, Coenzyme A, and the Origin of Coded Peptide Synthesis †

Hieronim Jakubowski [1,2]

[1] Department of Microbiology, Biochemistry and Molecular Genetics, New Jersey Medical School, Rutgers University, Newark, NJ 07103, USA; jakubows2@gmail.com or jakubows@rutgers.edu; Tel.: +1-973-972-8733
[2] Department of Biochemistry and Biotechnology, University of Life Sciences, Poznan 60-632, Poland
† Presented at the Banbury Center, Cold Spring Harbor Laboratory, NY meeting on "Evolution of the Translational Apparatus and implication for the origin of the Genetic Code", 13–16 November 2016.

Academic Editor: Koji Tamura
Received: 3 January 2017; Accepted: 3 February 2017; Published: 9 February 2017

Abstract: Aminoacyl-tRNA synthetases (AARSs) have evolved "quality control" mechanisms which prevent tRNA aminoacylation with non-protein amino acids, such as homocysteine, homoserine, and ornithine, and thus their access to the Genetic Code. Of the ten AARSs that possess editing function, five edit homocysteine: Class I MetRS, ValRS, IleRS, LeuRS, and Class II LysRS. Studies of their editing function reveal that catalytic modules of these AARSs have a thiol-binding site that confers the ability to catalyze the aminoacylation of coenzyme A, pantetheine, and other thiols. Other AARSs also catalyze aminoacyl-thioester synthesis. Amino acid selectivity of AARSs in the aminoacyl thioesters formation reaction is relaxed, characteristic of primitive amino acid activation systems that may have originated in the Thioester World. With homocysteine and cysteine as thiol substrates, AARSs support peptide bond synthesis. Evolutionary origin of these activities is revealed by genomic comparisons, which show that AARSs are structurally related to proteins involved in coenzyme A/sulfur metabolism and non-coded peptide bond synthesis. These findings suggest that the extant AARSs descended from ancestral forms that were involved in non-coded Thioester-dependent peptide synthesis, functionally similar to the present-day non-ribosomal peptide synthetases.

Keywords: aminoacyl-tRNA synthetase; homocysteine editing; thioester; coenzyme A; non-coded peptide synthesis; prebiotic chemistry; thioester world; evolution

1. Introduction

Each of the 20 aminoacyl-tRNA synthetases (AARSs) fulfils two important functions in the initial steps in the translation of the Genetic Code: Chemical Activation and Information Transfer. For example, methionyl-tRNA synthetase (MetRS) catalyzes Chemical Activation of the carboxyl group of its cognate amino acid methionine using ATP, which affords Met-AMP bound to the catalytic module of MetRS (Figure 1). The Information Transfer function involves attachment of the activated Met to the 3′ adenosine of tRNA(CAU)^Met according to the rules of the Genetic Code thereby matching Met with its anticodon CAU, which is read by the AUG codon in the mRNA on the ribosome (Figure 1).

AARSs belong to two structurally unrelated classes, of ten enzymes each, which have different catalytic domains indicating their independent evolutionary origin [1,2]. Class I AARSs usually have monomeric structure with a Rossman-fold catalytic domain characterized by the HIGH and KMSKS signature sequences. Class II AARSs have a two- or four-subunit quaternary structure and an antiparallel β sheet catalytic domain with class II-defining motifs. With the exception of LysRS enzymes, which exist as a Class I or Class II structure in different organisms, other AARSs have a class-specific structure in the three domains of life.

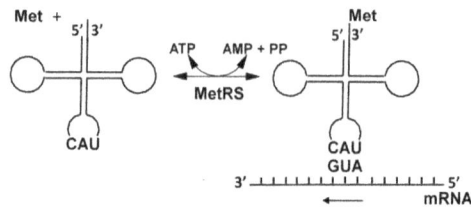

Figure 1. Chemical activation and information transfer by aminoacyl-tRNA synthetases (AARSs).

Catalytic domains of Class I and II AARSs exhibit pronounced differences in their modes of substrate binding. For example, class I AARSs bind ATP in an extended conformation, while class II AARSs bind ATP in a bent conformation with the γ-phosphate folding back over the adenine ring. While Class I AARSs bind tRNA via the minor groove side of its amino acid acceptor stem helix, Class II AARSs bind tRNA via the major groove side. Catalytic domains of Class I and II AARSs exhibit also *functional* differences. Specifically, Class II AARS, such as LysRS, PheRS, HisRS, SerRS, and AspRS, catalyze formation of diadenosine $5',5'''-P^1,P^4$-tetraphosphate (AppppA) [3–5], a signaling molecule that participates in transcriptional regulation of IgE-mediated immune response [6,7]. In contrast, Class I AARSs, such as ArgRS and TrpRS, do not possess the AppppA synthetase activity or have >10–100-fold lower activity (ValRS, MetRS, TyrRS) [4,5,8].

AARSs exhibit high selectivity for their cognate amino acid and tRNA substrates with error rates in the section of amino acids and tRNAs of 10^{-4} to 10^{-5} and 10^{-6}, respectively [9–12]. Although in general unambiguous translation according to the rules of the Genetic Code is crucial for cellular homeostasis, some species have adapted to grow optimally in the presence of ambiguous translation since, under stress conditions, higher error rates assure survival [13–15]. AARSs achieve unambiguous pairing of amino acids with their cognate tRNAs by preferential binding of cognate amino acids and a *quality control* [16] step, in which non-cognate amino acids are selectively edited [10,11,16,17]. The *quality control* step involves either *pre-transfer* or *post-transfer* mechanisms, or both [18]. The major *pre-transfer* mechanism involves hydrolysis of AA~AMP at the catalytic domain, first discovered for ValRS, IleRS, and MetRS [18,19], while *post-transfer* mechanism involves hydrolysis of AA-tRNA at a separate editing domain, originally discovered for IleRS [20] and PheRS [21]. Of the 20 extant AARSs, ten possess an editing function which corrects errors in amino acid selection [22]. For some AARS (IleRS, ValRS, or AlaRS), the editing function is conserved throughout the three domains of life, while editing function of other AARSs (LeuRS, ProRS, or PheRS) is phylogenetically restricted ([23] and references therein).

Editing by AARSs prevents access of *non-proteinogenic* amino acids such as homocysteine [24–28], ornithine [29], homoserine [10,16], or norvaline [30] to the Genetic Code and effectively partitions amino acids present in extant organisms into *proteinogenic* and *non-proteinogenic* amino acids. Natural non-proteinogenic amino acids vastly outnumber the 20 canonical proteinogenic amino acids found in all organisms, plus selenocysteine and pyrrolysine encoded in only some genomes [31]. Hundreds of non-proteinogenic amino acids are known in various species [32]: about 240 in plants, 75 in fungi, 50 in animals, and 50 in prokaryotes [33].

Of the 10 AARS that possess the editing function, five edit the thiol amino acid homocysteine (Hcy) at the catalytic domain: Class I MetRS, LeuRS, IleRS, ValRS, and Class II LysRS [10,11]. Other misactivated amino acids are edited at the catalytic domain, a dedicated editing domain, or both [11,31]. Phylogenetic analyses of structural domains present in proteomes [34] suggest that catalytic domains of AARSs belong to the oldest fold families and may have appeared about 3.7 billion years ago, while separate domains that edit misacylated tRNA appeared later, about 3.2 billion years ago [35] (these timelines are based on counting relative node, i.e., branch point, distance in the rooted trees and using a molecular clock of protein folds to convert relative age into geological time [36]).

Because of their crucial role in the translation and maintenance of the Genetic Code, analysis of AARSs structures and mechanisms of reactions catalyzed by AARSs can provide insights into the origin and evolution of the Genetic Code [35,37]. The present article examines the links between Hcy editing, thioester chemistry, and the origin of the amino acid activation for the coded protein synthesis. Available data suggest that ancestral AARSs were involved in the thiol (coenzyme A, pantetheine) aminoacylation reactions and thioester-based non-coded peptide synthesis before the emergence of the Genetic Code and the ribosomal protein biosynthetic machinery.

2. Homocysteine (Hcy) is Edited by Class I and Class II Aminoacyl-tRNA Synthetases (AARSs)

One of the selectivity problems in protein biosynthesis is discrimination against the non-proteinogenic thiol amino acid Hcy, a universal precursor of methionine. Hcy is misactivated (Reaction (1)) by Class I MetRS, IleRS, LeuRS [25,26], ValRS [18] and class II LysRS [10,11,29].

$$AARS + Hcy + ATP \rightleftharpoons AARS \bullet Hcy \sim AMP + PP_i \qquad (1)$$

Misactivated Hcy is edited by an intramolecular reaction between the side chain thiolate and the activated carboxyl of Hcy, affording the thioester Hcy-thiolactone (Reaction (2)) [18,38].

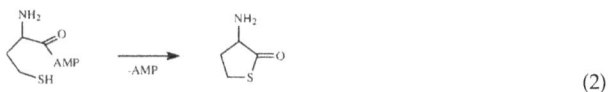

$$(2)$$

Hcy editing does not depend on tRNA [10,18], consumes one mole of ATP per mole Hcy-thiolactone [38], and prevents attachment of Hcy to tRNA, and thus Hcy access to the Genetic Code.

The energy of the anhydride bond of Hcy~AMP is conserved in the thioester bond of Hcy-thiolactone. Consequently, Hcy-thiolactone easily reacts with free amino acids forming Hcy-AA dipeptides [39,40] and with protein lysine residues forming *N*-Hcy-protein [39–41].

2.1. Hcy Editing is Universal

Hcy is an important intermediate in the metabolism of Met, Cys, and one-carbon units carried on folates in *Bacteria* and *Eukarya* [41,42]. Hcy has also been shown to be an intermediate in Met and Cys metabolism in *Archaea* methanogens [43]. Because Hcy is a non-coded amino acid, living organisms must have evolved the ability to prevent its access to the Genetic Code. Indeed, in bacteria (*E. coli*, *M. smegmatis*) [24,44], the yeast *Saccharomyces cerevisiae* [27], plants [45], mice [28], and humans [28,46], Hcy is edited and metabolized to Hcy-thiolactone by MetRS. In *E. coli* and *S. cerevisiae* Hcy thiolactone accumulation is proportional to the expression level of MetRS. In *S. cerevisiae*, both cytoplasmic and mitochondrial MetRSs edit Hcy [47]. Editing of endogenous Hcy by MetRS in cultured microbial and mammalian cells is prevented by supplementation with excess Met. In *E. coli* cultures supplemented with Hcy, two other AARSs LeuRS and IleRS, in addition to MetRS, catalyze Hcy-thiolactone formation [25,26]. As a result, Hcy-thiolactone formation is fully prevented only by simultaneous supplementation with excess Ile, Leu, and Met.

In all organisms, including human, genetic deficiencies in the Hcy/Cys/Met pathways or inadequate supply of cofactors of enzymes participating in Hcy metabolism (folate, cobalamin/vitamin B_{12}, pyridoxal phosphate/vitamin B_6) lead to the accumulation of Hcy and its metabolites, including Hcy-thiolactone, which are implicated in cardiovascular and neurodegenerative diseases [41] through mechanisms involving pro-atherogenic changes in gene expression [48], modification of protein structure [40] leading to amyloid formation [49], activation of mTORC1 signaling and inhibition of autophagy [50], and induction of inflammatory and autoimmune responses [51–53].

Structural similarities between Archaeal and Bacterial MetRSs suggest that Hcy can also be edited by Archaeal MetRSs. For example the catalytic domain of the Archeon *Pyrococcus abyssi* MetRS is very similar to catalytic domains of *E. coli* and *T. thermophilus* MetRSs and can be superimposed with a root mean square deviation (RMSD) values of 1.7 Å for 481 Cα atoms and 1.6 Å for 406 Cα atoms, respectively. Residues important for the synthetic and editing functions of the Bacterial MetRSs are conserved in the Archaeal MetRS and have similar positions in crystal structures [54], including a glutamic acid residue, E259, in *P. abyssi* MetRS homologous to aspartic acid D259 residue in *E. coli* MetRS, which participates as a mechanistic base in the Hcy editing reaction (discussed in Section 2.2.1 below).

2.2. Mechanism of Hcy Editing

Hcy editing is unique in that it involves an *intramolecular* Reaction (2), in which the side chain thiolate of Hcy molecule is a nucleophile, to accomplish editing. Editing reactions of all other amino acids, including a related thio-amino acid Cys, are *intermolecular* and use water hydroxide as a nucleophile [11]. Hcy is edited by the cyclization to Hcy-thiolactone at the synthetic/editing catalytic site in the Rossman-fold domain of class I AARS [55,56] and at the β sheet catalytic domain of Class II LysRS [29,38].

2.2.1. Methionyl-tRNA Synthetase (MetRS)

A model for pre-transfer Hcy editing explains how MetRS partitions Met and Hcy between the synthetic and editing pathways, respectively. The model is supported by the crystal structure [57], structure/function [55,56], and computational [58] studies of *E. coli* MetRS. In the synthetic pathway, the activated carboxyl of Met reacts with the 2′-hydroxyl of the 3′-terminus of tRNAMet, affording Met-tRNAMet. In the editing pathway, the activated carboxyl of Hcy reacts with the thiolate of its side chain, affording Hcy~ thiolactone.

Whether an amino acid substrate completes the synthetic or editing pathway is determined by the partitioning of its side chain between the specificity and thiol-binding subsites [55]. Met completes the synthetic pathway because its side chain is bound by the hydrophobic and hydrogen-bonding interactions with W305 and Y15 residues in the specificity sub-site (Figure 2).

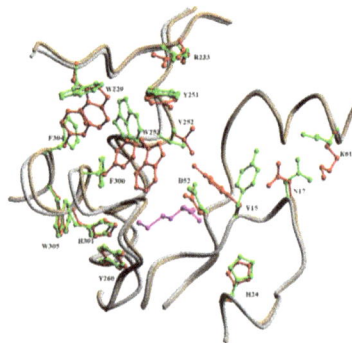

Figure 2. The synthetic/editing active site of *E. coli* methionyl-tRNA synthetase (MetRS): Hydrophobic and hydrogen-bonding interactions provide specificity for the cognate substrate L-methionine. Superimposition of Cα carbon atoms for the MetRS·Met complex (beige) and free MetRS (light grey), solved at 1.8 Å resolution, shows movements of active site residues upon binding of Met. Residue colors are red in the MetRS·Met complex and green in free MetRS, and L-methionine is magenta. Reprinted with permission from reference [57].

In contrast, the side chain of Hcy, missing the methyl group of Met, interacts weakly with the specificity sub-site. This allows the side chain of Hcy to interact with D259 [58] in the thiol-binding sub-site [55], which facilitates editing by cyclization to Hcy-thiolactone (Figure 3). Consistent with this model, mutations of W305 and Y15 residues, which form the hydrophobic Met-binding sub-site, reduce the Hcy/Met discrimination by the enzyme [56].

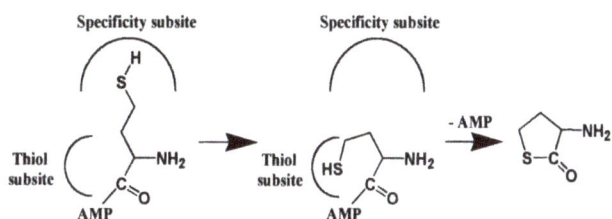

Figure 3. Editing of miscativated homocysteine (Hcy~AMP) at the catalytic module of an AARS: The MetRS-catalyzed cyclization of homocysteinyl adenylate to form Hcy-thiolactone and AMP, which are subsequently released from the synthetic/editing active site of MetRS.

Computational studies [58] suggest that D259 plays an essential role as a mechanistic base that deprotonates the side chain thiol in Hcy~AMP at the catalytic module of MetRS (Figure 4). In the initial MetRS·Hcy~AMP complex the distance $S_{Hcy} \cdots O_{Asp259}$ (5.30 Å) is markedly shorter than the distance $S_{Hcy} \cdots O_{phos}$ in Hcy~AMP (7.15 Å) (Table 1). The rate-limiting step in Hcy-thiolactone formation is the rotation about the $C_\beta - C_\gamma$ bond with 27.5 kJ·mol^{-1} energy barrier. This is more favorable than 98.25 kJ·mol^{-1} energy barrier for an alternative substrate-assisted mechanism in which the non-bridging oxygen of phosphate in Hcy~AMP acts as a base.

Relative Free Energy (kJ mol^{-1})

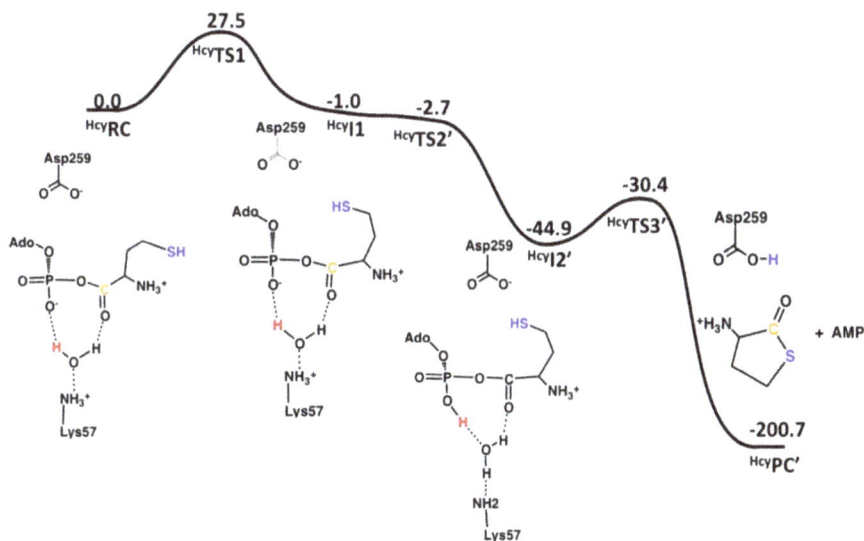

Figure 4. Energetics of Hcy~AMP editing at the catalytic module of MetRS. Asp259 is a mechanistic base that deprotonates the side chain thiol in Hcy~AMP. Reproduced with permission from reference [58].

Table 1. Average distances calculated from molecular dynamics simulation for Hcy~AMP bound in the active site of Class I AARSs.

	Average Distance, Å			
	MetRS	LeuRS	ValRS	IleRS
$S_{Hcy}\cdots O_{Asp/Glu}$	5.30	5.51	4.94	4.70
$S_{Hcy}\cdots C_{carb}$	3.91	5.19	4.14	4.66
$C_{carb}\cdots O_{Asp/Glu}$	4.25	4.39	5.26	4.10
$S_{Hcy}\cdots O_{phos}$	6.99	7.78	7.01	4.25

Cognate Met can also enter the editing pathway when the thiol-binding subsite is occupied by a thiol mimicking the side chain of Hcy [55]. Under these circumstances the activated carboxyl and thiol functions are on separate molecules and MetRS becomes a Met:thiol ligase that catalyzes synthesis of Met thioesters (Figure 5).

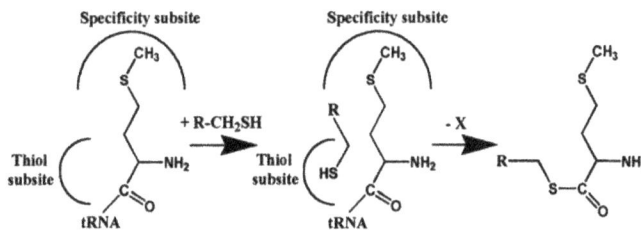

Figure 5. Amino acid:coenzyme A (CoA-SH) ligase activity of MetRS. When the active site is occupied by Met-tRNA or Met-AMP, and the thiol subsite is occupied by CoA-SH, MetRS catalyzes the formation of Met-*S*-CoA thioester. An R represents the bulk of CoA-SH or other thiol molecule.

2.2.2. Leucyl-tRNA Synthetase (LeuRS), Isoleucyl-tRNA Synthetase (IleRS), Valyl-tRNA Synthetase (ValRS)

A similar model explains Hcy editing [25,26] by related Class I AARS. ValRS, LeuRS, and IleRS have active sites, D490, E532, and E550, respectively, that correspond to D359 of MetRS and are similarly positioned with respect to the C_{carb} center of the substrate Hcy~AMP [58]. Computational analyses show that the Hcy~AMP substrate binds in their active sites in a linear conformation similar to that observed for MetRS. The average $C_{carb}\cdots O_{Asp/Glu}$ distances are within 1.16 Å of each other whereas all $S_{Hcy}\cdots C_{carb}$ distances are within 1.24 Å of each other (Table 1). More importantly, for MetRS, LeuRS, and ValRS, the average $S_{Hcy}\cdots O_{Asp/Glu}$ distance is significantly shorter than the average $S_{Hcy}\cdots O_{phos}$ distance by 2.15 Å, 2.27 Å, and 2.06 Å, respectively. Thus, each of these Asp/Glu residues can act as a mechanistic base that deprotonates the side chain thiol in Hcy~AMP and facilitate Hcy-thiolactone formation by MetRS, LeuRS, and ValRS. However, it is less clear whether Glu550 can act as a base in Hcy-thiolactone formation catalyzed by IleRS.

2.2.3. Lysyl-tRNA Synthetase (LysRS) Edits Homocysteine (Hcy), Ornithine (Orn), Homoserine (Hse), but Mischarges tRNA^{Lys} with Proteinogenic Amino Acids

Hcy editing occurs also at the β sheet catalytic domain of Class II *E. coli* LysRS [38], which also edits Orn and Hse. Editing by LysRS is not affected by tRNA^{Lys} and there is no tRNA^{Lys} mischarging with Hcy or Orn [29]. However, LysRS mischarges tRNA^{Lys} with several other amino acids (Arg, Thr, Met, Cys, Leu, Ala, or Ser) and does not deacylate mischarged Arg-tRNA^{Lys}, Thr-tRNA^{Lys}, and Met-tRNA^{Lys}. Recent data show that Met-tRNA^{Lys} (and other tRNAs mischarged with Met) are formed in *E. coli* and mammalian cells in response to stress conditions [13,14].

3. Expanding the Genetic Code: Decoding Methionine Codons by Homocysteine

By preventing the attachment of Hcy to tRNA, the editing reaction assures that Hcy is excluded from the Genetic Code. Weak interactions of the side chain of Hcy with the specificity subsite allow binding of the Hcy side chain to the thiol-binding subsite in the catalytic domain of MetRS [55,56]. Thus, modifications of the side chain of Hcy that increase binding to the specificity subsite should prevent editing and facilitate the transfer to tRNA^{Met}. This is achieved by *S*-nitrosylation of Hcy to *S*-nitroso-Hcy (*S*-NO-Hcy), which binds to the MetRS with affinity 10-fold greater than Hcy and is activated by MetRS to form *S*-NO-Hcy~AMP [59]. In contrast to Hcy-AMP which is edited, *S*-NO-Hcy~AMP is resistant to editing due to stronger binding to the specificity subsite which leads to the transfer of *S*-NO-Hcy to tRNA^{Met}, affording *S*-NO-Hcy~tRNA^{Met} (Figure 6).

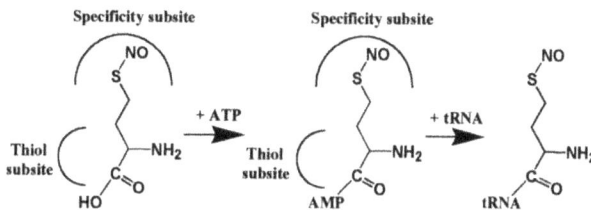

Figure 6. Aminoacylation of tRNA with *S*-NO-Hcy catalyzed by MetRS [59].

The *S*-NO-Hcy-tRNA^{Met} has a similar susceptibility to spontaneous deacylation as Met~tRNA^{Met} with a half-life of 28 min. De-nitrosylation of *S*-nitroso-Hcy~tRNA^{Met} affords Hcy~tRNA^{Met}, the least stable aminoacyl-tRNA known, which spontaneously deacylates with a half-life of 15 s to form Hcy-thiolactone and free tRNA^{Met}.

As expected, *S*-NO-Hcy~tRNA^{Met} is a substrate for protein synthesis on ribosomes, which allows translational incorporation of *S*-NO-Hcy into protein at positions normally occupied by Met [59]. For example, when cultures of *E. coli metE* mutant cells (unable to metabolize Hcy to Met) expressing mouse dihydrofolate reductase (DHFR) protein were supplemented with *S*-NO-Hcy, the DHFR protein was found to contain Hcy. Control experiments in which *E. coli metE* cultures were supplemented with Hcy or Hcy-thiolactone, instead of *S*-NO-Hcy, show that there is no incorporation of Hcy into bacterial proteins [59]. Globin and luciferase, produced in an in vitro mRNA-programmed rabbit reticulocyte protein synthesis system supplemented with *S*-NO-Hcy~tRNA^{Met} contain Hcy at positions normally occupied by Met.

Translationally incorporated Hcy has also been identified in protein from cultured human vascular endothelial cells (HUVECs) (Table 2), which endogenously produce nitric oxide and *S*-nitroso-Hcy [60]. Translationally incorporated Hcy is resistant to Edman degradation whereas post-translationally incorporated Hcy (by the reaction of Hcy-thiolactone with protein lysine residues [40]) is not (Table 2, last row) last, which allows to distinguish between the two mechanisms [61].

Table 2. Translational and post-translational incorporation of Hcy into HUVEC protein [60,62].

Cell Labeling Conditions	Translational		Post-Translational
	[^{35}S]Hcy-Protein	[^{35}S]Met-Protein	εN-[^{35}S]Hcy-Lys-Protein
	%	%	%
[^{35}S]Hcy (10 µM, 50 µCi/mL)	37	25	38
[^{35}S]Hcy + folic acid, 10 µM	<1	>98	<1
[^{35}S]Hcy + HDL, 1 mg/mL	68	25	7
[^{35}S]Hcy + Met, 20 µM	12	76	12
Control, εN-[^{35}S]Hcy-Lys-protein	<4	0	>96

Taken together, these findings show that Hcy can gain an access to the Genetic Code by nitric oxide-mediated invasion of the methionine-coding pathway [59].

4. AARSs Support the Aminoacylation of Thiols and Peptide Bond Synthesis

4.1. Methionyl-tRNA Synthetase (MetRS)

In the model of *pre-transfer* Hcy editing by MetRS, the side chain thiol of Hcy-AMP is bound to D259 at the thiol-binding subsite (Figure 4). When the cognate Met occupies the catalytic site of MetRS, the thiol-binding site is vacant. A prediction of this model is that activated Met (i.e., Met~tRNA or Met~AMP), which binds in the synthetic mode, will enter the editing pathway when the thiol-binding sub-site is filled by a thiol mimicking the side chain of Hcy (Figure 5).

This prediction is confirmed by findings showing that MetRS has the ability to catalyze the transfer of Met from Met~AMP or Met~tRNA to a variety of thiols with the formation of Met thioesters [55,63], i.e., has a Met:CoA-SH ligase activity (Scheme 1).

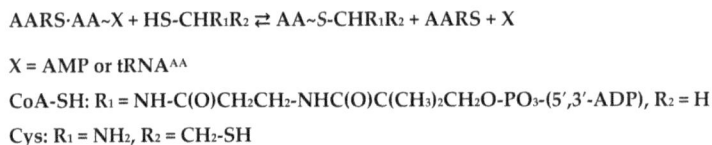

AARS·AA~X + HS-CHR₁R₂ ⇄ AA~S-CHR₁R₂ + AARS + X

X = AMP or tRNA^AA

CoA-SH: R_1 = NH-C(O)CH₂CH₂-NHC(O)C(CH₃)₂CH₂O-PO₃-(5′,3′-ADP), R_2 = H

Cys: R_1 = NH₂, R_2 = CH₂-SH

Scheme 1. Aminoacylation of thiols catalyzed by an AARS.

The thiol-binding site exhibits a remarkable selectivity for coenzyme A (CoA-SH) and cysteine, which are the preferred thiol substrates forming Met~S-CoA thioester and MetCys dipeptide, respectively. The formation of thioesters and dipeptides exhibits saturation kinetics with respect to thiols, characteristic of enzymatic reactions. Rates of the thiol aminoacylation reaction approach the rate of Hcy editing by MetRS (k_{cat} = 1 s⁻¹). The rate enhancement by MetRS of the aminoacyl thioester formation reaction is up to 1,000,000-fold [55]. Pantetheine is a 22-fold less catalytically efficient substrate than CoA-SH, indicating that the adenine nucleotide portion of CoA-SH structure is important for the Met~S-CoA thioester formation. DesulfoCoA, a CoA-SH analogue without the thiol, is not aminoacylated, indicating absolute requirement of the CoA-SH thiol [63].

With Cys as a thiol substrate and with Met~tRNA or Met-AMP in the active site, MetRS catalyzes the aminoacylation of Cys thiol with Met, forming Met~S-Cys thioester, which then rearranges to MetCys dipeptide. Facile intramolecular reaction results from the favorable geometric arrangement of the α-NH₂ of Cys with respect to the thioester bond in Met~S-Cys (Scheme 2). That Met~S-Cys thioester is an intermediate is supported by the finding that Met-S-mercaptopropionate thioester forms with 3-mercaptopropionic acid (an analogue of Cys missing the NH₂ group) as a thiol substrate.

With Hcy as a thiol substrate, MetHcy dipeptide is formed by a similar mechanism [55]. CysGly dipeptide is also aminoacylated by MetRS forming MetCysGly tripetide [55].

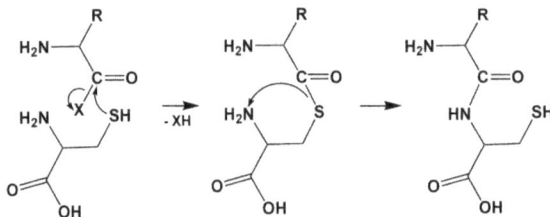

Scheme 2. X = tRNA^Met or AMP; R = side chain of Met or other amino acid.

Active site residues of MetRS that are important for catalysis of the synthetic reactions (Hcy~AMP, Met~AMP, and Met~tRNA formation) and Hcy editing (Hcy-thiolactone formation) are also important for the Met-thioester formation reactions.

That Met and thiols bind at different sites is supported by the non-competitive inhibition of the thioester formation reaction by Met, which affects the k_{cat} but not K_m for Cys [55].

4.2. IleRS, ValRS, LysRS

In addition to MetRS, other Hcy-editing AARSs have the ability to catalyze the synthesis of aminoacyl~*S*-thioesters and aminoacyl-Cys dipeptides (Table 3). For example, with CoA-SH as a thiol, substrate Class I IleRS [55,64] and ValRS [64], and Class II LysRS [64], catalyze the synthesis of Ile~*S*-CoA, Val~*S*-CoA, and Lys~*S*-CoA thioesters, respectively. With Cys as a thiol substrate, ValRS, IleRS, and LysRS catalyze the synthesis of ValCys [65], IleCys [65,66], LysCys [38] dipeptides, respectively (Scheme 2).

However, in contrast to their high selectivity in the tRNA aminoacylation reaction, AARSs exhibit relaxed amino acid selectivity in the CoA-SH aminoacylation reaction (Table 4). For example, IleRS aminoacylates CoA-SH with noncognate Val, Leu, Thr, Ala, and Ser in addition to the cognate isoleucine. The catalytic efficiencies for the non-cognate amino acids are only 10-fold (for Val) to 370-fold (for Thr) lower than the catalytic efficiency for the cognate isoleucine in the CoA-SH aminoacylation reaction. ValRS is even less selective and catalyzes aminoacylation of CoA-SH with threonine, alanine, serine, and isoleucine, in addition to valine (Table 4) with catalytic efficiencies only 2.4-fold (for Thr) to 28-fold (for Ile) lower than the catalytic efficiency for the cognate Val.

Table 3. Hcy editing, thiol aminoacylation, and peptide synthesis by AARSs [10,11].

AARS	Hcy Editing	Thiol-Binding Site Residue	Thiol Aminoacylation	Peptide Synthesis	References
			Class I		
MetRS	Yes	Asp359	Yes	Yes	[24–27,47,55,56,58,63,67]
LeuRS	Yes	Glu532	?	?	[25,26,58]
IleRS	Yes	Glu550	Yes	Yes	[25,26,58,63,64,66]
ValRS	Yes	Asp490	Yes	Yes	[58,64,65]
CysRS	No	?	Yes	?	[68]
ArgRS	No	?	Yes	Yes	[65]
			Class II		
SerRS	No	?	Yes	Yes	[38,63]
AspRS	No	?	Yes	Yes	[38,63]
LysRS	Yes	?	Yes	Yes	[29,38]

?—Not examined.

Table 4. Class I IleRS and ValRS, and Class II LysRS exhibit relaxed amino acid selectivity in the CoA-SH aminoacylation reaction.

Amino Acid	IleRS, k_{cat}/K_M (min$^{-1}\cdot$M^{-1})	ValRS, k_{cat}/K_M (min$^{-1}\cdot$M^{-1})	LysRS, v (μM/h)
Isoleucine	440,000	115	0.24
Valine	45,000	3300	0.67
Leucine	7700	<5	2.7
Threonine	1200	1400	2.7
Serine		170	<0.1
Alanine	1300	500	1.7
Lysine			1.9

LysRS is the least selective and aminoacylates CoA-SH with noncognate leucine, threonine, alanine, valine, and isoleucine, in addition to the cognate lysine. *The rates of transfer of noncognate leucine and threonine are faster than the rate of lysine transfer to CoA-SH, while other noncognate amino acids are transferred by LysRS at rates only 1.1- to 8-fold slower.* The aminoacylation of CoA-SH with noncognate amino acids is prevented by the cognate amino acid, indicating that the ability to aminoacylate CoA-SH with a noncognate amino acid is an inherent property of IleRS, ValRS, and LysRS [64]. Apparently, binding of CoA-SH relaxes the selectivity of the amino acid substrate-binding site, whereas binding of tRNA increases the selectivity towards the cognate amino acid. The relaxed amino acid selectivity is indicative of a more primitive aminoacylation system involving CoA-SH that may have originated in a Thioester World (Section 9).

4.3. Arginyl-tRNA Synthease (ArgRS), Aspartyl-tRNA Synthetase (AspRS), Seryl-tRNA Synthetases (SerRS)

Surprisingly, AARSs that do not misactivate/edit Hcy also have the ability to catalyze the thiol aminoacylation reactions (Table 3). For example, Class I ArgRS [65] catalyzes aminoacylation of Cys, cysteamine, Hcy, and other thiols. With Cys or Hcy as thiol substrates ArgRS catalyzes the synthesis of ArgCys and ArgHcy dipeptides with Arg~*S*-Cys and Arg~*S*-Hcy thioesters as intermediates formed by a mechanism analogous to that depicted in Scheme 2 for MetRS. The rate enhancement by ArgRS of the Arg-thioester formation reaction with Cys is >3300-fold. Remarkably, the aminoacylation of Cys exhibits some degree of stereospecificity with a 6-fold preference for L-Cys vs. D-Cys. Further, ArgRS has the ability to add Arg to *N*-terminal Cys of Cys-Gly dipeptide forming ArgCysGly tripeptide [65].

Class II AspRS catalyzes aminoacylation of CoA-SH, pantetheine, Cys, and Hcy with Asp, forming Asp~*S*-CoA and Asp~*S*-pantetheine [63] thioesters, and AspCys [38,63] and AspHcy [38] dipeptides, respectively. Although other thiols are aminoacylated with Asp by AspRS, CoA-SH, pantetheine, and cysteine are the preferred substrates [63]. Similar to ArgRS [65], the AspRS-catalyzed aminoacylation of Cys is stereospecific, with a 10-fold preference for L-Cys vs. D-Cys.

Class II SerRS [63] catalyzes aminoacylation of Cys and other thiols. Aminoacylation of Cys catalyzed by SerRS leads to the formation of SerCys dipeptide with Ser~*S*-Cys thioester as an intermediate (Scheme 2). The rate enhancements by SerRS, AspRS, and LysRS of the aminoacyl thioester formation reaction are up to 30,000-fold [38]. Thus, the ability to aminoacylate CoA-SH, Cys, Hcy, and other thiols is a general feature of the catalytic domains of both Class I and Class II AARSs [10,11,62].

The thioester chemistry of AARSs underlying their ability to support synthesis of dipeptides, is reminiscent of the adenylation domains of non-ribosomal thio-template peptide synthetases [69,70], which belong to the ANL superfamily that also includes firefly luciferase and acyl-CoA synthetases, but is structurally unrelated to AARS families. However, as discussed in Sections 5–7 below, genomic comparisons reveal that AARSs are *structurally related* to enzymes participating in coenzyme A/sulfur metabolism and peptide bond synthesis, which has important evolutionary implications and sheds a light on the origin of coded peptide synthesis, discussed in Section 9.

5. Class I AARS Are Related to Proteins Involved in Sulfur/CoA Metabolism

The ribosomal translation apparatus is the most conserved component of cellular metabolism [1,37,71]. Its core components, including Class I and Class II AARSs and ribosomes, belong to the oldest fold families that can be traced back to the last universal common ancestor (LUCA) of life and were the earliest systems to evolve into a form close to that of the present day before the divergence of the three kingdoms of life from the LUCA [1,34,35,72]. Phylogenetic analyses of structural domains present in proteomes [34] and of substructures of RNA molecules [35] suggest that catalytic domains of AARSs have appeared before the ribosome [34,35]. The evolution of AARS from their respective common ancestors must have predated the LUCA [73,74].

The catalytic domain of Class I AARSs includes a three-layered $\alpha/\beta/\alpha$ domain with five core β-strands in the order 3-2-1-4-5 surrounded by four α-helices (Figure 7). The signature HIGH motif is located in the loop between strand 1 and helix 1. Class I AARSs also have an extension that consists of a loop, containing the KMSKS motif, followed by two helices C-terminal to the $\alpha/\beta/\alpha$ domain. The HIGH and KMSKS motifs are also present in a diverse family of nuclotidyltransferases. Genomic and structural comparisons show that catalytic domains of Class I AARSs and nucleotidyltransferases form a distinct HIGH superfamily that is related to two other superfamilies: PP-ATPases, a diverse superfamily of domains that catalyze reactions involving hydrolysis of the α-β pyrophosphate bond in ATP, and USPA-like group, that includes USPA domains, electron transport flavoproteins, and photolyases. Together these superfamilies comprise a distinct class of $\alpha/\beta/\alpha$ domains designated the HUP (HIGH-signature proteins, UspA, and PP-ATPase) domain [73] (Figure 7). Similarities between these protein families were initially detected using structural comparisons but can also be detected at the sequence level. A structure-based multiple sequence alignment shows that specific features of the HUP domain include a core of five β-strand sheet in the 3-2-1-4-5 order and four α-helices, horizontal depression of β-strand 3 with respect to the rest of the sheet, crossover of β-strand 4 and 5 or their extensions, a sequence motif corresponding to β-strand 4 with a conserved small amino acid residue at its C-terminus, and ATP ligand (Figure 7). These features distinguish the HUP family from other domains with a Rossman-fold-like geometry (Figure 8) and suggest a monophyletic origin of the HUP domains [73].

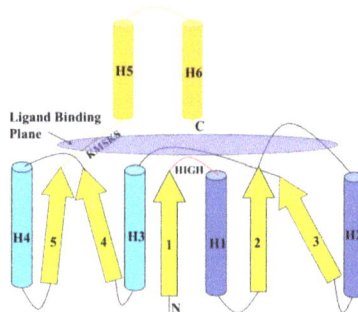

Figure 7. Topological display of the catalytic core of the HUP (HIGH-signature proteins, UspA, and PP-ATPase) domain proteins. Adapted with permission from reference [73].

In addition to Class I AARSs, the HIGH superfamily family includes protein families involved in CoA-SH biosynthesis such as panthotenate synthase (pantoate-β-alanine ligase) PanC [75], phosphopantetheine adenosyltransferase PPAT [76], and sulfate metabolism (sulfate adenylyltransferase SAT1) [73]. The HIGH superfamily is a distinct lineage within the HUP family (Figure 8), characterized by the HIGH motif between β-strand 1 and α-helix 1, and a two-helix extension at the C-terminus of the core domain via a loop with a KMSKS or related sequence involved in nucleotide binding (Figure 7). The second lysine of the KMSKS motif is almost always present in Class I AARSs but absent in other members of the HIGH superfamily, consistent with different modes of interaction with ATP phosphates.

An evolutionary link between Class I AARSs and enzymes of CoA metabolism is revealed by a structure-based alignment of amino acid sequences of PanC, PPAT, GluRS, GlnRS, and TyrRS, which shows that sequences corresponding to the HIGH and KMSKS signature motifs are present, although the KMSKS motif is only evident in structural alignment [75]. Serine at position three and basic or hydroxyl function at positions four and five of the KMSKS motif are conserved in all structures. The superposition of crystallographic structures of CoA-SH biosynthetic enzymes (pantothenate synthtase PanC, phosphopantetheine adenylyltransferase PPAT) and Class IAARSs (GluRS, GlnRS, TyrRS) shows that the highest ranking structural matches to the Cα coordinates of the N-terminal domain of *E. coli* PanC are the ATP-binding domains of *T. thermophilus* GluRS, GlnRS, and TyrRS, *E. coli*

GlnRS, *B. stearothermphilus* TyrRS, and *E. coli* PPAT [75]. Although sequence identities are low (10.8% to 14.5%), these structures are very similar with a root mean square deviation (RMSD) of Cα atoms of 1.5 to 2.2 Å over 62–76 residues (Table 5).

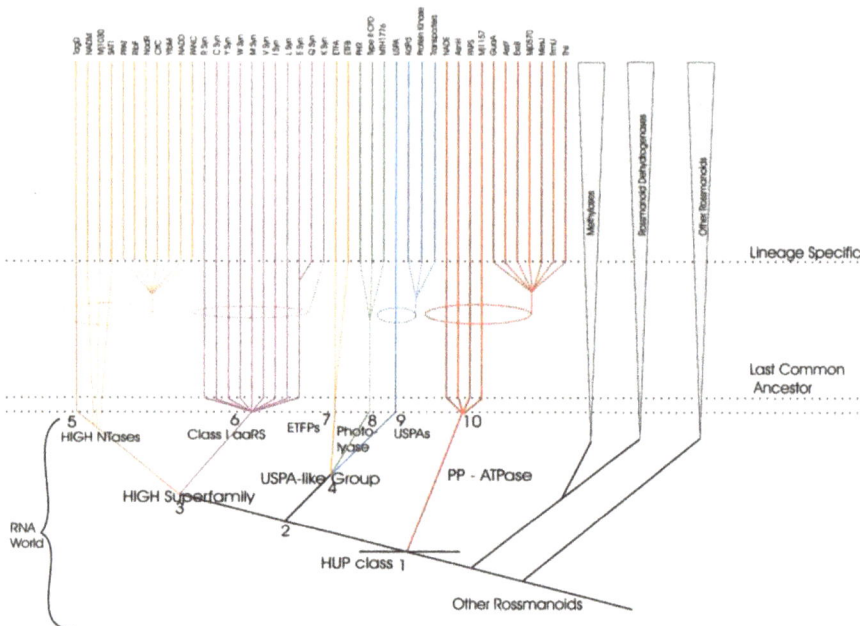

Figure 8. Evolutionary scenario (cladogram) for the HUP domain class (HIGH-signature proteins, UspA, and PP-ATPase). The specific characters associated with each of the nodes as derived through the cladistic analysis are as follows. **Node 1** (the HUP domain): presence of a core 5-strand sheet in the 3–2–1–4–5 order; **Node 2**: configuration of the region between the last helix and strand of the HUP domain, strand 4 and 5 hydrogen-bonded through most of their lengths; **Node 3**: HIGH motif in the loop between strand 1 and helix 1, bihelical extension C-terminal to the core HUP domain, with a secondary role in nucleotide-binding; **Node 4**: loop between strand 3 and helix 4 tends to face outward, helix 4 tends to be placed behind the central sheet, no strongly conserved motif between strand-1 and helix-1, possible loss of ability to hydrolyze α–β bond in ATP; **Node 5**: nucleotidyl transferase activity, two small residues in the KMSKS loop; **Node 6**: AARS activity, classical KMSKS motif; **Node 7**: insertion of a β-hairpin between helix 4 and strand 5; **Node 8**: fusion with large, α-helical C-terminal domain; **Node 9**: distinct loop between strand 4 and helix 4; **Node 10**: SXGXDS motif between strand-1 and helix-1, conserved helix N-terminal to strand 1. The conserved families and their probable temporal points of origin are shown for each of the six major lineages. Dotted ellipses and lines leading to a particular lineage indicate uncertainty regarding its emergence. Reproduced with permission from reference [73].

Table 5. Structural similarity of pantothenate synthetase Pan C to Class I AARSs and PPAT [75].

	GlnRS	GluRS	TyrRS	PPAT
Sequence identity to Pan C (%) *	10.8	14.5	11.2	14.5
RMSD of Cα atoms (Å)	1.8	1.5	2.2	1.8
Number of Cα aligned for RMSD	65	62	68	76

* Calculated for those residues used for RMSD calculation.

Catalytic domains of class I AARSs are also similar to proteins involved in sulfate assimilation for cysteine biosynthesis such as sulfate adenosyltransferase [77] in the HIGH superfamily and 3'-phosphoadenosine-5'-phosphosulfate reductase [78] in the PP-LOOP superfamily [73]. These findings suggest an evolutionary link between class I AARSs and sulfur metabolism.

Analysis of patters of phyletic distribution of distinct families within these major lineages (Figure 8) suggests that the Last Universal Common Ancestor (LUCA) of modern life encoded 15–18 α/β ATPases and nucleotide-binding proteins of the HUP domain. This points to an extensive radiation of HUP domains before the LUCA, during which class I AARSs emerged at a later stage [73]. This also suggests that substantial evolutionary diversification of protein domains occurred well before the present-day version of protein-dependent translation apparatus was established, possibly in the Thioester World (Section 9).

6. AARSs Are Related to Proteins Involved in Peptide Bond Synthesis

6.1. Class I AARSs

Two peptide bond-forming systems related to Class I AARSs have been identified. The first system includes mycothiol synthase MshC, a Class I CysRS paralogue that functions independently of tRNA. MshC activates Cys with ATP, forming Cys~AMP, and then transfers the activated Cys to an amino group of glucosamine in the mycothiol biosynthetic pathway [79,80]. MshC shares 36.1% primary sequence identity to CysRS. Crystallographic structure of MshC is similar to that of CysRS and other Class I AARSs with the catalytic Rossman-fold domain of five-stranded parallel β-sheet surrounded by α-helices [81]. A superposition of MshC and CysRS structures, excluding the anticodon-binding domain of CysRS, shows an RMSD of 2.70 Å for overlapping α-carbon atoms [80].

The second system includes cyclic dipeptide synthetases (CDPs) [82], a group of enzymes belonging to the HUP superfamily of Rosmanoids folds, related to catalytic domains of Class I TyrRS and TrpRS. The CDPSs participate in biosynthesis of biologically active diketopiperazines such as albonoursin (Alb) in *Streptomyces noursei*, pulcherrimin in *Bacillus subtilis*, and mycocyclosin in *Mycobacterium tuberculosis* [82,83]. CDPs do not activate amino acids, but use aminoacyl~tRNAs, synthesized by canonical AARSs, to form cyclodipeptides. The synthesis of the cyclo(Phe-Leu) dipeptide, the first step of the albonoursin [cyclo(α,β-dehydroPhe-α,β-dehydroLeu)] biosynthetic pathway is catalyzed by AlbC, which transfers Phe from Phe~tRNA[Phe] to the conserved Ser37. The Phe-AlbC intermediate reacts with Leu~tRNA[Leu], forming a dipeptidyl-AlbC, which undergoes intramolecular cyclization to generate the cyclo(Phe-Leu) dipeptide [84]. AlbC can also incorporate Tyr and Met from the corresponding AA-tRNAs into cyclodipeptides [82,83].

In some actinomycetes (*Actinosynnema mirum* and *Streptomyces* sp. *AA4*), the AlbC-like CDP genes are associated with genes encoding an acyl-CoA ligase [85]. The acyl-CoA ligase family of enzymes activates carboxyl groups with ATP and then transfers them to CoA-SH, forming acyl~S-CoA thioesters. Thus, this neighborhood association suggests that these CDPS might use aminoacyl~S-CoA (generated by an acyl:CoA-SH ligase) as substrate for dipeptide synthesis, rather than aminoacyl~ tRNAs.

Genomic comparisons also identified bacterial Class I MetRS paralogues [85] that are proposed to participate in the synthesis of a dipeptide through the condensation of Met~AMP with the α-NH$_2$ group of a lysine derivative acetylated at the ϵ-NH$_2$. These MetRS paralogues are often found in operons containing non-ribosomal peptide synthetases and acyl-CoA ligases, which could charge CoA-SH with amino acids for use by the peptide synthetases [85].

6.2. Class II AARSs

Two other peptide bond-forming systems related to Class II AARSs have also been identified. The first system uses aminoacyl~tRNA while the second uses aminoacyl~AMP for the peptide bond formation. The first system includes a SerRS paralogue VlmL while the second system includes biotin-protein ligase BirA, a SerRS paralogue; PoxA, a LysRS paralogue; and an AspRS/AsnRS paralogue AsnA.

VlmL is a dedicated SerRS that produces Ser~tRNA^{Ser} that is used by VlmA in transferring serine to isobutylhydroxylamine in biosynthesis of the antibiotic valanimycin by *Streptomyces viridifaciens* [86]. VlmL is essential and cannot be substituted by the canonical SerRS, suggesting that VlmL-VlmA complex formation is required for the VlmA function.

BirA is a protein ligase that attaches biotin to $N\varepsilon$-amino group of lysine residues of metabolic proteins involved in caboxylation or decarboxylation, e.g., acyl-CoA carboxylase [87]. Biotin is activated with ATP, forming BirA•biotin~AMP. The catalytic domain of BirA (residues 60–270) includes seven-stranded β-sheet and five α-helices with topology identical [87] to that observed in the catalytic domain of SerRS [88]. The RMSD for the superposition of 31 α-carbon atoms in the sheet of both structures is very low, 1.17 Å, indicating that the twist and curvature of the β-sheet are very similar. Two α-helices on sides of the sheet are also topologically equivalent and occur in the same sequence order. The binding sites for biotin in BirA and Ser~AMP analogue in SerRS occupy equivalent positions with respect to the β-sheet. Structural and functional similarities between BirA and SerRS suggest that their catalytic domains diverged from a common ancestor [87].

PoxA is a LysRS paralogue, the first known AARS paralogue that modifies a protein with an amino acid, transfers β-lysine to the ε-NH_2 group of a conserved Lys34 residue of translation elongation factor P [89,90]. β-Lysine is first activated with ATP, forming β-Lys~AMP.

AsnA is an AspRS/AsnRS paralogue [91] that activates β-carboxyl of aspartic acid with ATP to form β-Asp~AMP and then transfers β-aspartate to ammonia forming asparagine.

7. Thioester Chemistry of SerRS Homologues from Methanogenic Archaea

Genome sequences of methanogenic archaea encode AA:CP ligases homologous to the catalytic domain of Class II SerRS [92], which aminoacylate the thiol of phosphopantetheine prosthetic group of a carrier protein (CP) with Gly or Ala, but do not aminoacylate tRNA. CP is encoded in the same operon that encodes the AA:CP ligase in the bacterial chromosome. These AA:CP ligases support the aminoacyl-thioester formation reaction similar to the CoA-SH/pantetheine aminoacylation reactions catalyzed by canonical AARSs (Scheme 1). Crystallographic studies show the pantetheine thiol arm enters the AA:CP ligase active site from the opposite direction relative to the entry of the 3'-end of tRNA in the canonical SerRS, a mode of interaction predicted for CoA-SH and tRNA within the active sites of AARSs. Further, similar to AARSs, AA:CP ligases catalyze aminoacylation of free CoA-SH and Cys with Ala and Gly, forming AA~S-CoA (Scheme 1) and AA-Cys dipeptides (Scheme 2), respectively [93], indicating that AA:CP ligases are both structurally *and* functionally related to canonical class II SerRS. These similarities suggest that catalytic domains of present-day AARSs may have evolved from ancestral forms that functioned as AA:CoA-SH ligases and represent molecular fossils that originate from an ancient catalytic domain that utilized thioester chemistry to activate amino acids for non-coded peptide bond synthesis, before acquiring the ability to aminoacylate tRNA. The AA:CoA-SH ligase activities of extant AARSs and CPs appear to be vestiges of an evolutionary link between thioester-dependent and RNA-dependent peptide synthesis [63,64].

8. Acyl~S-CoA Thioesters and Aminoacyl~tRNA Esters are Used in Peptide Bond Synthesis Catalyzed the Gcn5-related N-acetyltransferases (GNAT) Fold Enzymes

The Gcn5-related *N*-acetyltransferases (GNAT) are a superfamily of enzymes that are universally distributed in nature and use acyl~S-CoA to acylate their cognate substrates [94]. These include histone acetyltransferases, ribosomal protein S18 Nα-acetyltransferase (RimI), protein *N*-myristoyltransferase, aminoglycoside *N*-acetyltransferases, serotonin *N*-acetyltransferase, glucosamine-6-phosphate *N*-acetyltransferase, mycothiol synthase (MshD), and the Fem family of amino acyl transferases. MshD catalyzes the final step in mycothiol biosynthesis, the acetylation of the α-amino group of cysteine attached to glucosamine part of mycothiol using acetyl~S-CoA as a substrate [95].

Crystallographic studies of the GNAT fold proteins with bound acetyl-*S*-CoA or CoA-SH show that the GNAT fold is a phosphopantetheine-binding domain [95], which contains a central five-β-stranded mixed polarity sheet with four α-helices (Figure 9) [94]. Two of the α-helices lie on top of the β-sheet aligned in parallel with the β-strands. The other two helices are stacked on the bottom of the β-sheet with one α-helix aligned in parallel with the β-strands and the other at a 60° angle. The GNAT fold has two distinct binding sites: one for the pyrophosphate moiety of CoA-SH with a signature motif Q/RxxGxG/A and another site for the pantetheine arm. The pyrophosphate-binding is located at a loop between β-strand 4 and α-helix 3, while the pantetheine arm-binding site is located between β-strands 4 and 5, which splay apart to allow the pantetheine arm to make pseudo β-sheet interactions with the exposed backbone atoms of β-strand 4 (Figure 9).

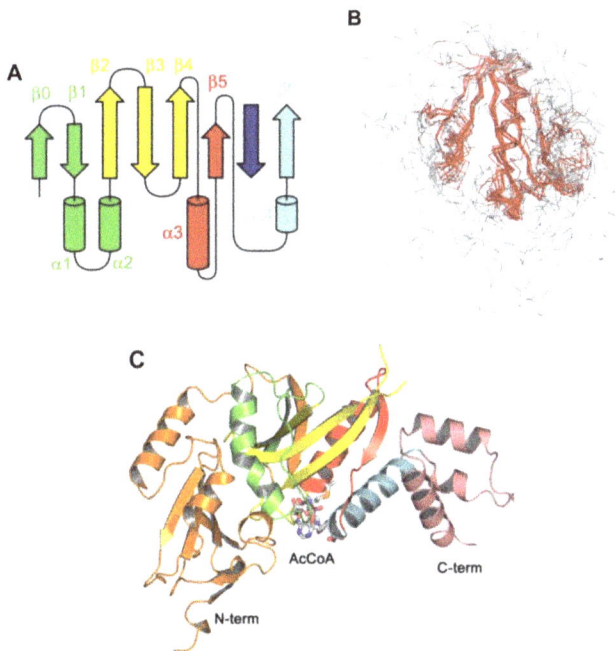

Figure 9. (**A**) Topology of the GNAT fold. From the *N*-terminus, secondary structural elements are colored green (β1, α1, α2), yellow (β2–4), red (α3, β5) and blue (α4, β6). The dark green (β0) *N*-terminal strand is not completely conserved and the deep blue *C*-terminal strand may be from the same monomer, or contributed by another; (**B**) Superposition of 15 GNAT structures. Residues in which the RMSD is <2.7 Å are highlighted in red; (**C**) Yeast Hat1 histone acetyltransferase in complex with Ac-*S*-CoA (1BOB.pdb). Reproduced with permission from reference [94].

However, the GNAT fold of some proteins binds tRNA, instead of CoA-SH [96,97]. For example, FemABX ligases of the GNAT fold (Figure 10) use aminoacyl~tRNAs as substrates for peptide ligation in the cell wall peptidoglycan synthesis in the Gram-positive bacteria. The Fem ligases catalyze the addition of L-amino acids and glycine from aminoacyl-tRNAs synthesized by AARSs to the ε-amino group of lysine in the pentapeptide (L-Ala-D-γ-Glu-L-Lys-D-Ala-D-Ala) of a peptidoglycan. Unfortunately, details of tRNA-GNAT fold interaction are not known, because the structures of any of those proteins in complexes with tRNA are not yet available.

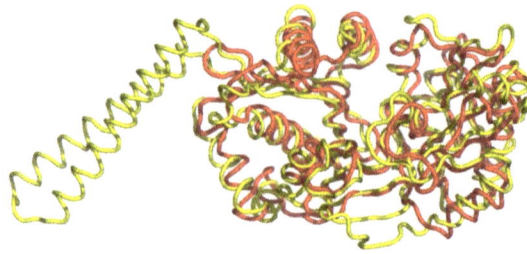

Figure 10. Superposition of FemX (red) with *S. aureus* FemA (yellow, PDB 1LRZ). The overall structures are similar: RMSD is 2.8 Å for the 309 Cα atoms). The major structural difference is the absence in FemX of the coiled-coil region constituted by two helices (residues 246–307; found in SerRS) inserted in FemA domain 2. Reproduced with permission from reference [97].

Another member of the FemABX family, Phe/Leu transferase, uses aminoacyl-tRNA to transfer these amino acids to the *N*-terminal amino acid of proteins, usually lysine or arginine [98,99]. A structure-based search shows that FemX and Phe/Leu transferases have similar structures (Figure 11).

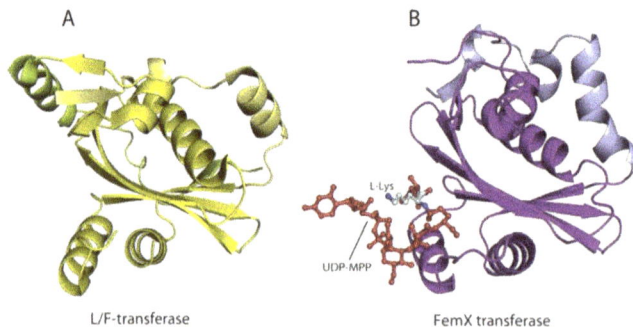

Figure 11. The GNAT superfamily fold in Leu/Phe-tRNA-protein transferase (L/F-transferase) (**A**) and FemX transferase (**B**). Uridine 5′-diphosphate-*N*-acetylmuramoylpentapeptide (UDP-MPP) bound in the FemX transferase structure is colored red and depicted as a ball-and-stick model. The L-Lys of UDP-MPP is shown in pale cyan. The two structures superimpose with RMSD of 2.7 Å. Reproduced with permission from reference [99].

Despite their low sequence identity, 12%, the *C*-terminal domain of Phe/Leu transferase has a core region, formed by six β-strands and four α-helices, found in the GNAT family proteins, which is involved in CoA-SH binding. *C*-terminal domains of Phe/Leu transferase and FemX superimpose with an RMSD of 2.7 Å for the conserved GNAT fold.

The GNAT superfamily proteins provide the second example of a protein fold, the first being the Rossman-fold in AARSs, which can evolve to bind both CoA-SH and tRNA.

9. Evolutionary Implications

As discussed in Section 4, Class I ArgRS and CysRS as well as class II AspRS and SerRS do not possess the Hcy editing function but nevertheless have the ability to catalyze the thiol aminoacylation reaction. This suggests that these AARSs possess a vestigial thiol-binding site, functionally similar to the thiol binding sites of Class I MetRS, ValRS, LeuRS, and IleRS. The ability of extant AARSs to catalyze the synthesis of aminoacyl thioesters and peptides suggests a vestigial thiol-dependent peptide synthesizing function, reminiscent of non-ribosomal peptide synthesis involving thioesters [69,70,100].

The AA:thiol ligase activity of the present-day AARSs suggests an evolutionary link between thioester-dependent and RNA-dependent peptide synthesis. Another example of a link between thioester-dependent and RNA-dependent peptide synthesis is provided by the GNAT fold superfamily of enzymes, discussed in Section 8. Taken together, these findings suggest that ancestral AARSs may have functioned as AA:pantetheine or AA:CoA-SH ligases before acquiring the ability to aminoacylate tRNA. CoA-SH itself appears to be a primitive analogue of tRNA [101] in that both molecules can carry amino acids for peptide bond synthesis [63]. Consistent with this scenario are the findings that IleRS, ValRS, and LysRS exhibit relaxed amino acid selectivity in the CoA-SH aminoacylation reaction, expected of a more primitive amino acid activating system in contrast to essentially absolute selectivity of IleRS and ValRS in the tRNA or RNA-minihelix aminoacylation reactions [64].

9.1. Prebiotic Synthesis of Amino Acids and Organosulfur Compounds

Classical spark discharge experiments have established that amino acids are produced under possible primitive Earth conditions [102]. Organosulfur compounds are also formed under simulated prebiotic conditions [103] (Figure 12). For instance, methionine is formed on a simulated primitive Earth atmosphere containing methane, nitrogen, ammonia, water, and H_2S or CH_3SH subjected to spark discharges [104]. Cysteine forms in gas mixtures containing methane, ethane, ammonia, water, and hydrogen sulfide irradiated with long-wavelength ultraviolet light [105]. Amino acids can also form on pyrite or other metal sulfides under simulated volcanic conditions [106]. However, organic sulfur-containing amino acids are formed more readily in H_2S-containing primitive atmospheres [107].

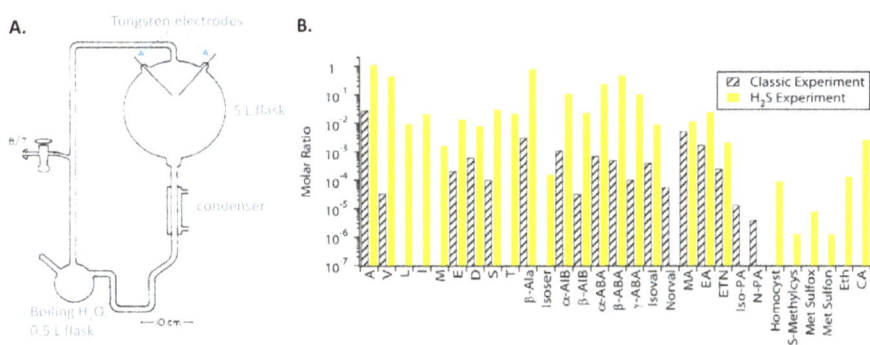

Figure 12. Amino acids and organosulfur compounds produced in the spark discharge experiments. (**A**) Diagram of the spark discharge apparatus (adapted from reference [108]); (**B**) Abundances of amino acids and organosulfur compounds relative to glycine. Letter amino acid abbreviations are used. Reproduced from reference [103] with permission.

Acrolein is also a product of the spark discharge and is thought to be an intermediate in the prebiotic synthesis of methionine and Hcy [104], in addition to homoserine, glutamic acid and α,γ-diaminobutyric acid. Analysis of samples from an unreported 1958 Stanley Miller experiment shows that homocysteic acid, cysteamine, S-methyl-cysteine, ethionine, methionine sulfoxide and sulfone, in addition to methionine, form under prebiotic conditions [109]. Components of CoA-SH (β-alanine, pantoyl lactone cysteamine, and adenosine) are known to be likely prebiotic compounds. The pantetheine moiety of CoA-SH is obtained by mildly heating (40 °C) solutions of pantoyl lactone, β-alanine and cysteamine [110]. These findings support the suggestion that organosulfur compounds and thiols played important roles in prebiotic evolution [111].

Another approach to prebiotic chemistry, called "cyanosulfidic" chemical homologation, and relying on hydrogen cyanide as the sole carbon and nitrogen source, H_2S as a reductant, and ultraviolet light and Cu(I)/Cu(II) catalysis of photo-redox cycling, can generate with high efficiency the building blocks of protein, RNA, and lipids [112] (Figure 13).

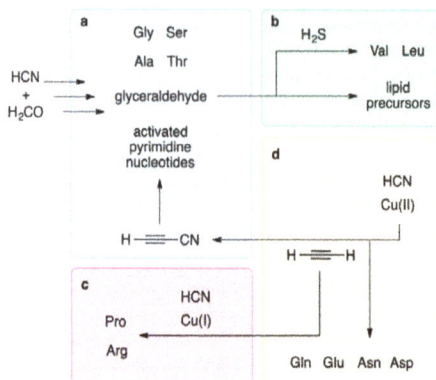

Figure 13. Prebiotic syntheses of amino acid, ribonucelotide, and lipid precursors. Reproduced from reference [113] with permission.

Reductive homologation of HCN (**a**) provides C2 and C3 sugar precursors of amino acids Gly, Ala, Ser, and Thr, as well as of ribonucleotides. Reductive homologation of the products of glyceraldehyde isomerization and reduction leads to lipid precursors, as well as amino acids Val and Leu (**b**). Cu(I) catalyzed cross-coupling followed by reductive homologation gives precursors of Pro and Arg (**c**), while Cu(II)-driven oxidative cross-coupling leads to precursors of Gln, Glu, Asn, and Asp (**d**) [112].

This simple approach inexorably leads to the very set of molecules used by modern biology. Different components may be delivered at different times and places via pools and streams, rainfall, and evaporite basins. Not only does this approach suggest that the set of proteinogenic amino acids might be preordained for life, it also overcomes perceived incompatibilities between the key subsystems and suggests that they could have developed together rather than sequentially [113].

9.2. The Thioester World

Sulfur is an important component of the present-day metabolic pathways. Several coenzymes, including CoA-SH and pantetheine, and two of the 20 proteinogenic amino acids, methionine and cysteine, contain sulfur. CoA-SH is thought to be a molecular fossil from an early metabolic state [114], *"a surviving representative of a group of thiol-amino acid-nucleotide compounds which contained within their structures the potential for both peptide synthesis and nucleotide extension"* and *"a primitive analogue of tRNA"* [101]. Extant AARSs exhibit a vestigial ability to catalyze aminoacylation of CoA-SH [64]. Both tRNA and CoA-SH, or its homologue pantetheine, can carry amino acids for peptide bond synthesis [63]. A more recent finding points to a close metabolic relationship between CoA-SH and RNA. For example, a chemical screen reveals a number of small molecule-RNA conjugates, including 3'-dephospho-CoA-SH and its acetyl, succinyl, and methylmalonyl-thioester derivatives [115]. These CoA species are attached at the 5'-terminus of small (<200 nucleotide) bacterial RNA that have yet to be identified. There are ~100 CoA-RNA molecules per *E. coli* cell [115], which suggests that these species are 10-fold less abundant than Phe-tRNA^{Phe} [116].

Thioester chemistry underlies the participation of CoA-SH in acyl group activation in many biological processes in the three kingdoms of life. As the thioester bond is crucial in biochemical processes in extant organisms, then, by the principle of congruence, it must also have been important in the origin of life [101,111]. It is thought that CoA-thioesters originated very early in the development of life, dating back to the LUCA [117]. Particularly relevant in this context is the role of pantetheine, a precursor of CoA-SH, in the activation of amino acids for non-coded peptide synthesis, which forms the basis for Lipmann's proposal that the thioester-dependent mechanism of peptide bond formation may have preceded the RNA-dependent mechanism of protein synthesis in the development of life [118]. This proposal was taken further by De Duve who pointed out that peptides form spontaneously from Aminoacyl Thioesters in aqueous solution [119] and suggested that Thioesters, which would have provided proto-metabolism with *catalysis* and *energy*, were among early organic molecules that seeded the development of life on the prebiotic Earth, i.e., in the Thioester World that preceded the RNA World [111] (Scheme 3).

Since the original proposal [111], new findings providing strong support to the Thioester World hypothesis have been generated [120]. Comparative genomic, structural, and biochemical studies revealed that *the present-day AARSs contain telltale traces pointing to a crucial role of sulfur and Thioesters in the origin of amino acid activation and peptide bond synthesis.* Hydrogen sulfide was a predominant form of sulfur in the prebiotic world and, as shown by the spark discharge experiments in a simulated prebiotic Earth environment (Figure 12), facilitated prebiotic formation of amino acids and a variety of organosulfur compounds [103], including S-methyl-Cys, homocysteic acid, methionine, Met-sulfoxide, Met-sulfone, ethionine (Eth), and cysteamine (CA) [107].

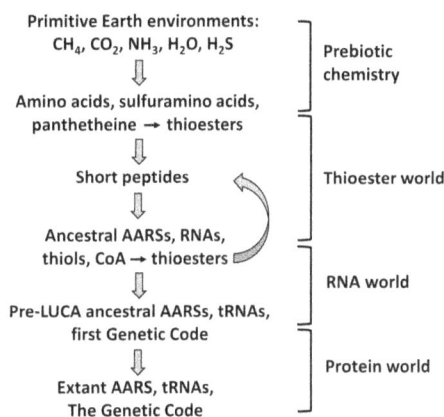

Scheme 3. Major stages in the evolution of life.

Because it forms easily by demethylation of methionine in the presence of iron or copper [121,122] or under acidic conditions [41], Hcy must also have been present in prebiotic environments, as well as the thioester Hcy-thiolactone, which forms easily from Hcy under acidic conditions [123]. Due to its high chemical reactivity towards amine nucleophiles, Hcy-thiolactone could have participated in the peptide bond formation [40]. Although amino acids can form under simulated hot volcanic conditions in an alkaline aqueous $Ni(OH)_2/KCN/CO$ system [124], organosulfur compounds such as cysteine [105], homocysteic acid, cysteamine, and methionine are produced more readily from H_2S-containing primitive atmospheres [107]. Pantetheine, a precursor to CoA-SH, is also formed under simulated prebiotic conditions [110].

Thioesters form easily under acidic conditions [41] and can spontaneously react with amino acids to form peptides [119]. The thioester bond [125] is a high-energy bond and thus is highly reactive towards amino groups [39,41]. Peptides can also form in other plausible prebiotic environments containing carbon monoxide, H_2S or CH_3SH on (Fe,Ni)S or pyrite FeS_2 surfaces [126], or by using a plausible prebiotic condensing agent cyanamide [127]. However, peptide synthesis from thioesters is much more robust [119,128] than from pyrite/methyl sulfide- or cyanamide-dependent reactions.

The thioester chemistry of extant Class I and Class II AARSs (Table 4), and structural/ functional similarities of their catalytic domains to enzymes participating in CoA-SH and sulfur metabolism point to the origin of the amino acid activation and peptide bond synthesis functions in the Thioester World (Scheme 3). These findings strongly suggest that *before the development of coded protein synthesis, catalytic domains of ancestral AARSs may have facilitated the formation of aminoacyl thioestersupported non-coded peptide synthesis.* With Cys and Hcy as thiol substrates, AARSs support the synthesis of AA-Cys and AA-Hcy dipeptides. With Cys-Gly as a thiol substrate tripeptide AA-Cys-Gly forms in reactions catalyzed by AARSs.

Peptides containing *N*-terminal Cys can be extended to longer peptides by condensing with aminoacyl-thioesters in reactions analogous to native chemical ligation exploited in chemical synthesis of proteins [128]. Hcy-(AA)$_n$ peptides are easily formed in reactions of Hcy~thiolactone with amino acids or dipeptides [39]. The $\alpha/\beta/\alpha$-layered and antiparallel β-sheet structures that gave rise to catalytic domains of primordial Class I and Class II AARSs, respectively, may have been assembled from short peptide hairpins [35,37] formed in non-coded thioester-dependent peptide synthesis. Cys-containing peptides must have played an important role in peptide -> protein evolution: Cys residues endow extant proteins with the ability to modify their structure by splicing out specific peptides [129].

With the advent of RNA, which may have been synthesized in catalytic pockets of ancestral AARSs [35] (as suggested by the ability of extant AARSs to catalyze the formation of AppppA and related dinucleoside oligophosphates [4,5]), the function of thiols (CoA-SH, panthethiene) in the non-coded peptide synthesis was expanded to RNA minihelices [64,130]. Before the emergence of the Genetic Code, ancestral AARSs could have facilitated formation of aminoacyl-S-CoA/panthetheine thioesters and aminoacyl-RNA esters for non-coded peptide assembly [63]. The thioester-dependent and RNA-dependent peptide synthesizing systems could have been developing in parallel, possibly via interrelated stages. Vestiges of this stage of prebiotic evolution, i.e., CoA-SH-dependent and RNA-dependent peptide synthesizing activities, are still visible in catalytic domains of extant AARSs [63,64] and the CDPs [82], respectively. The traces of the transition from the Thioester World to RNA World can also be found in the GNAT fold superfamily, which can use acyl~*S*-CoA thioesters or aminoacyl~tRNA esters as substrates for amide or peptide bond synthesis.

10. Conclusions

- Coded peptide synthesis has been preceded by a prebiotic stage, a Thioester World, in which thioesters played key roles.
- Remnants of the Thioester World can be found in extant aminoacyl-tRNA synthetases (AARSs) and related proteins:
- A thiol-binding site at the catalytic domain of AARSs confers the amino acid:thiol ligase activity;
- AARSs are structurally related to proteins involved in sulfur/CoA-SH metabolism and peptide bond synthesis;
- Protein folds that bind pantetheine/CoA-SH can also bind tRNA.

Acknowledgments: I thank L. Aravind, John Blanchard, and Claudine Mayer for providing high resolution images of Figures 7, 9A and 10, respectively. Supported in part by grants from the National Science Center, Poland (2011/02/A/NZ1/00010, 2012/07/B/NZ7/01178, 2013/09/B/NZ5/02794, 2013/11/B/NZ1/00091) and the American Heart Association (12GRNT9420014, 17GRNT32910002).

Conflicts of Interest: The authors declare no conflict of interest. The founding sponsors had no role in the design of the study; in the collection, analyses, or interpretation of data; in the writing of the manuscript, and in the decision to publish the results.

References

1. Woese, C.R.; Olsen, G.J.; Ibba, M.; Soll, D. Aminoacyl-tRNA synthetases, the genetic code, and the evolutionary process. *Microbiol. Mol. Biol. Rev.* **2000**, *64*, 202–236. [CrossRef] [PubMed]

2. Giege, R.; Springer, M. Aminoacyl-tRNA Synthetases in the Bacterial World. *EcoSal Plus* **2016**, *7*. [CrossRef] [PubMed]

3. Plateau, P.; Mayaux, J.F.; Blanquet, S. Zinc(II)-dependent synthesis of diadenosine $5',5'''$-P^1,P^4-tetraphosphate by *Escherichia coli* and yeast phenylalanyl transfer ribonucleic acid synthetases. *Biochemistry* **1981**, *20*, 4654–4662. [CrossRef] [PubMed]

4. Jakubowski, H. Synthesis of diadenosine $5',5'''$-P^1,P^4-tetraphosphate and related compounds by plant (*Lupinus luteus*) seryl-tRNA and phenylalanyl-tRNA synthetases. *Acta Biochim. Pol.* **1983**, *30*, 51–69. [PubMed]

5. Goerlich, O.; Foeckler, R.; Holler, E. Mechanism of synthesis of adenosine(5')tetraphospho(5')adenosine (AppppA) by aminoacyl-tRNA synthetases. *Eur. J. Biochem.* **1982**, *126*, 135–142. [CrossRef] [PubMed]

6. Guo, M.; Schimmel, P. Essential nontranslational functions of tRNA synthetases. *Nat. Chem Biol.* **2013**, *9*, 145–153. [PubMed]

7. Ofir-Birin, Y.; Fang, P.; Bennett, S.P.; Zhang, H.M.; Wang, J.; Rachmin, I.; Shapiro, R.; Song, J.; Dagan, A.; Pozo, J.; et al. Structural switch of lysyl-tRNA synthetase between translation and transcription. *Mol. Cell* **2013**, *49*, 30–42. [CrossRef] [PubMed]

8. Brevet, A.; Chen, J.; Leveque, F.; Plateau, P.; Blanquet, S. In vivo synthesis of adenylylated bis(5'-nucleosidyl) tetraphosphates (Ap4N) by *Escherichia coli* aminoacyl-tRNA synthetases. *Proc. Natl. Acad. Sci. USA* **1989**, *86*, 8275–8279. [CrossRef] [PubMed]

9. Yamane, T.; Hopfield, J.J. Experimental evidence for kinetic proofreading in the aminoacylation of tRNA by synthetase. *Proc. Natl. Acad. Sci. USA* **1977**, *74*, 2246–2250. [CrossRef] [PubMed]

10. Jakubowski, H. Quality control in tRNA charging—Editing of homocysteine. *Acta Biochim. Pol.* **2011**, *58*, 149–163. [PubMed]

11. Jakubowski, H. Quality control in tRNA charging. *Wiley Interdiscip. Rev. RNA* **2012**, *3*, 295–310. [CrossRef] [PubMed]

12. Schulman, L.H. Recognition of tRNAs by aminoacyl-tRNA synthetases. *Prog. Nucleic Acid Res. Mol. Biol.* **1991**, *41*, 23–87. [PubMed]

13. Netzer, N.; Goodenbour, J.M.; David, A.; Dittmar, K.A.; Jones, R.B.; Schneider, J.R.; Boone, D.; Eves, E.M.; Rosner, M.R.; Gibbs, J.S.; et al. Innate immune and chemically triggered oxidative stress modifies translational fidelity. *Nature* **2009**, *462*, 522–526. [CrossRef] [PubMed]

14. Schwartz, M.H.; Waldbauer, J.R.; Zhang, L.; Pan, T. Global tRNA misacylation induced by anaerobiosis and antibiotic exposure broadly increases stress resistance in *Escherichia coli*. *Nucleic Acids Res.* **2016**, *44*, 10292–10303. [PubMed]

15. Ling, J.; Soll, D. Severe oxidative stress induces protein mistranslation through impairment of an aminoacyl-tRNA synthetase editing site. *Proc. Natl. Acad Sci. USA* **2010**, *107*, 4028–4033. [CrossRef] [PubMed]

16. Jakubowski, H.; Goldman, E. Editing of errors in selection of amino acids for protein synthesis. *Microbiol. Rev.* **1992**, *56*, 412–429. [PubMed]

17. Baldwin, A.N.; Berg, P. Transfer ribonucleic acid-induced hydrolysis of valyladenylate bound to isoleucyl ribonucleic acid synthetase. *J. Biol. Chem.* **1966**, *241*, 839–845. [PubMed]

18. Jakubowski, H.; Fersht, A.R. Alternative pathways for editing non-cognate amino acids by aminoacyl-tRNA synthetases. *Nucleic Acids Res.* **1981**, *9*, 3105–3117. [CrossRef] [PubMed]

19. Jakubowski, H. Valyl-tRNA synthetase form yellow lupin seeds: Hydrolysis of the enzyme-bound noncognate aminoacyl adenylate as a possible mechanism of increasing specificity of the aminoacyl-tRNA synthetase. *Biochemistry* **1980**, *19*, 5071–5078. [CrossRef] [PubMed]

20. Eldred, E.W.; Schimmel, P.R. Rapid deacylation by isoleucyl transfer ribonucleic acid synthetase of isoleucine-specific transfer ribonucleic acid aminoacylated with valine. *J. Biol. Chem.* **1972**, *247*, 2961–2964. [PubMed]

21. Yarus, M. Phenylalanyl-tRNA synthetase and isoleucyl-tRNA Phe: A possible verification mechanism for aminoacyl-tRNA. *Proc. Natl. Acad. Sci. USA* **1972**, *69*, 1915–1919. [CrossRef] [PubMed]

22. Perona, J.J.; Gruic-Sovulj, I. Synthetic and editing mechanisms of aminoacyl-tRNA synthetases. *Top. Curr. Chem.* **2014**, *344*, 1–41. [PubMed]

23. SternJohn, J.; Hati, S.; Siliciano, P.G.; Musier-Forsyth, K. Restoring species-specific posttransfer editing activity to a synthetase with a defunct editing domain. *Proc. Natl. Acad. Sci. USA* **2007**, *104*, 2127–2132. [CrossRef] [PubMed]

24. Jakubowski, H. Proofreading in vivo: Editing of homocysteine by methionyl-tRNA synthetase in *Escherichia coli*. *Proc. Natl. Acad. Sci. USA* **1990**, *87*, 4504–4508. [CrossRef] [PubMed]

25. Jakubowski, H. Proofreading in vivo. Editing of homocysteine by aminoacyl-tRNA synthetases in *Escherichia coli*. *J. Biol. Chem.* **1995**, *270*, 17672–17673. [PubMed]

26. Sikora, M.; Jakubowski, H. Homocysteine editing and growth inhibition in *Escherichia coli*. *Microbiology* **2009**, *155*, 1858–1865. [CrossRef]

27. Jakubowski, H. Proofreading in vivo: Editing of homocysteine by methionyl-tRNA synthetase in the yeast Saccharomyces cerevisiae. *EMBO J.* **1991**, *10*, 593–598. [PubMed]

28. Chwatko, G.; Boers, G.H.; Strauss, K.A.; Shih, D.M.; Jakubowski, H. Mutations in methylenetetrahydrofolate reductase or cystathionine beta-synthase gene, or a high-methionine diet, increase homocysteine thiolactone levels in humans and mice. *FASEB J.* **2007**, *21*, 1707–1713. [CrossRef] [PubMed]

29. Jakubowski, H. Misacylation of tRNALys with noncognate amino acids by lysyl-tRNA synthetase. *Biochemistry* **1999**, *38*, 8088–8093. [CrossRef] [PubMed]

30. Cvetesic, N.; Palencia, A.; Halasz, I.; Cusack, S.; Gruic-Sovulj, I. The physiological target for LeuRS translational quality control is norvaline. *EMBO J.* **2014**, *33*, 1639–1653. [CrossRef] [PubMed]

31. Bullwinkle, T.; Lazazzera, B.; Ibba, M. Quality control and infiltration of translation by amino acids outside of the genetic code. *Annu. Rev. Genet.* **2014**, *48*, 149–166. [CrossRef] [PubMed]

32. Hunt, S. The Non-Protein Amino Acids. In *Chemistry and Biochemistry of the Amino Acids*; Springer: Wien, Austria, 1985; pp. 55–138.

33. Rosenthal, G.A. *Plant Nonprotein Amino and Imino Acids*; Academic Press: New York, NY, USA, 1982; p. 272.

34. Caetano-Anolles, G.; Kim, K.M.; Caetano-Anolles, D. The phylogenomic roots of modern biochemistry: Origins of proteins, cofactors and protein biosynthesis. *J. Mol. Evol.* **2012**, *74*, 1–34. [CrossRef]

35. Caetano-Anolles, D.; Caetano-Anolles, G. Piecemeal Buildup of the Genetic Code, Ribosomes, and Genomes from Primordial tRNA Building Blocks. *Life* **2016**, *6*. [CrossRef] [PubMed]

36. Wang, M.; Jiang, Y.Y.; Kim, K.M.; Qu, G.; Ji, H.F.; Mittenthal, J.E.; Zhang, H.Y.; Caetano-Anolles, G. A universal molecular clock of protein folds and its power in tracing the early history of aerobic metabolism and planet oxygenation. *Mol. Biol. Evol.* **2011**, *28*, 567–582. [CrossRef] [PubMed]

37. Smith, T.F.; Hartman, H. The evolution of Class II Aminoacyl-tRNA synthetases and the first code. *FEBS Lett.* **2015**, *589*, 3499–3507. [CrossRef] [PubMed]

38. Jakubowski, H. Aminoacyl thioester chemistry of class II aminoacyl-tRNA synthetases. *Biochemistry* **1997**, *36*, 11077–11085. [CrossRef] [PubMed]

39. Jakubowski, H. Metabolism of homocysteine thiolactone in human cell cultures. Possible mechanism for pathological consequences of elevated homocysteine levels. *J. Biol. Chem.* **1997**, *272*, 1935–1942. [PubMed]

40. Jakubowski, H. Protein homocysteinylation: Possible mechanism underlying pathological consequences of elevated homocysteine levels. *FASEB J.* **1999**, *13*, 2277–2283. [PubMed]

41. Jakubowski, H. *Homocysteine in Protein Structure/Function and Human Disease—Chemical Biology of Homocysteine-containing Proteins*; Springer: Wien, Austria, 2013.

42. Guzzo, M.B.; Nguyen, H.T.; Pham, T.H.; Wyszczelska-Rokiel, M.; Jakubowski, H.; Wolff, K.A.; Ogwang, S.; Timpona, J.L.; Gogula, S.; Jacobs, M.R.; et al. Methylfolate Trap Promotes Bacterial Thymineless Death by Sulfa Drugs. *PLoS Pathog.* **2016**, *12*, e1005949. [CrossRef] [PubMed]

43. Rauch, B.J.; Perona, J.J. Efficient Sulfide Assimilation in Methanosarcina acetivorans Is Mediated by the MA1715 Protein. *J. Bacteriol.* **2016**, *198*, 1974–1983. [CrossRef] [PubMed]

44. Jakubowski, H. The determination of homocysteine-thiolactone in biological samples. *Anal. Biochem.* **2002**, *308*, 112–119. [CrossRef]

45. Jakubowski, H.; Guranowski, A. Metabolism of homocysteine-thiolactone in plants. *J. Biol. Chem.* **2003**, *278*, 6765–6770. [CrossRef] [PubMed]

46. Chwatko, G.; Jakubowski, H. Urinary excretion of homocysteine-thiolactone in humans. *Clin. Chem.* **2005**, *51*, 408–415. [CrossRef] [PubMed]

47. Senger, B.; Despons, L.; Walter, P.; Jakubowski, H.; Fasiolo, F. Yeast cytoplasmic and mitochondrial methionyl-tRNA synthetases: Two structural frameworks for identical functions. *J. Mol. Biol.* **2001**, *311*, 205–216. [CrossRef] [PubMed]

48. Gurda, D.; Handschuh, L.; Kotkowiak, W.; Jakubowski, H. Homocysteine thiolactone and N-homocysteinylated protein induce pro-atherogenic changes in gene expression in human vascular endothelial cells. *Amino Acids* **2015**, *47*, 1319–1339. [CrossRef] [PubMed]

49. Paoli, P.; Sbrana, F.; Tiribilli, B.; Caselli, A.; Pantera, B.; Cirri, P.; De Donatis, A.; Formigli, L.; Nosi, D.; Manao, G.; et al. Protein N-homocysteinylation induces the formation of toxic amyloid-like protofibrils. *J. Mol. Biol.* **2010**, *400*, 889–907. [CrossRef] [PubMed]

50. Khayati, K.; Antikainen, H.; Bonder, E.M.; Weber, G.F.; Kruger, W.D.; Jakubowski, H.; Dobrowolski, R. The amino acid metabolite homocysteine activates mTORC1 to inhibit autophagy and form abnormal proteins in human neurons and mice. *FASEB J.* **2017**, *31*, 598–609. [CrossRef] [PubMed]

51. Keating, A.K.; Freehauf, C.; Jiang, H.; Brodsky, G.L.; Stabler, S.P.; Allen, R.H.; Graham, D.K.; Thomas, J.A.; van Hove, J.L.; Maclean, K.N. Constitutive induction of pro-inflammatory and chemotactic cytokines in cystathionine beta-synthase deficient homocystinuria. *Mol. Genet. Metab.* **2011**, *103*, 330–337. [CrossRef] [PubMed]

52. Undas, A.; Perla, J.; Lacinski, M.; Trzeciak, W.; Kazmierski, R.; Jakubowski, H. Autoantibodies against N-homocysteinylated proteins in humans: Implications for atherosclerosis. *Stroke* **2004**, *35*, 1299–1304. [CrossRef] [PubMed]

53. Fang, P.; Zhang, D.; Cheng, Z.; Yan, C.; Jiang, X.; Kruger, W.D.; Meng, S.; Arning, E.; Bottiglieri, T.; Choi, E.T.; et al. Hyperhomocysteinemia potentiates hyperglycemia-induced inflammatory monocyte differentiation and atherosclerosis. *Diabetes* **2014**, *63*, 4275–4290. [CrossRef] [PubMed]

54. Crepin, T.; Schmitt, E.; Blanquet, S.; Mechulam, Y. Three-dimensional structure of methionyl-tRNA synthetase from Pyrococcus abyssi. *Biochemistry* **2004**, *43*, 2635–2644. [CrossRef] [PubMed]

55. Jakubowski, H. The synthetic/editing active site of an aminoacyl-tRNA synthetase: Evidence for binding of thiols in the editing subsite. *Biochemistry* **1996**, *35*, 8252–8259. [CrossRef] [PubMed]

56. Kim, H.Y.; Ghosh, G.; Schulman, L.H.; Brunie, S.; Jakubowski, H. The relationship between synthetic and editing functions of the active site of an aminoacyl-tRNA synthetase. *Proc. Natl. Acad. Sci. USA* **1993**, *90*, 11553–11557. [CrossRef] [PubMed]

57. Serre, L.; Verdon, G.; Choinowski, T.; Hervouet, N.; Risler, J.L.; Zelwer, C. How methionyl-tRNA synthetase creates its amino acid recognition pocket upon L-methionine binding. *J. Mol. Biol.* **2001**, *306*, 863–876. [CrossRef] [PubMed]

58. Fortowsky, G.B.; Simard, D.J.; Aboelnga, M.M.; Gauld, J.W. Substrate-Assisted and Enzymatic Pretransfer Editing of Nonstandard Amino Acids by Methionyl-tRNA Synthetase. *Biochemistry* **2015**, *54*, 5757–5765. [CrossRef] [PubMed]

59. Jakubowski, H. Translational incorporation of S-nitrosohomocysteine into protein. *J. Biol. Chem.* **2000**, *275*, 21813–21816. [CrossRef] [PubMed]

60. Jakubowski, H. Translational accuracy of aminoacyl-tRNA synthetases: Implications for atherosclerosis. *J. Nutr.* **2001**, *131*, 2983S–2987S. [PubMed]

61. Jakubowski, H.; Zhang, L.; Bardeguez, A.; Aviv, A. Homocysteine thiolactone and protein homocysteinylation in human endothelial cells: Implications for atherosclerosis. *Circ. Res.* **2000**, *87*, 45–51. [CrossRef] [PubMed]

62. Jakubowski, H. Accuracy of Aminoacyl-tRNA Synthetases: Proofreading of Amino Acids. In *The Aminoacyl-tRNA Synthetases*; Ibba, M., Francklyn, C., Cusack, S., Eds.; Landes Bioscience/Eurekah.com: Georgetown, TX, USA, 2005; pp. 384–396.

63. Jakubowski, H. Aminoacylation of coenzyme A and pantetheine by aminoacyl-tRNA synthetases: Possible link between noncoded and coded peptide synthesis. *Biochemistry* **1998**, *37*, 5147–5153. [CrossRef] [PubMed]

64. Jakubowski, H. Amino acid selectivity in the aminoacylation of coenzyme A and RNA minihelices by aminoacyl-tRNA synthetases. *J. Biol. Chem.* **2000**, *275*, 34845–34848. [CrossRef] [PubMed]

65. Jakubowski, H. Synthesis of cysteine-containing dipeptides by aminoacyl-tRNA synthetases. *Nucleic Acids Res.* **1995**, *23*, 4608–4615. [CrossRef] [PubMed]

66. Jakubowski, H. Proofreading in trans by an aminoacyl-tRNA synthetase: A model for single site editing by isoleucyl-tRNA synthetase. *Nucleic Acids Res.* **1996**, *24*, 2505–2510. [CrossRef] [PubMed]

67. Jakubowski, H.; Goldman, E. Synthesis of homocysteine thiolactone by methionyl-tRNA synthetase in cultured mammalian cells. *FEBS Lett.* **1993**, *317*, 237–240. [CrossRef]

68. Jakubowski, H. Editing function of *Escherichia coli* cysteinyl-tRNA synthetase: Cyclization of cysteine to cysteine thiolactone. *Nucleic Acids Res.* **1994**, *22*, 1155–1160. [CrossRef] [PubMed]

69. Gulick, A.M. Conformational dynamics in the Acyl-CoA synthetases, adenylation domains of non-ribosomal peptide synthetases, and firefly luciferase. *ACS Chem. Biol.* **2009**, *4*, 811–827. [CrossRef] [PubMed]

70. Finking, R.; Marahiel, M.A. Biosynthesis of nonribosomal peptides. *Annu. Rev. Microbiol.* **2004**, *58*, 453–488. [CrossRef] [PubMed]

71. Hartman, H.; Smith, T.F. The evolution of the ribosome and the genetic code. *Life* **2014**, *4*, 227–249. [CrossRef] [PubMed]

72. Wolf, Y.I.; Aravind, L.; Grishin, N.V.; Koonin, E.V. Evolution of aminoacyl-tRNA synthetases—Analysis of unique domain architectures and phylogenetic trees reveals a complex history of horizontal gene transfer events. *Genome Res.* **1999**, *9*, 689–710. [PubMed]

73. Aravind, L.; Anantharaman, V.; Koonin, E.V. Monophyly of class I aminoacyl tRNA synthetase, USPA, ETFP, photolyase, and PP-ATPase nucleotide-binding domains: Implications for protein evolution in the RNA. *Proteins* **2002**, *48*, 1–14. [CrossRef] [PubMed]

74. Fournier, G.P.; Alm, E.J. Ancestral Reconstruction of a Pre-LUCA Aminoacyl-tRNA Synthetase Ancestor Supports the Late Addition of Trp to the Genetic Code. *J. Mol. Evol.* **2015**, *80*, 171–185. [CrossRef] [PubMed]

75. Von Delft, F.; Lewendon, A.; Dhanaraj, V.; Blundell, T.L.; Abell, C.; Smith, A.G. The crystal structure of *E. coli* pantothenate synthetase confirms it as a member of the cytidylyltransferase superfamily. *Structure* **2001**, *9*, 439–450. [CrossRef]

76. Izard, T. A novel adenylate binding site confers phosphopantetheine adenylyltransferase interactions with coenzyme A. *J. Bacteriol.* **2003**, *185*, 4074–4080. [CrossRef] [PubMed]

77. Yu, Z.; Lansdon, E.B.; Segel, I.H.; Fisher, A.J. Crystal structure of the bifunctional ATP sulfurylase-APS kinase from the chemolithotrophic thermophile Aquifex aeolicus. *J. Mol. Biol.* **2007**, *365*, 732–743. [CrossRef] [PubMed]

78. Yu, Z.; Lemongello, D.; Segel, I.H.; Fisher, A.J. Crystal structure of Saccharomyces cerevisiae 3′-phosphoadenosine-5′-phosphosulfate reductase complexed with adenosine 3′,5′-bisphosphate. *Biochemistry* **2008**, *47*, 12777–12786. [CrossRef] [PubMed]

79. Sareen, D.; Steffek, M.; Newton, G.L.; Fahey, R.C. ATP-dependent L-cysteine:1D-myo-inosityl 2-amino-2-deoxy-alpha-D-glucopyranoside ligase, mycothiol biosynthesis enzyme MshC, is related to class I cysteinyl-tRNA synthetases. *Biochemistry* **2002**, *41*, 6885–6890. [CrossRef] [PubMed]

80. Tremblay, L.W.; Fan, F.; Vetting, M.W.; Blanchard, J.S. The 1.6 A crystal structure of Mycobacterium smegmatis MshC: The penultimate enzyme in the mycothiol biosynthetic pathway. *Biochemistry* **2008**, *47*, 13326–13335. [CrossRef] [PubMed]

81. Newberry, K.J.; Hou, Y.M.; Perona, J.J. Structural origins of amino acid selection without editing by cysteinyl-tRNA synthetase. *EMBO J.* **2002**, *21*, 2778–2787. [CrossRef] [PubMed]

82. Gondry, M.; Sauguet, L.; Belin, P.; Thai, R.; Amouroux, R.; Tellier, C.; Tuphile, K.; Jacquet, M.; Braud, S.; Courcon, M.; et al. Cyclodipeptide synthases are a family of tRNA-dependent peptide bond-forming enzymes. *Nat. Chem. Biol.* **2009**, *5*, 414–420. [CrossRef] [PubMed]

83. Belin, P.; Moutiez, M.; Lautru, S.; Seguin, J.; Pernodet, J.L.; Gondry, M. The nonribosomal synthesis of diketopiperazines in tRNA-dependent cyclodipeptide synthase pathways. *Nat. Prod. Rep.* **2012**, *29*, 961–979. [CrossRef] [PubMed]

84. Moutiez, M.; Schmitt, E.; Seguin, J.; Thai, R.; Favry, E.; Belin, P.; Mechulam, Y.; Gondry, M. Unravelling the mechanism of non-ribosomal peptide synthesis by cyclodipeptide synthases. *Nat. Commun.* **2014**, *5*, 5141. [CrossRef] [PubMed]

85. Aravind, L.; de Souza, R.F.; Iyer, L.M. Predicted class-I aminoacyl tRNA synthetase-like proteins in non-ribosomal peptide synthesis. *Biol. Direct* **2010**, *5*, 48. [CrossRef] [PubMed]

86. Garg, R.P.; Qian, X.L.; Alemany, L.B.; Moran, S.; Parry, R.J. Investigations of valanimycin biosynthesis: Elucidation of the role of seryl-tRNA. *Proc. Natl. Acad. Sci. USA* **2008**, *105*, 6543–6547. [CrossRef] [PubMed]

87. Artymiuk, P.J.; Rice, D.W.; Poirrette, A.R.; Willet, P. A tale of two synthetases. *Nat. Struct. Biol.* **1994**, *1*, 758–760. [CrossRef] [PubMed]

88. Cusack, S.; Berthet-Colominas, C.; Hartlein, M.; Nassar, N.; Leberman, R. A second class of synthetase structure revealed by X-ray analysis of *Escherichia coli* seryl-tRNA synthetase at 2.5 A. *Nature* **1990**, *347*, 249–255. [CrossRef] [PubMed]

89. Yanagisawa, T.; Sumida, T.; Ishii, R.; Takemoto, C.; Yokoyama, S. A paralog of lysyl-tRNA synthetase aminoacylates a conserved lysine residue in translation elongation factor P. *Nat. Struct. Mol. Biol.* **2010**, *17*, 1136–1143. [CrossRef] [PubMed]

90. Zou, S.B.; Roy, H.; Ibba, M.; Navarre, W.W. Elongation factor P mediates a novel post-transcriptional regulatory pathway critical for bacterial virulence. *Virulence* **2011**, *2*, 147–151. [CrossRef] [PubMed]

91. Blaise, M.; Frechin, M.; Olieric, V.; Charron, C.; Sauter, C.; Lorber, B.; Roy, H.; Kern, D. Crystal structure of the archaeal asparagine synthetase: Interrelation with aspartyl-tRNA and asparaginyl-tRNA synthetases. *J. Mol. Biol.* **2011**, *412*, 437–452. [CrossRef] [PubMed]

92. Mocibob, M.; Ivic, N.; Bilokapic, S.; Maier, T.; Luic, M.; Ban, N.; Weygand-Durasevic, I. Homologs of aminoacyl-tRNA synthetases acylate carrier proteins and provide a link between ribosomal and nonribosomal peptide synthesis. *Proc. Natl. Acad. Sci. USA* **2010**, *107*, 14585–14590. [CrossRef] [PubMed]

93. Mocibob, M.; Ivic, N.; Luic, M.; Weygand-Durasevic, I. Adaptation of aminoacyl-tRNA synthetase catalytic core to carrier protein aminoacylation. *Structure* **2013**, *21*, 614–626. [CrossRef] [PubMed]

94. Vetting, M.W.; LP, S.d.C.; Yu, M.; Hegde, S.S.; Magnet, S.; Roderick, S.L.; Blanchard, J.S. Structure and functions of the GNAT superfamily of acetyltransferases. *Arch. Biochem. Biophys.* **2005**, *433*, 212–226. [CrossRef] [PubMed]

95. Vetting, M.W.; Roderick, S.L.; Yu, M.; Blanchard, J.S. Crystal structure of mycothiol synthase (Rv0819) from Mycobacterium tuberculosis shows structural homology to the GNAT family of *N*-acetyltransferases. *Protein Sci.* **2003**, *12*, 1954–1959. [CrossRef] [PubMed]

96. Benson, T.E.; Prince, D.B.; Mutchler, V.T.; Curry, K.A.; Ho, A.M.; Sarver, R.W.; Hagadorn, J.C.; Choi, G.H.; Garlick, R.L. X-ray crystal structure of Staphylococcus aureus FemA. *Structure* **2002**, *10*, 1107–1115. [CrossRef]

97. Biarrotte-Sorin, S.; Maillard, A.P.; Delettre, J.; Sougakoff, W.; Arthur, M.; Mayer, C. Crystal structures of Weissella viridescens FemX and its complex with UDP-MurNAc-pentapeptide: Insights into FemABX family substrates recognition. *Structure* **2004**, *12*, 257–267. [CrossRef] [PubMed]

98. Watanabe, K.; Toh, Y.; Suto, K.; Shimizu, Y.; Oka, N.; Wada, T.; Tomita, K. Protein-based peptide-bond formation by aminoacyl-tRNA protein transferase. *Nature* **2007**, *449*, 867–871. [CrossRef]

99. Dong, X.; Kato-Murayama, M.; Muramatsu, T.; Mori, H.; Shirouzu, M.; Bessho, Y.; Yokoyama, S. The crystal structure of leucyl/phenylalanyl-tRNA-protein transferase from *Escherichia coli*. *Protein Sci.* **2007**, *16*, 528–534. [PubMed]

100. Linne, U.; Schafer, A.; Stubbs, M.T.; Marahiel, M.A. Aminoacyl-coenzyme A synthesis catalyzed by adenylation domains. *FEBS Lett.* **2007**, *581*, 905–910. [CrossRef] [PubMed]

101. Reanney, D.C. Aminoacyl thiol esters and the origins of genetic specificity. *J. Theor. Biol.* **1977**, *65*, 555–569. [CrossRef]

102. Miller, S.L. A production of amino acids under possible primitive earth conditions. *Science* **1953**, *117*, 528–529. [CrossRef] [PubMed]

103. Bada, J.L. New insights into prebiotic chemistry from Stanley Miller's spark discharge experiments. *Chem. Soc. Rev.* **2013**, *42*, 2186–2196. [CrossRef] [PubMed]

104. Van Trump, J.E.; Miller, S.L. Prebiotic synthesis of methionine. *Science* **1972**, *178*, 859–860. [CrossRef] [PubMed]

105. Khare, B.N.; Sagan, C. Synthesis of cystine in simulated primitive conditions. *Nature* **1971**, *232*, 577–579. [CrossRef] [PubMed]

106. Huber, C.; Eisenreich, W.; Wachtershauser, G. Synthesis of α-amino and α-hydroxy acids under volcanic conditions: Implications for the origin of life. *Tetrahedron Lett.* **2010**, *51*, 1069–1071. [CrossRef]

107. Parker, E.T.; Cleaves, H.J.; Callahan, M.P.; Dworkin, J.P.; Glavin, D.P.; Lazcano, A.; Bada, J.L. Prebiotic synthesis of methionine and other sulfur-containing organic compounds on the primitive Earth: A contemporary reassessment based on an unpublished 1958 Stanley Miller experiment. *Orig. Life Evol. Biosph.* **2011**, *41*, 201–212. [CrossRef] [PubMed]

108. Miller, S.L. Production of some organic compounds under possible primitive Earth conditions. *J. Am. Chem. Soc.* **1955**, *77*, 2351–2361. [CrossRef]

109. Parker, E.T.; Cleaves, H.J.; Dworkin, J.P.; Glavin, D.P.; Callahan, M.; Aubrey, A.; Lazcano, A.; Bada, J.L. Primordial synthesis of amines and amino acids in a 1958 Miller H2S-rich spark discharge experiment. *Proc. Natl. Acad. Sci. USA* **2011**, *108*, 5526–5531. [CrossRef] [PubMed]

110. Keefe, A.D.; Newton, G.L.; Miller, S.L. A possible prebiotic synthesis of pantetheine, a precursor to coenzyme A. *Nature* **1995**, *373*, 683–685. [CrossRef] [PubMed]

111. De Duve, C. *Vital Dust*; Basic Books: New York, NY, USA, 1994.

112. Patel, B.H.; Percivalle, C.; Ritson, D.J.; Duffy, C.D.; Sutherland, J.D. Common origins of RNA, protein and lipid precursors in a cyanosulfidic protometabolism. *Nat. Chem.* **2015**, *7*, 301–307. [CrossRef] [PubMed]

113. Wagner, A.J.; Blackmond, D.G. The Future of Prebiotic Chemistry. *ACS Cent. Sci.* **2016**, *2*, 775–777. [CrossRef] [PubMed]

114. White, H.B., 3rd. Coenzymes as fossils of an earlier metabolic state. *J. Mol. Evol.* **1976**, *7*, 101–104. [CrossRef] [PubMed]

115. Kowtoniuk, W.E.; Shen, Y.; Heemstra, J.M.; Agarwal, I.; Liu, D.R. A chemical screen for biological small molecule-RNA conjugates reveals CoA-linked RNA. *Proc. Natl. Acad. Sci. USA* **2009**, *106*, 7768–7773. [CrossRef] [PubMed]

116. Jakubowski, H.; Goldman, E. Quantities of individual aminoacyl-tRNA families and their turnover in *Escherichia coli*. *J. Bacteriol.* **1984**, *158*, 769–776. [PubMed]

117. Benner, S.A.; Ellington, A.D.; Tauer, A. Modern metabolism as a palimpsest of the RNA world. *Proc. Natl. Acad. Sci. USA* **1989**, *86*, 7054–7058. [CrossRef] [PubMed]

118. Lipmann, F. Attempts to map a process evolution of peptide biosynthesis. *Science* **1971**, *173*, 875–884. [CrossRef] [PubMed]

119. Wieland, T.; Bokelmann, E.; Bauer, L.; Lang, H.U.; Lau, H. Uber Peptidsynthesen. 8. Mitteilung. Bildung von S-haltigen Peptiden durch intramolekulare Wanderung Von Aminoacylresten. *Eur. J. Org. Chem.* **1953**, *583*, 129–149.

120. Jakubowski, H. Aminoacyl-tRNA synthetases and the evolution of coded peptide synthesis: The Thioester World. *FEBS Lett.* **2016**, *590*, 469–481. [CrossRef] [PubMed]

121. Baggott, J.E.; Tamura, T. Iron-dependent formation of homocysteine from methionine and other thioethers. *Eur. J. Clin. Nutr.* **2007**, *61*, 1359–1363. [CrossRef] [PubMed]

122. Lieberman, M.; Kunishi, A.T. Ethylene production from methionine. *Biochem. J.* **1965**, *97*, 449–459. [CrossRef] [PubMed]

123. Jakubowski, H. Facile syntheses of [^{35}S]homocysteine-thiolactone, [^{35}S]homocystine, [^{35}S]homocysteine, and [S-nitroso-^{35}S]homocysteine. *Anal. Biochem.* **2007**, *370*, 124–126. [CrossRef] [PubMed]

124. Huber, C.; Wachtershauser, G. α-Hydroxy and α-amino acids under possible Hadean, volcanic origin-of-life conditions. *Science* **2006**, *314*, 630–632. [CrossRef] [PubMed]

125. Racker, E. Metabolism of thiolesters of glutathione. *Fed. Proc.* **1953**, *12*, 711–714. [PubMed]

126. Huber, C.; Eisenreich, W.; Hecht, S.; Wachtershauser, G. A possible primordial peptide cycle. *Science* **2003**, *301*, 938–940. [CrossRef] [PubMed]

127. Parker, E.T.; Zhou, M.; Burton, A.S.; Glavin, D.P.; Dworkin, J.P.; Krishnamurthy, R.; Fernandez, F.M.; Bada, J.L. A plausible simultaneous synthesis of amino acids and simple peptides on the primordial Earth. *Angew. Chem. Int. Ed. Engl.* **2014**, *53*, 8132–8136. [CrossRef] [PubMed]

128. Dawson, P.E.; Muir, T.W.; Clark-Lewis, I.; Kent, S.B. Synthesis of proteins by native chemical ligation. *Science* **1994**, *266*, 776–779. [CrossRef] [PubMed]

129. Novikova, O.; Topilina, N.; Belfort, M. Enigmatic distribution, evolution, and function of inteins. *J. Biol. Chem.* **2014**, *289*, 14490–14497. [CrossRef] [PubMed]

130. Schimmel, P.; Kelley, S.O. Exiting an RNA world. *Nat. Struct. Biol.* **2000**, *7*, 5–7. [CrossRef] [PubMed]

Article

On the Uniqueness of the Standard Genetic Code

Gabriel S. Zamudio and Marco V. José *

Theoretical Biology Group, Instituto de Investigaciones Biomédicas,
Universidad Nacional Autónoma de México, México D.F. 04510, Mexico; gazaso92@gmail.com
* Correspondence: marcojose@biomedicas.unam.mx; Tel.: +52-5562-3894

Academic Editor: Koji Tamura
Received: 15 December 2016; Accepted: 8 February 2017; Published: 13 February 2017

Abstract: In this work, we determine the biological and mathematical properties that are sufficient and necessary to uniquely determine both the primeval RNY (purine-any base-pyrimidine) code and the standard genetic code (SGC). These properties are: the evolution of the SGC from the RNY code; the degeneracy of both codes, and the non-degeneracy of the assignments of aminoacyl-tRNA synthetases (aaRSs) to amino acids; the wobbling property; the consideration that glycine was the first amino acid; the topological and symmetrical properties of both codes.

Keywords: RNY code; Standard genetic code; evolution of the genetic code; frozen code; degeneracy; aminoacyl-tRNA synthetases; symmetry

---

1. Introduction

A fundamental feature of all life forms existing on Earth is that, with several minor exceptions, they share the same standard genetic code (SGC). This universality led Francis Crick to propose the frozen accident hypothesis [1], i.e., the SGC does not change. According to Crick [1], the SGC code remained universal because any change would be lethal, or would have been very strongly selected against and extinguished.

The astonishing diversity of living beings in the history of the biosphere has not been halted by a frozen SGC. The inherent structure of the frozen SGC, in concert with environmental influences, has unleashed life from determinism.

It is widely accepted that there was an age in the origin of life in which RNA played the role of both genetic material and the main agent of catalytic activity [1–3]. This period is known as the RNA World [4,5].

The reign of the RNA World on Earth probably began no more than about 4.2 billion years ago, and ended no less than about 3.6 billion years ago [6]. Eigen and coworkers (1968) [7] revealed kinship relations by alignments of tRNA sequences and they concluded that the genetic code is not older than but almost as old as our planet. There is an enormous leap from the RNA World to the complexity of DNA replication, protein manufacture and biochemical pathways. Code stability since its formation on the early Earth has contributed to preserving evidence of the transition from an RNA World to a protein-dependent world.

The transfer RNA (tRNA) is perhaps the most important molecule in the origin and evolution of the genetic code. Just two years after the discovery of the double-helix structure of DNA, Crick [8,9] proposed the existence of small adaptor RNA molecules that would act as decoders carrying their own amino acids and interacting with the messenger RNA (mRNA) template in a position for polymerization to take place.

The SGC is written in an alphabet of four letters (C, A, U, G), grouped into words three letters long, called triplets or codons. Crick represented the genetic code in a two-dimensional table arranged in such a way that it is possible to readily find any amino acid from the three letters, written in the 5′

to 3′ direction of the codon [1]. Each of the 64 codons specifies one of the 20 amino acids or else serves as a punctuation mark signaling the end of a message.

Crick proposed the wobble hypothesis [10,11], which accounts for the degeneracy of the SGC: the third position in each codon is said to wobble because it is much less specific than the first and second positions.

Given 64 codons and 20 amino acids plus a punctuation mark, there are $21^{64} \approx 4 \times 10^{84}$ possible genetic codes. This staggering number is beyond any imaginable astronomical number, the total count of electrons in the universe being well below this number. Note, however, that this calculation tacitly ignores the evolution of the SGC. If we assume two sets of 32 complementary triplets where each set codes for 10 amino acids, we would have $10^{32} \times 10^{32} = 10^{64}$ possible codes. Then we have a reduction of the order of 4×10^{20}. Albeit this is a significant reduction, it is still a very large number. Many more biological constraints are necessary. The result that only one in every million random alternative codes is more efficient than the SGC [12] implies that there could be $\sim 4 \times 10^{78}$ genetic codes as efficient as the SGC. This calculation does not offer deeper insights concerning the origin and structure of the SGC, particularly the frozen accident.

Crick [1] argued that the SGC need not be special at all; it could be nothing more than a "frozen accident". This concept is not far away from the idea that there was an age of miracles. However, as we show in this article, there are indeed several features that are special about the SGC: first, it can be partitioned into two classes of aminoacyl-tRNA synthetases (aaRSs) [13]; secondly, the SGC can be broken down into a product of simpler groups reflecting the pattern of degeneracy observed [14,15]; third, it has symmetrical properties, and evolution did not erase its own evolutionary footsteps [16].

Several models on the origin of the genetic code from prebiotic constituents have been proposed [17–21]. Among the 20 canonical amino acids of the biological coding system, the amino acid glycine is one of the most abundant in prebiotic experiments that simulate the conditions of the primitive planet, either by electrical discharges or simulations of volcanic activity [22–24], and this amino acid is also abundant in the analysis of meteorites [25]. Bernhardt and Patrick (2014), and Tamura (2015) [26,27] also suggested that glycine was the first amino acid incorporated into the genetic code according to an internal analysis of its corresponding tRNA and its crucial importance in the structure and function of proteins. Part of this abundance can be ascribed to its structural simplicity when compared with the structure of the remaining 19 canonical amino acids. Several models for the origin of the coding system mirror glycine as one of the initial amino acids in this system [26–29].

The SGC was theoretically derived from a primeval RNY (R means purine, Y pyrimidine, and N any of them) genetic code under a model of sequential symmetry breakings [14,15], and vestiges of this primeval RNY genetic code were found in current genomes of both Eubacteria and Archaea [16]. All distance series of codons showed critical-scale invariance not only in RNY sequences (all ORFs (Open reading frames) concatenated after discarding the non-RNY triplets), but also in all codons of two intermediate steps of the genetic code and in all kind of codons in the current genomes [16]. Such scale invariance has been preserved for at least 3.5 billion years, beginning with an RNY genetic code to the SGC throughout two evolutionary pathways. These two likely evolutionary paths of the genetic code were also analyzed algebraically and can be clearly visualized in three, four and six dimensions [15,30,31].

The RNY subcode is widely considered as the primeval genetic code [32]. It comprises 16 triplets and eight amino acids, where each amino acid is encoded by two codons. The abiotic support of the RNY primeval code is in agreement with observations on abundant amino acids in Miller's sets [33] and in the chronology of the appearance of amino acids according to Trifonov's review [34]. It has been shown that once the primeval genetic code reached the RNY code, the elimination of any amino acid at this stage would be strongly selected against and therefore the genetic code was already frozen [35].

There are 20 aaRSs which are divided into two 10-member, non-overlapping classes, I and II, and they provide virtually errorless aminoacylation of tRNAs [36,37]. Therefore, this operational code is non- degenerate [36,37].

In this work, we pose the following question: What are the minimum necessary and sufficient biological and mathematical properties to uniquely determine the primeval RNY code and the SGC?

2. Mathematical Model of the RNY Code

The RNY code consists of codons where the first base is a purine (R), the third is a pyrimidine (Y) and the second is any of them (Table 1). In this code, the wobble position is strictly present on the third base of the triplet. The number of possible RNY codes is $8^{16} = 2.81 \times 10^{14}$.

Table 1. RNY code. Amino acids that pertain to class I are in red, and those that correspond to class II are in black.

Amino Acid	Codons	Amino Acid	Codons
Asn	AAC, AAU	Thr	ACC, ACU
Asp	GAC, GAU	Ala	GCC, GCU
Ser	AGC, AGU	Ile	AUC, AUU
Gly	GGC, GGU	Val	GUC, GUU

The SGC has been represented in a six-dimensional hypercube [30,38]. Observing that 64 is equal not only to 4^3 but also to 2^6, the codon table can be organized as a six-dimensional hypercube [30]. In such a model, the set of codons are treated as the 64 vertices of the hypercube, and they are joined by edges which connect codons that differ by a single nucleotide. Each dimension describes a type of mutation, transition or transversion acting on each of three bases of any codon. Consequently, we obtain the six dimensions.

This symmetrical model [38] can be partitioned *exactly* into two classes of aaRSs in six dimensions; it displays symmetry groups when the polar requirement is used, and the SGC can be broken down into a product of simpler groups reflecting the pattern of degeneracy observed, and the salient fact that evolution did not erase its own evolutionary footsteps. The symmetrical model and the Rodin-Ohno model [13] are one and the same [38].

Similarly, the RNY subcode can be represented in a four-dimensional hypercube (Figure 1). This hypercube will be employed to reduce the possible number of mappings, by considering its topology and neighborhood properties. Codons that codify the same amino acid are neighbors. Note from Figure 1 that codons for the same amino acid are next to each other, due to the fact that they differ in only the third base and therefore they are at distance of one. A detailed description of the 6D hypercube representing the SGC can be found in Reference [30].

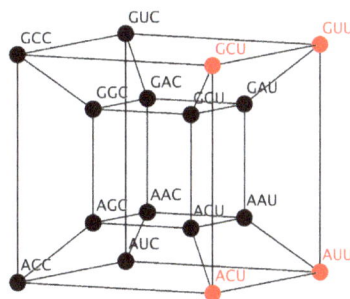

Figure 1. Four-dimensional hypercube that represents the RNY code. Codons for amino acids of class I are in red and those for class II are in black.

3. Combinatorics of the RNY Code

We have noted above that the number of possible codes composed by eight amino acids and 16 triplets is $8^{16} = 2.81 \times 10^{14}$. This number includes codes completely redundant (all codons assigned to the same amino acid) or codes in which all amino acids share the same degeneration, as in the present RNY code. Also, there may not be restrictions between the two classes of aaRS and their corresponding amino acids. First, we consider the restriction in which all amino acids are coded by two triplets, and such codes are given by the multinomial coefficient $(2, 2, 2, 2, 2, 2, 2, 2)! = \frac{16!}{2^{18}} \simeq 8.17 \times 10^{10}$. The present RNY code arranges the triplets so that two codons for the same amino acid are neighbors in the four-dimensional cube. With such a restriction, there are $\begin{pmatrix} 4 \\ 1 \end{pmatrix} 8! = 161,280$ possible RNY codes, since there are four possible configurations in which amino acids can be arranged in the 4-dimensional hypercube hypercube. This neighborhood property preserves the degeneracy irrrespective of the particular wobbling nucleotide, not necessarily the third position. The number 8! accounts for the fact that all the permutations in the assignation of amino acids maintain the property that the two codons that encode the same amino acid must be neighbors.

Considering the third base as the source of variability in the code, the number of posibilities is reduced to $8! = 40,320$. If we consider only the first two bases that determine the amino acid, it is possible to reduce the four-dimensional cube to a three-dimensional cube in which the vertices represent the first two nucleotides (Figure 2a). If the vertices are relabeled to show the codified amino acid, we obtain a phenotypic cube (Figure 2b).

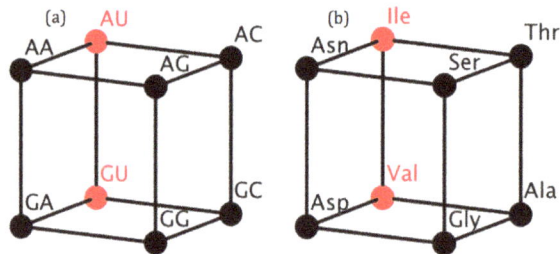

Figure 2. (**a**) Cube of RNY dinucleotides according to the four-dimensional model of the code. Dinucleotides for class I amino acids are in red; and those for class II are in black; (**b**) Phenotypic cube of amino acids according to the four-dimensional model of the RNY code. Class I amino acids are in red and those of class II are in black.

If we consider that there are two amino acids that belong to class I and six amino acids that correspond to class II, then there are $2 \begin{pmatrix} 8 \\ 2 \end{pmatrix} 6! = 37,440$ possible codes. This calculation comes from taking two out of the eight amino acids and assigning them to class I, considering its permutations and also the permutations of the amino acids of class II. To maintain the topological properties of the RNY model, four triplets of class I must form a square in the four-dimensional model, or similarly, the dinucleotides must be neighbors, i.e., they are connected by an edge in the cube representation. In this case, there are $2(12)6! = 17,280$ different codes that preserve the aaRSs distribution in the code and in the model. This number arises from the 12 edges available in the cube to join classs I amino acids and the permutations of classs II amino acids.

In order to maintain the topological properties of the three-dimensional cube, the amino acids of a code must share the neighboring properties of the current RNY code. In other words, if two amino acids are next to each other in the current model, then they are also adjacent in a model constructed by such a code. This property is manifested by the fact that such codes are built by the symmetries of the present model, so that there are 48 different codes that keep the topology of the current code intact.

The ocurrence of glycine as the first amino acid and its assignment to the triplets GGC and GGU as a fixed starting point in the evolution of the SGC impose another restriction, particularly when contrasted with the topology of the four- and three-dimensional cubes, since it fixates isoleucine to AUC and AUU in order to keep the adjacency properties. In this case, there are as many as $\binom{3}{1}2 = 6$ possible codes, due to the fact that there are three possible positions for valine that maintain its adjacency to isoleucine, and there are two symmetrical configurations (given by a reflection) that maintain the rest of the topology.

In the actual code, all triplets where the middle base is uracile codify for amino acids of class I, and this pattern forces the triplets of valine to be GUC and GUU, which in turn also fixes AGC and AGU for serine. This results in two possible RNY codes, which here and further on will be denoted by ○RNY and ∅RNY. The ○RNY denotes the actual and original RNY code, whereas ∅RNY represents an alternative code in which the codons for threonine and alanine are simultaneously interchanged with the ones of aspartic acid and asparagine, respectively. The fixation of another amino acid would completely constraint the number of RNY possible codes to only one!

4. Evolution of the RNY Code by Means of Frame-Shifts and Transversions

Two genetic codes from which the primeval RNA code could have originated the SGC were derived [14–16]. The primeval RNA code consists of 16 codons that specify eight amino acids (then this code shows a slight degeneration). The extended RNA code type I consists of all codons of the RNY type plus codons obtained by considering the RNA code, but in the second (NYR-type) and third (YRN-type) reading frames. The extended RNA code type II comprises all codons of the RNY type plus codons that arise from transversions of the RNA code in the first (YNY type) and third (RNR) nucleotide bases. Then, by allowing frame-reading mistranslations, we arrived at 48 codons that specify 17 amino acids and the three stop codons. If transversions in the first or third nucleotide bases of the RNY pattern are permitted, then there are also 48 codons that encode for 18 amino acids but no stop codons.

In the context of the frozen concept, it was concluded that considering the symmetries of both extended RNA codes, the primeval RNY code was already frozen and it evolved like a replicating and growing icicle [14]. The composition of both extended codes eventually leads to the actual SGC.

As the RNY is described mathematically as a four-dimensional cube, each extended code comprises a duplication of the RNY cube in order to determine a five-dimensional prism as an intermediate step towards the final six-dimensional cube for the SGC. Supposing one of the two alternative RNY codes as the initial code, the number of possible extended codes can be calculated. Then, assuming, as before, that wobbling occurs principally at the third base, the current degeneration of the code and the topology given by the mathematical model shall be maintained.

If the ○RNY is used as a cornerstone for the formation of the genetic code, then, regardless of the evolutionary path chosen, there are two SGCs which are compatible with all the assumptions. These are the actual SGC and a second one in which the codifications of AUG and UGG are interchanged with the ones of AUA and UGA, respectively. These modifications make it so that methionine is codified by AUA and tryptophane by UGA, while AUG codes for isoleucine and UGG is a stop signal. The rest of the code remains unaltered.

On the other hand, if ∅RNY is used as an initial condition, then there are no possible codes on any evolutionary path which meet all hypotheses. In other words, it is not possible to derive the SGC from ∅RNY without violating at least one of the considered properties. This is due to the fact that the mathematical model forbids the possible extended codes that would keep biological properties such as wobbling and the binary division of aaRSs.

5. Discussion

It is possible to gradually add properties to the RNY code to reduce the number of possible codes from 2.81×10^4 to only one. This is done when considering the current properties of degeneracy of the RNY code and the wobble, the aaRSs distribution in the RNY and in the SGC, and finally the mathematical model to represent the genetic code and its induced property of adjacency. The mathematical model plays an important role in the reduction of the possible number of codes. The $37,440$ possible RNY codes were obtained by considering the degeneration in the third base and by assuming that the distribution of aaRRs classes is the same as in the current RNY code. Further reductions, up to one code, were only accomplished by the use of our mathematical model. Both evolutionary paths majorly reduce the number of possible genetic codes from the staggering number of 4.18×10^{84} to only two, which consists of the current code and an alternative code with a subtle modification. The alternative RNY code, $\varnothing$RNY, cannot lead to an SGC that is compatible with all the hypotheses by means of the transversions and frame-shift reading mistranslations. Hence, the SGC evolved from the $\bigcirc$RNY code.

Novozhilov et al. [39] found that the SGC is a suboptimal random code in regard to robustness to error of translations. Thus, the SGC appears to be a point on an evolutionary trajectory from a random code about halfway to the summit (or to the valley) of the local peak in a rugged fitness landscape.

So far, all we know is terrestrial biology. If life is to be found somewhere else in the universe, and even if its ancestry can be traced back to primitive organisms, the rules of the assignments of codons to amino acids may not necessarily be the same and the amino acids may be even chemically different to those found in known terrestrial life. Different environments and different evolutionary paths on different worlds could result in completely different genetic codes and patterns of evolution.

In conclusion, the SGC is certainly ubiquitous in Earth, and what we would expect to find in living beings on other planets is, precisely, this universal biological property: a genetic coding system.

Acknowledgments: Gabriel S. Zamudio is a doctoral student from Programa de Doctorado en Ciencias Biomédicas, Universidad Nacional Autónoma de México (UNAM) and a fellowship recipient from Consejo Nacional de Ciencia y Tecnología (CONACYT) (number: 737920); Marco V. José was financially supported by PAPIIT-IN224015, UNAM, México.

Author Contributions: Gabriel S. Zamudio performed the calculations and figures, wrote a draft of the manuscript; Gabriel S. Zamudio and Marco V. José conceived the work, contributed to ideas, performed the analyses; Marco V. José wrote the manuscript, and prepared the paper.

Conflicts of Interest: The authors declare no conflict of interest.

References

1. Crick, F.H.C. The origin of the genetic code. *J. Mol. Biol.* **1968**, *38*, 367–379. [CrossRef]
2. Woese, C. *The Genetic Code*; Harper and Row: New York, NY, USA, 1967; Chapter 7.
3. Kenneth, D.J.; Ellington, A.D. The search for missing links between self-replicating nucleic acids and the RNA world. *Orig. Life Evol. Biosph.* **1995**, *25*, 515–530.
4. Gilbert, W. The RNA World. *Nature* **1986**, *319*, 618. [CrossRef]
5. Gesteland, R.F.; Cech, T.R.; Atkins, J.F. *The RNA World*; Cold Spring Harbor Laboratory Press: New York, NY, USA, 1999.
6. Joyce, G.F. The antiquity of RNA-based evolution. *Nature* **2002**, *418*, 214–221. [CrossRef] [PubMed]
7. Eigen, M.; Lindemann, B.F.; Tietze, M.; Winkler-Oswatitsch, R.; Dress, A.; Haeseler, A. How old is the genetic code? Statistical geometry of tRNA provides an answer. *Science* **1968**, *244*, 673–679. [CrossRef]
8. Crick, F.H.C. *On Degenerate Templates and Adaptor Hypothesis Draft*; CSHL Archives Repository: Long Island, NY, USA, 1955.
9. Crick, F.H.C. On Degenerate Templates and the Adaptor Hypothesis: A Note for the RNA Tie Club; unpublished but cited by M B Hoagland (1960). In *The Nucleic Acids*; Chargaff, E., Davidson, J.N., Eds.; Academic Press: New York, NY, USA, 1955; Volume 3, p. 349.
10. Crick, F.H.C. On protein synthesis. *Symp. Soc. Exp. Biol.* **1958**, *12*, 138–163. [PubMed]

11. Crick, F.H.C.; Brenner, S.; Klug, A.; Pieczenik, G. A speculation on the origin of protein synthesis. *Orig. Life* **1976**, *7*, 389–397. [CrossRef] [PubMed]

12. Freeland, S.J.; Hurst, L.D. The genetic code is one in a million. *J. Mol. Evol.* **1998**, *47*, 238–248. [CrossRef] [PubMed]

13. Rodin, S.N.; Rodin, S.A. Partitioning of aminoacyl-tRNA synthetases in two classes could have been encoded in a strand-symmetric RNA World. *DNA Cell Biol.* **2006**, *25*, 617–626. [CrossRef] [PubMed]

14. José, M.V.; Morgado, E.R.; Govezensky, T. An extended RNA code and its relationship to the standard genetic code: An algebraic and geometrical approach. *Bull. Math. Biol.* **2007**, *69*, 215–243. [CrossRef] [PubMed]

15. José, M.V.; Morgado, E.R.; Guimarães, R.C.; Zamudio, G.S.; Farías, S.T.; Bobadilla, J.R.; Sosa, D. Three-dimensional algebraic models of the tRNA code and the 12 graphs for representing the amino acids. *Life* **2014**, *4*, 341–373. [CrossRef] [PubMed]

16. José, M.V.; Govezensky, T.; García, J.A.; Bobadilla, J.R. On the evolution of the standard genetic code: Vestiges of scale invariance from the RNA World in current prokaryote genomes. *PLoS ONE* **2009**, *4*, e4340. [CrossRef] [PubMed]

17. Wong, J.T. Evolution of the genetic code. *Microbiol. Sci.* **1988**, *5*, 174–181. [PubMed]

18. Wong, J.T. Coevolution theory of the genetic code at age thirty. *BioEssays* **2005**, *27*, 416–425. [CrossRef] [PubMed]

19. Bandhu, A.V.; Aggarwal, N.; Sengupta, S. Revisiting the physico-chemical hypothesis of code origin: An analysis based on code-sequence coevolution in a finite population. *Orig. Life Evol. Biosph.* **2013**, *43*, 465–489. [CrossRef] [PubMed]

20. Di Giulio, M. The origin of the genetic code: Matter of metabolism or physicochemical determinism? *J. Mol. Evol.* **2013**, *77*, 131–133. [CrossRef] [PubMed]

21. Rouch, D.A. Evolution of the first genetic cells and the universal genetic code: A hypothesis based on macromolecular coevolution of RNA and proteins. *J. Theor. Biol.* **2014**, *357*, 220–244. [CrossRef] [PubMed]

22. Miller, S.L. A production of amino acids under possible primitive earth conditions. *Science* **1953**, *15*, 528–529. [CrossRef]

23. Parker, E.T.; Zhou, M.; Burton, A.S.; Glavin, D.P.; Dworkin, J.P.; Krishnamurthy, R.; Fernández, F.M.; Bada, J.L. A plausible simultaneous synthesis of amino acids and simple peptides on the primordial Earth. *Angew. Chem. Int. Ed. Engl.* **2014**, *28*, 8270–8274. [CrossRef]

24. Bada, J.L. New insights into prebiotic chemistry from Stanley Miller's spark discharge experiments. *Chem. Soc. Rev.* **2013**, *7*, 2186–2196. [CrossRef] [PubMed]

25. Callahan, M.P.; Martin, M.G.; Burton, A.S.; Glavin, D.P.; Dworkin, J. Amino acid analysis in micrograms of meteorite sample by nanoliquid chromatography-high-resolution mass spectrometry. *J. Chromatogr. A* **2014**, *1332*, 30–34. [CrossRef] [PubMed]

26. Bernhardt, H.S.; Patrick, W.M. Genetic code evolution started with the incorporation of glycine, followed by other small hydrophilic amino acids. *J. Mol. Evol.* **2014**, *78*, 307–309. [CrossRef] [PubMed]

27. Tamura, K. Beyond the Frozen Accident: Glycine Assignment in the Genetic Code. *J. Mol. Evol.* **2015**, *81*, 69–71. [CrossRef] [PubMed]

28. Bernhardt, H.S.; Tate, W.P. Evidence from glycine transfer RNA of a frozen accident at the dawn of the genetic code. *Biol. Direct* **2008**, *3*. [CrossRef] [PubMed]

29. Parker, E.T.; Cleaves, H.J.; Dworkin, J.P.; Glavin, D.P.; Callahan, M.; Aubrey, A.; Lazcano, A.; Bada, J.L. Primordial synthesis of amines and amino acids in a 1958 Miller H_2S-rich spark discharge experiment. *Proc. Natl. Acad. Sci. USA* **2011**, *5*, 5526–5531. [CrossRef] [PubMed]

30. José, M.V.; Morgado, E.R.; Sánchez, R.; Govezensky, T. The 24 possible algebraic representations of the standard genetic code in six and three dimensions. *Adv. Stud. Biol.* **2012**, *4*, 119–152.

31. José, M.V.; Morgado, E.R.; Govezensky, T. Genetic hotels for the standard genetic code: Evolutionary analysis based upon novel three-dimensional algebraic models. *Bull. Math. Biol.* **2011**, *73*, 1443–1476. [CrossRef] [PubMed]

32. Eigen, M.; Winkler-Oswatitsch, R. Transfer-RNA: An early gene? *Naturwissenschaften* **1981**, *68*, 282–292. [CrossRef] [PubMed]

33. Miller, S.L.; Urey, H.C.; Oró, J. Origin of organic compounds on the primitive earth and in meteorites. *J. Mol. Evol.* **1976**, *9*, 59–72. [CrossRef] [PubMed]

34. Trifonov, E.N. Consensus temporal order of amino acids and evolution of the triplet code. *Gene* **2000**, *261*, 139–151. [CrossRef]

35. José, M.V.; Zamudio, G.S.; Palacios-Pérez, M.; Bobadilla, J.R.; Farías, S.T. Symmetrical and thermodynamic properties of phenotypic graphs of amino acids encoded by the primeval RNY code. *Orig. Life Evol. Biosph.* **2015**, *45*, 77–83. [CrossRef] [PubMed]

36. de Pouplana, L.R.; Schimmel, P. Aminoacyl-tRNA synthetases: Potential markers of genetic code development. *Trends Biochem. Sci.* **2001**, *26*, 591–596. [CrossRef]

37. Schimmel, P.; Giégé, R.; Moras, D.; Yokoyama, S. An operational RNA code for amino acids and possible relationship to genetic code. *Proc. Natl. Acad. Sci. USA* **1993**, *90*, 8763–8768. [CrossRef] [PubMed]

38. José, M.V.; Zamudio, G.S.; Morgado, E.R. A unified model of the standard genetic code. *R. Soc. Open Sci.* **2017**, *4*, 160908. [CrossRef]

39. Novozhilov, A.S.; Wolf, Y.I.; Koonin, E. Evolution of the genetic code: Partial optimization of a random code for robustness to translation error in a rugged fitness landscape. *Biol. Direct* **2007**, *2*, 1–24. [CrossRef] [PubMed]

Article

Bioinformatic Analysis Reveals Archaeal tRNA^Tyr and tRNA^Trp Identities in Bacteria

Takahito Mukai [1], Noah M. Reynolds [1], Ana Crnković [1] and Dieter Söll [1,2,*]

[1] Department of Molecular Biophysics and Biochemistry, Yale University, New Haven, CT 06520, USA; takahito.mukai@yale.edu (T.M.); noah.reynolds@yale.edu (N.M.R.); ana.crnkovic@yale.edu (A.C.)
[2] Department of Chemistry, Yale University, New Haven, CT 06520, USA
* Correspondence: dieter.soll@yale.edu; Tel.: +1-203-432-6200; Fax: +1-203-432-6202

Academic Editor: Koji Tamura
Received: 16 January 2017; Accepted: 17 February 2017; Published: 21 February 2017

Abstract: The tRNA identity elements for some amino acids are distinct between the bacterial and archaeal domains. Searching in recent genomic and metagenomic sequence data, we found some candidate phyla radiation (CPR) bacteria with archaeal tRNA identity for Tyr-tRNA and Trp-tRNA synthesis. These bacteria possess genes for tyrosyl-tRNA synthetase (TyrRS) and tryptophanyl-tRNA synthetase (TrpRS) predicted to be derived from DPANN superphylum archaea, while the cognate tRNA^Tyr and tRNA^Trp genes reveal bacterial or archaeal origins. We identified a trace of domain fusion and swapping in the archaeal-type TyrRS gene of a bacterial lineage, suggesting that CPR bacteria may have used this mechanism to create diverse proteins. Archaeal-type TrpRS of bacteria and a few TrpRS species of DPANN archaea represent a new phylogenetic clade (named TrpRS-A). The TrpRS-A open reading frames (ORFs) are always associated with another ORF (named ORF1) encoding an unknown protein without global sequence identity to any known protein. However, our protein structure prediction identified a putative HIGH-motif and KMSKS-motif as well as many α-helices that are characteristic of class I aminoacyl-tRNA synthetase (aaRS) homologs. These results provide another example of the diversity of molecular components that implement the genetic code and provide a clue to the early evolution of life and the genetic code.

Keywords: tRNA; aaRS; genetic code; evolution; lateral gene transfer

---

1. Introduction

Bacteria, archaea and eukarya share the standard genetic code, which suggests that they share a universal common ancestor (LUCA). However, the molecular systems underlying the standard genetic code are not completely conserved between all domains of life. In aminoacyl-tRNA synthesis, several elements of tRNA such as the anticodon sequence, other nucleotide residues, post-transcriptional modifications, and local and global tertiary structures are recognized by the cognate aminoacyl-tRNA synthetase (aaRS) [1–3]. While it is known that in all domains of life the anticodon sequences of tRNA^Tyr and tRNA^Trp are recognized by their cognate aaRSs, the other major identity elements of tRNA^Tyr and tRNA^Trp are distinct between the bacterial domain and the archaeal and eukaryotic domains [1,2,4,5] (Figure 1A). In bacteria, tRNA^Tyr contains a G1-C72 base pair and a variable arm (V-arm) that is recognized by the additional C-terminal S4-like domain of bacteria-type TyrRS (Figure 1A,B) [6]. On the other hand, archaea and eukaryotes encode a tRNA^Tyr lacking the V-arm and containing a C1-G72 base pair (Figure 1A) [6]. For tRNA^Trp, bacteria encode a tRNA^Trp with G73, whereas archaea and eukaryotes have a tRNA^Trp with A73 and a G1-C72 base pair (Figure 1A) [7]. Thus, unlike the other aaRS species, archaeal and eukaryotic TyrRS and TrpRS have not been found in the bacterial domain [6,8–15].

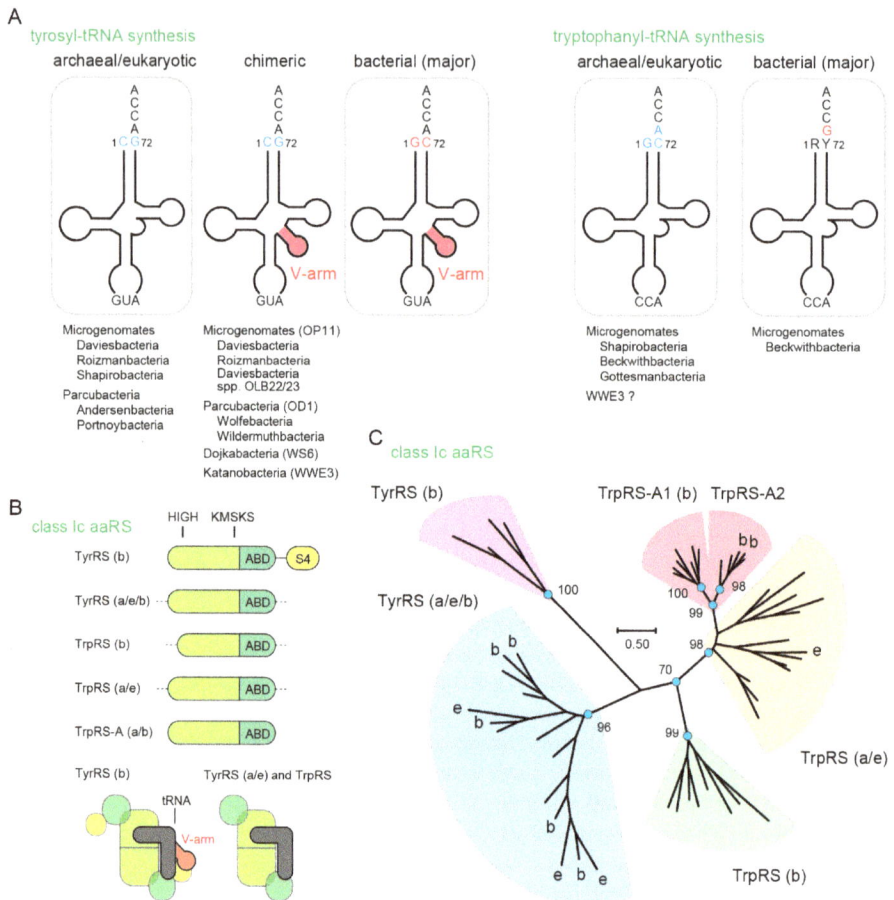

Figure 1. Bacteria with an archaeal tRNA^{Tyr} or tRNA^{Trp} identity. (**A**) The major identity elements for tyrosyl-tRNA synthetase (TyrRS) and tryptophanyl-tRNA synthetase (TrpRS) are shown in blue and red. Diverse subgroups of the Candidate Phyla Radiation (CPR) have archaea-like tRNA^{Tyr} and tRNA^{Trp} genes, as shown below the tRNA structures. Ca. Beckwithbacteria bacterium RBG_13_42_9 has both bacterial and archaeal pairs of TrpRS•tRNA^{Trp}; (**B**) Domain structures of the class Ic aminoacyl-tRNA synthetase (aaRS) family (homodimer). The S4-like domain binds to the V-arm of bacterial tRNA^{Tyr}. ABD denotes anticodon-binding domain. TrpRS-A is a newly identified TrpRS homolog. Class Ic aaRS is known to form a homodimer (in a few cases pseudo-homodimer) and binds to one tRNA molecule at a time (half-of-the-sites). Bacterial, archaeal and eukaryotic origins are indicated with b, a and e, respectively; (**C**) Phylogenetic analysis of the class Ic aaRS family. Maximum likelihood bootstrap values (100 replicates) are shown. The TrpRS-A species are split into two clades. The TrpRS-A2 proteins may chelate a [4Fe–4S] cluster.

A fundamental question in understanding the evolution of the genetic code is whether tRNA identities were established at the time of LUCA, and if so, which tRNA identity set was used [13,16]. The discovery of bacteria with an archaeal tRNA identity would provide support for the hypothesis that archaeal tRNA identity sets may have been used in LUCA. A clue to the answer to this question was provided by two synthetic biology studies [17,18]. The artificial gene transfer of an archaeal or eukaryotic TyrRS or TrpRS gene to *Escherichia coli* was successfully achieved by the simultaneous transfer of an archaeal or eukaryotic tRNA gene [17,18]. These heterologous aaRS•tRNA pairs functionally replaced the endogenous aaRS•tRNA pairs in *E. coli*. Thus, it can be hypothesized that the archaeal tRNA^{Tyr} and tRNA^{Trp} identities might have been used in LUCA.

Inspired by these studies, we carefully re-investigated the phylogenetic distribution of TyrRS and TrpRS. In the present study, archaea-, eukarya-, and bacteria-type is used to indicate the canonical archaea-, eukarya-, or bacteria-type aaRS, respectively, independent of the organism in which the enzyme is identified. Surprisingly, we found a putative bacterial species annotated to have a eukarya-type TyrRS gene (EKE14628.1) [19]. This Ca. Roizmanbacteria bacterium belongs to the candidate phyla radiation (CPR) composed of diverse uncultured bacteria which are often symbiotic with DPANN archaea [20–24]. Although composite genomes of CPR bacteria and DPANN archaea are sometimes contaminated by DPANN archaeal genomes and CPR bacterial genomes, respectively, a recent study was able to identify archaea-like form II/III RubisCO genes in CPR bacteria [25]. These findings prompted us to search for archaea/eukarya-type Tyr- and Trp-encoding systems in bacteria.

2. Materials and Methods

Bioinformatics

Archaeal and eukaryotic TyrRS genes and non-canonical bacteria-type TyrRS genes were collected in three steps. First, TyrRS genes of representative archaea and eukaryotes were collected by a keyword search (tyrosyl/tryptophanyl-tRNA synthetase; tyrosine/tryptophan–tRNA ligase) and a BLASTp search in the National Center for Biotechnology Information (NCBI) database. Next, archaea/eukarya-type TyrRS genes in the bacterial domain were collected by a BLASTp search in the NCBI database and manually curated. Lastly, TyrRS amino acid sequences which showed about $\geq$40% similarity with a query sequence (GenBank: KKM02188.1) were collected from all genome, metagenome and metatranscriptome protein sequence datasets in the Integrated Microbial Genomes (IMG) system [26] (last update September 2016). The reason for employing KKM02188.1 was to find bacteria whose TyrRSs resemble opisthokontal (fungal and animal) TyrRSs. The query protein belongs to an unknown fungus in a marine sediment metagenome and shows good similarity (41%–51%) to both opisthokontal and Daviesbacteria GW2011_GWA1_38_7 TyrRS species.

Non-canonical TrpRS genes in the bacterial domain were first identified by a BLASTp search in the NCBI database using the *Pyrococcus horikoshii* TrpRS (UniProtKB: O59584.2) as query. Non-canonical TrpRS genes which showed >40% amino acid similarity with Ca. Beckwithbacteria bacterium RBG_13_42_9 TrpRS-A were collected by BLASTp searches in the NCBI database and IMG's groundwater metagenome datasets. The obtained protein sequences were analyzed by Clustal X 2.1 [27] (for rough alignment), SeaView ver 4.0 [28] (for manual curation), MEGA7 [29] with the default settings (Maximum Likelihood, JTT model, Uniform rates, Use all Gaps/Missing sites, for phylogeny estimation), and BoxShade Server ver. 3.21 (for visualization). Multiple sequence alignment analyses by Clustal X were followed by a manual curation based on the reported structure-based alignments of TyrRS and TrpRS [30–35]. For the phylogenetic analyses of class Ic aaRSs, *N*- and *C*-terminal protein sequences were trimmed and nonconserved insertion sequences removed; up to 13 residues upstream of the HIGH motif and to the end of the anticodon binding domain were included in the analyses. Protein two- and three-dimentional structural prediction was performed using JPred 4 [36] and SWISS-MODEL [37], respectively. tRNA^{Tyr} and tRNA^{Trp} sequences were identified by a BLASTn search using automatically annotated tRNA^{Tyr} and tRNA^{Trp} sequences as queries.

3. Results

3.1. Identification of Non-Canonical Class Ic aaRS Sequences

We found tRNA^{Tyr} with C1-G72 and archaea/eukarya-type TyrRS genes in diverse subgroups of the Parcubacteria (OD1), Microgenomates (OP11), Dojkabacteria (WS6) and Katanobacteria (WWE3) phyla in CPR [22,25] (Figure 1A). In many cases, the CPR tRNA^{Tyr} species with C1-G72 contain a V-arm, indicating that these non-canonical tRNA^{Tyr} species are derived from bacterial tRNA^{Tyr} with a V-arm. Both a V-arm-containing and a V-arm-lacking tRNA^{Tyr} with C1-G72 are found in Ca. Roizmanbacteria bacterium GW2011_GWC2_34_23. On the other hand, tRNA^{Trp} with A73 and archaea/eukarya-type TrpRS genes are found in a few Microgenomates bacteria (Figure 1A). Both a bacterial tRNA^{Trp} with G73 and a tRNA^{Trp} with A73 exist in Ca. Bechwithbacteria bacterium RBG_13_42_9. We named these non-canonical TrpRS species as TrpRS-A (Figure 1B). Interestingly, TrpRS-A is slightly different from the canonical archaeal TrpRS species, but highly similar to minor DPANN archaeal TrpRS species (also named TrpRS-A) found only in the groundwater metagenomes (Figure 1C). We identified only 12 instances of TrpRS-A genes in total, suggesting an infrequent distribution of these genes in nature. These bacterial and archaeal TrpRS-A species form a terminal clade within the archaeal/eukaryotic TrpRS clade and can be grouped into two sub-clades (TrpRS-A1 and TrpRS-A2) (Figure 1C). The TrpRS-A1 proteins appear to be restricted within CPR bacteria. The TrpRS-A2 proteins are predicted to chelate a [4Fe–4S] cluster through their four cysteine residues, like some bacterial and archaeal TrpRS proteins having a C-x22-C-x6-C-x2-C motif [38].

3.2. Archaea/Eukarya-Type TyrRS in the Bacterial Domain

We then investigated the collected TyrRS sequences. A high-resolution phylogeny for archaea/eukarya-type TyrRS suggests that several lineages of CPR bacteria independently obtained an archaea/eukarya-type TyrRS gene from archaea (Figure 2). Alternatively, lateral gene transfer (LGT) might have occurred from bacteria to archaea and other groups of bacteria. Interestingly, one bacterial TyrRS sequence (3300007427.a:Ga0100483_102719) is highly similar to the TyrRS sequences of *Acanthamoeba castellanii* and Pandoraviruses (the Eukarya domain) (see the Ultra-High resolution region in Figure 2). Thus, LGT between bacteria and amoeba or giant viruses can be reasonably assumed. This Pandoravirus-like TyrRS, as well as two other TyrRSs derived from WS6 bacterium GW2011 GWA2_37_6 [24,39] and an active sludge metagenome, possess a B2 domain of bacterial phenylalanyl-tRNA synthetase β-subunit (PheRSβ) [40]. This domain is fused to the N-terminus of the TyrRS by a long α-helix (Figure 3A). The B2 domain belongs to the RNA-binding OB folds, but it is missing in many CPR-bacterial PheRSβ (for example, OGE14653.1). We also found that a few Microgenomates lysyl-tRNA synthetases (KKR67068.1 & KKQ91124.1) have an additional C-terminal domain that is very similar to this B2 domain and predicted α-helix. It is known that aaRS proteins are often fused with an OB domain [41]. The two Klosterneuburg active sludge metagenomic contigs containing the B2-TyrRS fusion genes showed almost identical gene organization and gene sequences, indicating that these two contigs belong to two closely-related bacteria. However, only the TyrRS 'domain' sequences are different in terms of sequence similarity (Figure 3A,B). While one is Pandoravirus-like, as previously mentioned, the other is most similar to the TyrRS 'domain' sequence of the B2-TyrRS species of WS6 bacterium GW2011 GWA2_37_6 (Figures 2 and 3A). A possible explanation for this is that the TyrRS 'domain' sequence was replaced with the Pandoravirus-like sequence in a bacterial lineage (Figure 3B, the upper contig). This finding will help us understand aaRS evolution through domain fusion and swapping.

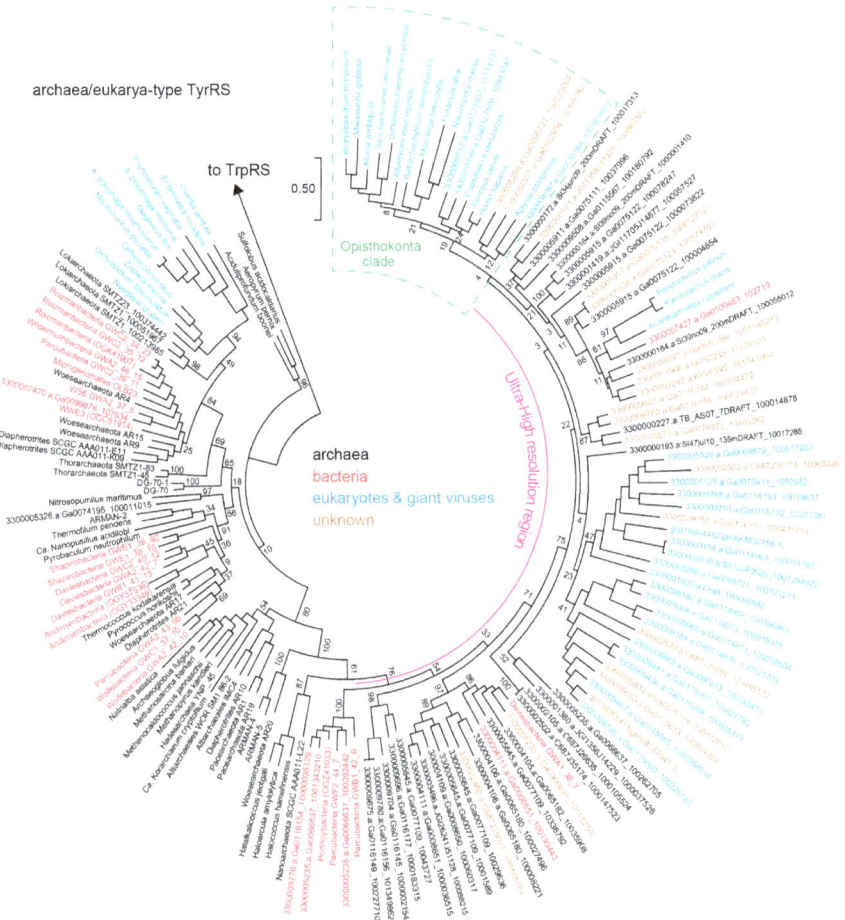

Figure 2. A phylogenetic tree for archaea/eukarya-type TyrRS. Bootstrap values (%) are shown for the rooted Maximum Likelihood tree made with 100 replicates using MEGA7. The 'Ultra-High resolution' region shows almost all TyrRS sequences identified by the comprehensive genome/metagenome/metatranscriptome analysis using gi | 816604452 | gb | KKM02188.1 | as query for BLASTp. The archaeal species in the Ultra-High resolution region may belong to the DPANN superphylum according to the Joint Genome Institute's annotation pipeline and our manual annotation. The opisthokontal (fungal and animal) TyrRS clade is marked with a green box. We chose a few representative TyrRS sequences for each major bacterial group (Roizmanbacteria, Daviesbacteria, Shapirobacteria, Wolfebacteria and Andersenbacteria) after confirming the sequence similarity within the same group. In contrast, we identified three orphan TyrRS genes belonging to bacteria in the Ultra-High resolution region. TrpRS sequences of *Thermus thermophilus* (bacteria) and *Pyrococcus horikoshii* (archaea) were used as an outgroup.

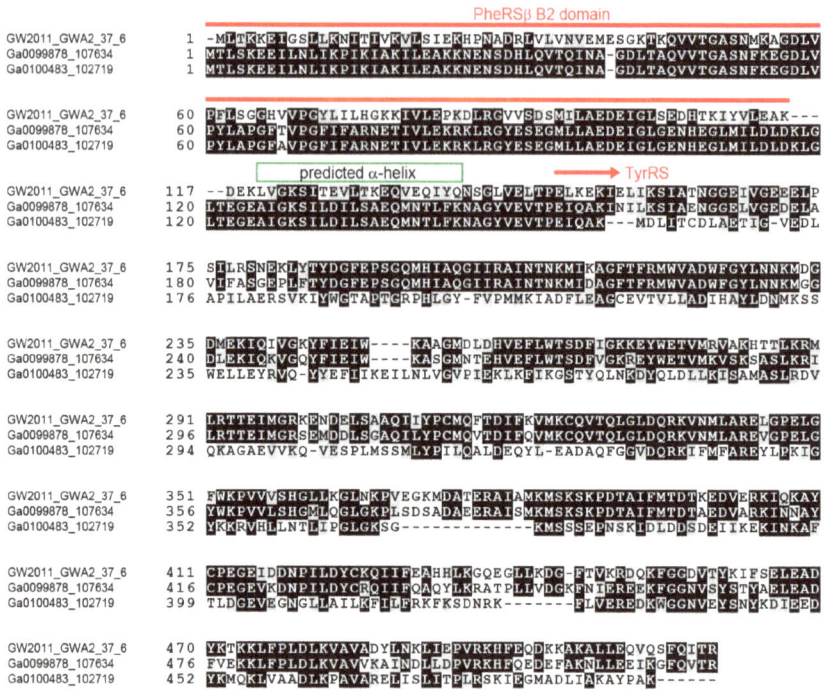

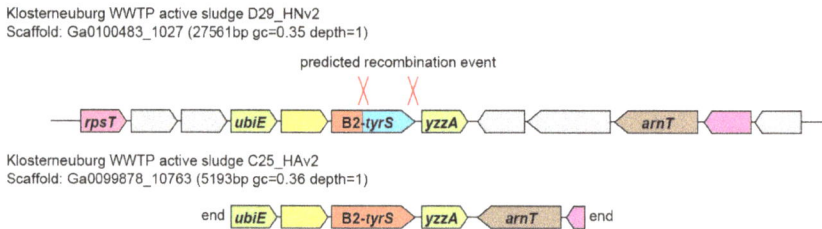

Figure 3. The TyrRS species with a *N*-terminal B2 domain fusion. (**A**) Multiple sequence alignment of the three B2-TyrRS proteins; (**B**) The genomic loci of the two B2-TyrRS genes in active sludge metagenomes. Predicted recombination sites are indicated.

3.3. Non-Canonical TrpRS Species and Their Associating Proteins

We then investigated the genetic loci of the TrpRS-A genes (Figure 4A). In bacteria, tRNA^{Trp} (A73) and TrpRS-A comprise an operon. Interestingly, this operon is either followed by a bacteria-type TrpRS gene or headed by a *trp* repressor gene (Figure 4A). This is consistent with Ca. Beckwithbacteria bacterium RBG_13_42_9 possessing a bacterial tRNA^{Trp} (G73) somewhere else in the genome. The observed operon structures indicate that the tRNA^{Trp} (A73) and TrpRS-A1 genes may be regulated by tryptophan availability and could coexist with a canonical bacterial TrpRS•tRNA^{Trp} pair. In archaea, one or two tRNA^{Trp} (A73) genes, one or two TrpRS-A2 genes and another gene encoding a TrpRS-A homolog (named TrpRS-A-like) are found in addition to a single canonical archaeal TrpRS gene. TrpRS-A-like has several insertions and deletions compared to archaea/eukarya-type TrpRS and TrpRS-A. Thus, TrpRS-A and TrpRS-A-like coexist with a canonical archaeal TrpRS in archaea.

Interestingly, TrpRS-A2 and TrpRS-A-like genes are also found in putative bacterial metagenomic contigs (Figure 4A), implying multiple LGT events.

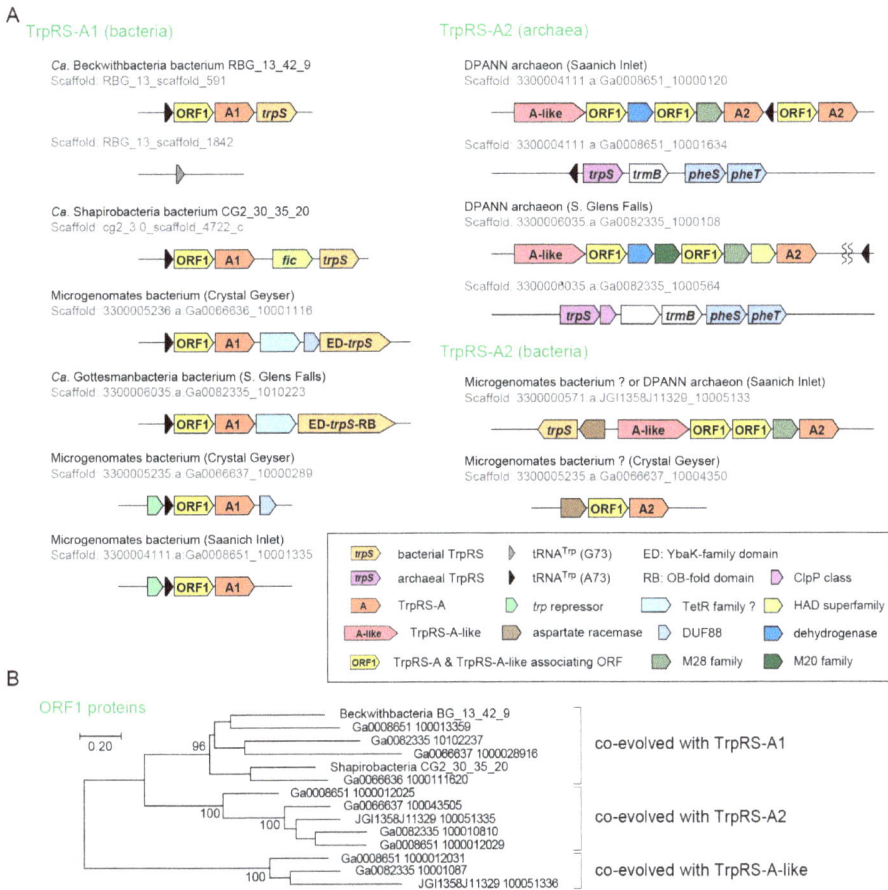

Figure 4. Non-canonical TrpRS species. (**A**) The genetic loci and the operon structures of TrpRS-A genes. The origins of these uncultured organisms are described in the parentheses and indicated with "CG" (Crystal Geyser groundwater) and "RBG" (Rifle BackGround groundwater). In a few Microgenomates species, TrpRS is fused with small proteins; (**B**) Co-evolution of the ORF1 genes with the TrpRS-A and TrpRS-A-like genes. Bootstrap values (%) are shown for the unrooted Maximum Likelihood tree made with 100 replicates using MEGA7.

We found an unusual open reading frame (named as ORF1) between the tRNA^{Trp} (A73) and TrpRS-A1 genes in the bacterial operons and in the gene clusters of the TrpRS-A2 and TrpRS-A-like in archaea (Figure 4A). The ORF1 is completely conserved and appears to have co-evolved with the TrpRS-A1, TrpRS-A2 and TrpRS-A-like ORFs (Figure 4B). ORF1 sequences show no significant similarity with any known protein (less than score 40 in NCBI BLASTp searches). However, the predicted structure of the ORF1 proteins includes HIGH-motif-like and KMSKS-motif-like motifs [2], as well as many α-helices, thereby suggesting that these proteins might be structural homologs of class I aaRSs (Figure 5). Seventy residues from the putative HIGH motif of ORF1 matched the corresponding region of cysteinyl-tRNA synthetase (a class Ia aaRS) with 16.9% sequence identity in our SWISS-MODEL prediction. However, in the ORF1 structure prediction, the Rossmann-fold domain is missing [42], indicating a function distinct from class I aaRS or class I aaRS homolog that synthesizes cyclo(L-Trp-L-Trp) using tryptophanyl-tRNA^{Trp} [43]. Interestingly, both TrpRS-A1 and ORF1 pairs and bacteria-type TrpRSs of these bacteria lack any tryptophan residues (see Figure 5 for the ORF1 cases). This might imply that tryptophan may be limited for these bacteria living in groundwater environments.

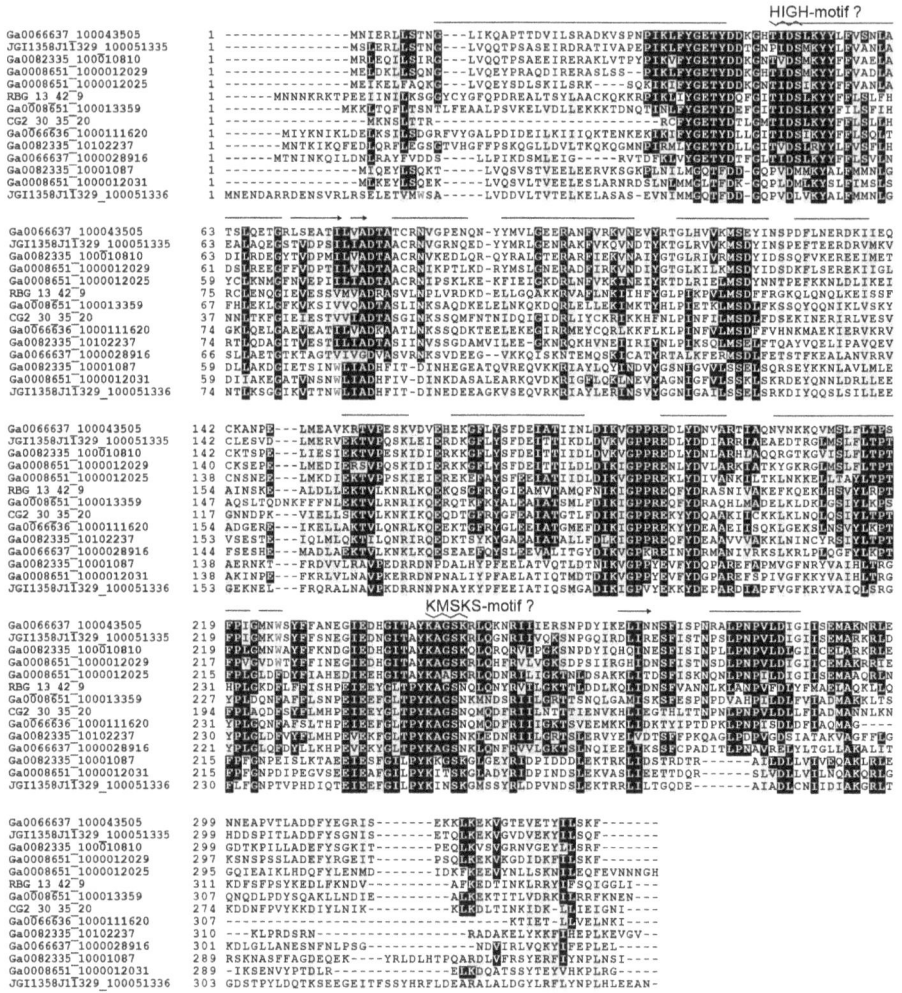

Figure 5. Multiple pairwise alignment of the ORF1 proteins and their structural prediction by SWISS-MODEL. Bars and arrows above the amino acid sequences represent predicted α-helices and β-strands, respectively, whereas zigzags indicate predicted HIGH and KMSKS motifs. The predicted overall structure suggests that the ORF1 protein might be a homolog of class I aaRS. Tryptophan (W) residues are shown as red letters.

4. Discussion

It was thought that LGT of some archaeal/eukaryotic aaRS genes to bacteria would be prevented by the difference in tRNA identity rules. A previous bioinformatic study detected a case of eukaryotic-like histidine (His) tRNA identity (i.e., A73) in certain α-proteobacteria [12]. These α-proteobacterial tRNA^{His} species have A73 and lack G-1 [44]. However, a subsequent biochemical study revealed A73 to be a minor identity element of the *Caulobacter crescentus* histidyl-tRNA synthetase (HisRS) [45]. A recent comprehensive bioinformatic study did not support any LGT event of HisRS from eukaryotes to α-proteobacteria [8], suggesting that the eukaryotic-like tRNA^{His} identity in certain α-proteobacteria might be a result of convergent evolution. Similarly, it is known that some mitochondrial aaRSs violate the bacterial identity rules. For example, human mitochondrial TyrRS charges mitochondrial tRNA^{Tyr} species with G1-C72 (wild-type) or C1-G72 (mutant) with the same efficiency [46]. The V-arm is missing in the human mitochondrial tRNA^{Tyr}, whereas the human mitochondrial TyrRS retains the S4-like domain [47].

In contrast to these previous findings, our results provide the first evidence that archaeal or eukaryotic TyrRS and TrpRS genes exist in a few lineages of bacteria that have tRNA^{Tyr} or tRNA^{Trp} species with archaeal and eukaryotic identity elements. Intimate relationships between CPR bacteria and DPANN archaea may have facilitated LGT of TyrRS and TrpRS genes. Furthermore, our high-resolution TyrRS phylogeny and another comprehensive study [8] revealed that the LGT of TyrRS may have occurred several times among DPANN archaea, CPR bacteria, eukaryotes and giant viruses (Figure 2). Taken together, our data make clear that tRNA identities may not be hardwired to each domain of life.

The functions of TrpRS-A, TrpRS-A-like and the ORF1 proteins remain unclear. Their operon structures suggest that they are involved in tryptophan metabolism and encoding, rather than producing complex antibiotics, plant toxins and peptidoglycans [43,48]. Since TrpRS-A and TrpRS-A-like genes appear to occur in addition to the canonical TrpRS gene in both bacteria and archaea, TrpRS-A and TrpRS-A-like genes might encode additional copies of TrpRS that display a higher or lower specificity for tryptophan than the canonical TrpRS in order to confer antibiotic resistance [49] or to cope with stress [50]. The ORF1 proteins might bind to the homodimers of TrpRS-A and TrpRS-A-like in order to be stabilized in a complex [51]. Another possibility is that TrpRS-A and TrpRS-A-like proteins might form heterodimers with the partner ORF1 protein. It is known that some eukaryotes have a double-length TyrRS species forming a pseudo-dimer in which one of the "subunits" is catalytic but has lost the affinity for the tRNA^{Tyr} anticodon, whereas the other is non-catalytic but still recognizes the anticodon [52]. In addition, trans-oligomerization of duplicated threonyl-tRNA synthetases is known [50]. Future studies should elucidate the biochemical properties and the biological functions of these proteins.

Acknowledgments: This work was supported by grants from the National Institute for General Medical Sciences (GM22854 to D.S.) and from the Division of Chemical Sciences, Geosciences, and Biosciences, Office of Basic Energy Sciences of the Department of Energy (DE-FG02-98ER20311 to D.S.; for funding the genetic experiments). T.M. is a Japan Society for the Promotion of Science Postdoctoral Fellow for Research Abroad. We thank Jill Banfield, Michael Wagner, Steven Hallam, Eugene Madsen and many others for permission to use sequence data produced through the DOE-JGI's community sequencing program, and Jill Banfield for depositing genome sequences of CPR bacteria and DPANN archaea in NCBI. We are grateful to Eugene V. Koonin, Oscar Vargas-Rodriguez, Anastasia Sevostiyanova, and Takuya Umehara for enlightened discussions.

Author Contributions: Takahito Mukai and Dieter Söll designed the project; Takahito Mukai performed bioinformatics; Takahito Mukai, Noah M. Reynolds and Ana Crnković investigated TrpRS-A1 and ORF1 proteins. All authors wrote the manuscript.

Conflicts of Interest: The authors declare no conflict of interest.

Abbreviations

The following abbreviations are used in this manuscript:

CPR	candidate phyla radiation
aaRS	aminoacyl-tRNA synthetase
TyrRS	tyrosyl-tRNA synthetase
TrpRS	tryptophanyl-tRNA synthetase
PheRS	phenylalanyl-tRNA synthetase
HisRS	histidyl-tRNA synthetase
ORF	open reading frame
LGT	lateral gene transfer

References

1. Giegé, R.; Sissler, M.; Florentz, C. Universal rules and idiosyncratic features in tRNA identity. *Nucleic Acids Res.* **1998**, *26*, 5017–5035. [CrossRef] [PubMed]
2. Ibba, M.; Francklyn, C.; Cusack, S.E. *The Aminoacyl-tRNA Synthetases*; Landes Biosciences: Georgetown, TX, USA, 2005.
3. Chaliotis, A.; Vlastaridis, P.; Mossialos, D.; Ibba, M.; Becker, H.D.; Stathopoulos, C.; Amoutzias, G.D. The complex evolutionary history of aminoacyl-tRNA synthetases. *Nucleic Acids Res.* **2016**, *45*, 1059–1068. [CrossRef] [PubMed]
4. Carter, C.W., Jr. Tryptophanyl-tRNA synthetases. In *The Aminoacyl-tRNA Synthetases*; Ibba, M., Francklyn, C., Cusack, S.E., Eds.; Landes Biosciences: Georgetown, TX, USA, 2005.
5. Bedouelle, H. Tyrosyl-tRNA synthetases. In *The Aminoacyl-tRNA Synthetases*; Ibba, M., Francklyn, C., Cusack, S.E., Eds.; Landes Biosciences: Georgetown, TX, USA, 2005.
6. Bonnefond, L.; Giegé, R.; Rudinger-Thirion, J. Evolution of the tRNA^{Tyr}/TyrRS aminoacylation systems. *Biochimie* **2005**, *87*, 873–883. [CrossRef] [PubMed]
7. Xue, H.; Shen, W.; Giegé, R.; Wong, J.T. Identity elements of tRNA^{Trp}. Identification and evolutionary conservation. *J. Biol. Chem.* **1993**, *268*, 9316–9322. [PubMed]
8. Furukawa, R.; Nakagawa, M.; Kuroyanagi, T.; Yokobori, S.I.; Yamagishi, A. Quest for Ancestors of Eukaryal Cells Based on Phylogenetic Analyses of Aminoacyl-tRNA Synthetases. *J. Mol. Evol.* **2017**, *84*, 51–66. [CrossRef] [PubMed]
9. Andam, C.P.; Gogarten, J.P. Biased gene transfer in microbial evolution. *Nat. Rev. Microbiol.* **2011**, *9*, 543–555. [CrossRef] [PubMed]
10. Shiba, K.; Motegi, H.; Schimmel, P. Maintaining genetic code through adaptations of tRNA synthetases to taxonomic domains. *Trends Biochem. Sci.* **1997**, *22*, 453–457. [CrossRef]
11. Sassanfar, M.; Kranz, J.E.; Gallant, P.; Schimmel, P.; Shiba, K. A eubacterial *Mycobacterium tuberculosis* tRNA synthetase is eukaryote-like and resistant to a eubacterial-specific antisynthetase drug. *Biochemistry* **1996**, *35*, 9995–10003. [CrossRef] [PubMed]
12. Ardell, D.H.; Andersson, S.G. TFAM detects co-evolution of tRNA identity rules with lateral transfer of histidyl-tRNA synthetase. *Nucleic Acids Res.* **2006**, *34*, 893–904. [CrossRef] [PubMed]
13. Fournier, G.P.; Alm, E.J. Ancestral Reconstruction of a Pre-LUCA Aminoacyl-tRNA Synthetase Ancestor Supports the Late Addition of Trp to the Genetic Code. *J. Mol. Evol.* **2015**, *80*, 171–185. [CrossRef] [PubMed]
14. Woese, C.R.; Olsen, G.J.; Ibba, M.; Söll, D. Aminoacyl-tRNA synthetases, the genetic code, and the evolutionary process. *Microbiol. Mol. Biol. Rev.* **2000**, *64*, 202–236. [CrossRef] [PubMed]
15. Wolf, Y.I.; Aravind, L.; Grishin, N.V.; Koonin, E.V. Evolution of aminoacyl-tRNA synthetases—Analysis of unique domain architectures and phylogenetic trees reveals a complex history of horizontal gene transfer events. *Genome Res.* **1999**, *9*, 689–710. [PubMed]
16. Ribas de Pouplana, L.; Frugier, M.; Quinn, C.L.; Schimmel, P. Evidence that two present-day components needed for the genetic code appeared after nucleated cells separated from eubacteria. *Proc. Natl. Acad. Sci. USA* **1996**, *93*, 166–170. [CrossRef] [PubMed]

17. Iraha, F.; Oki, K.; Kobayashi, T.; Ohno, S.; Yokogawa, T.; Nishikawa, K.; Yokoyama, S.; Sakamoto, K. Functional replacement of the endogenous tyrosyl-tRNA synthetase-tRNA^{Tyr} pair by the archaeal tyrosine pair in *Escherichia coli* for genetic code expansion. *Nucleic Acids Res.* **2010**, *38*, 3682–3691. [CrossRef] [PubMed]

18. Italia, J.S.; Addy, P.S.; Wrobel, C.J.J.; Crawford, L.A.; Lajoie, M.J.; Zheng, Y.; Chatterjee, A. An orthogonalized platform for genetic code expansion in both bacteria and eukaryotes. *Nat. Chem. Biol.* **2017**. [CrossRef] [PubMed]

19. Yutin, N.; Wolf, Y.I.; Koonin, E.V. Origin of giant viruses from smaller DNA viruses not from a fourth domain of cellular life. *Virology* **2014**, *466–467*, 38–52. [CrossRef] [PubMed]

20. Hug, L.A.; Baker, B.J.; Anantharaman, K.; Brown, C.T.; Probst, A.J.; Castelle, C.J.; Butterfield, C.N.; Hernsdorf, A.W.; Amano, Y.; Ise, K.; et al. A new view of the tree of life. *Nat. Microbiol.* **2016**, *1*, 16048. [CrossRef] [PubMed]

21. Eloe-Fadrosh, E.A.; Ivanova, N.N.; Woyke, T.; Kyrpides, N.C. Metagenomics uncovers gaps in amplicon-based detection of microbial diversity. *Nat. Microbiol.* **2016**, *1*, 15032. [CrossRef] [PubMed]

22. Anantharaman, K.; Brown, C.T.; Hug, L.A.; Sharon, I.; Castelle, C.J.; Probst, A.J.; Thomas, B.C.; Singh, A.; Wilkins, M.J.; Karaoz, U. Thousands of microbial genomes shed light on interconnected biogeochemical processes in an aquifer system. *Nat. Commun.* **2016**, *7*, 13219. [CrossRef] [PubMed]

23. Jaffe, A.L.; Corel, E.; Pathmanathan, J.S.; Lopez, P.; Bapteste, E. Bipartite graph analyses reveal interdomain LGT involving ultrasmall prokaryotes and their divergent, membrane-related proteins. *Environ. Microbiol.* **2016**, *18*, 5072–5081. [CrossRef] [PubMed]

24. Brown, C.T.; Hug, L.A.; Thomas, B.C.; Sharon, I.; Castelle, C.J.; Singh, A.; Wilkins, M.J.; Wrighton, K.C.; Williams, K.H.; Banfield, J.F. Unusual biology across a group comprising more than 15% of domain bacteria. *Nature* **2015**, *523*, 208–211. [CrossRef] [PubMed]

25. Wrighton, K.C.; Castelle, C.J.; Varaljay, V.A.; Satagopan, S.; Brown, C.T.; Wilkins, M.J.; Thomas, B.C.; Sharon, I.; Williams, K.H.; Tabita, F.R.; et al. RubisCO of a nucleoside pathway known from Archaea is found in diverse uncultivated phyla in bacteria. *ISME J.* **2016**, *10*, 2702–2714. [CrossRef] [PubMed]

26. Markowitz, V.M.; Chen, I.M.; Chu, K.; Szeto, E.; Palaniappan, K.; Pillay, M.; Ratner, A.; Huang, J.; Pagani, I.; Tringe, S.; et al. IMG/M 4 version of the integrated metagenome comparative analysis system. *Nucleic Acids Res.* **2014**, *42*, D568–D573. [CrossRef] [PubMed]

27. Larkin, M.A.; Blackshields, G.; Brown, N.P.; Chenna, R.; McGettigan, P.A.; McWilliam, H.; Valentin, F.; Wallace, I.M.; Wilm, A.; Lopez, R.; et al. Clustal W and Clustal X version 2.0. *Bioinformatics* **2007**, *23*, 2947–2948. [CrossRef] [PubMed]

28. Gouy, M.; Guindon, S.; Gascuel, O. SeaView version 4: A multiplatform graphical user interface for sequence alignment and phylogenetic tree building. *Mol. Biol. Evol.* **2010**, *27*, 221–224. [CrossRef] [PubMed]

29. Kumar, S.; Stecher, G.; Tamura, K. MEGA7: Molecular Evolutionary Genetics Analysis Version 7.0 for Bigger Datasets. *Mol. Biol. Evol.* **2016**, *33*, 1870–1874. [CrossRef] [PubMed]

30. Jia, J.; Chen, X.L.; Guo, L.T.; Yu, Y.D.; Ding, J.P.; Jin, Y.X. Residues Lys-149 and Glu-153 switch the aminoacylation of tRNA^{Trp} in *Bacillus subtilis*. *J. Biol. Chem.* **2004**, *279*, 41960–41965. [CrossRef] [PubMed]

31. Xu, F.; Chen, X.; Xin, L.; Chen, L.; Jin, Y.; Wang, D. Species-specific differences in the operational RNA code for aminoacylation of tRNA^{Trp}. *Nucleic Acids Res.* **2001**, *29*, 4125–4133. [CrossRef] [PubMed]

32. Yang, X.L.; Otero, F.J.; Skene, R.J.; McRee, D.E.; Schimmel, P.; Ribas de Pouplana, L. Crystal structures that suggest late development of genetic code components for differentiating aromatic side chains. *Proc. Natl. Acad. Sci. USA* **2003**, *100*, 15376–15380. [CrossRef] [PubMed]

33. Kobayashi, T.; Nureki, O.; Ishitani, R.; Yaremchuk, A.; Tukalo, M.; Cusack, S.; Sakamoto, K.; Yokoyama, S. Structural basis for orthogonal tRNA specificities of tyrosyl-tRNA synthetases for genetic code expansion. *Nat. Struct. Biol.* **2003**, *10*, 425–432. [CrossRef] [PubMed]

34. Stiebritz, M.T. A role for [Fe$_4$S$_4$] clusters in tRNA recognition—A theoretical study. *Nucleic Acids Res.* **2014**, *42*, 5426–5435. [CrossRef] [PubMed]

35. Abergel, C.; Rudinger-Thirion, J.; Giegé, R.; Claverie, J.M. Virus-encoded aminoacyl-tRNA synthetases: Structural and functional characterization of mimivirus TyrRS and MetRS. *J. Virol.* **2007**, *81*, 12406–12417. [CrossRef] [PubMed]

36. Drozdetskiy, A.; Cole, C.; Procter, J.; Barton, G.J. JPred4: A protein secondary structure prediction server. *Nucleic Acids Res.* **2015**, *43*, W389–W394. [CrossRef] [PubMed]

37. Biasini, M.; Bienert, S.; Waterhouse, A.; Arnold, K.; Studer, G.; Schmidt, T.; Kiefer, F.; Cassarino, T.G.; Bertoni, M.; Bordoli, L.; et al. SWISS-MODEL: Modelling protein tertiary and quaternary structure using evolutionary information. *Nucleic Acids Res.* **2014**, *42*, W252–W258. [CrossRef] [PubMed]
38. Han, G.W.; Yang, X.L.; McMullan, D.; Chong, Y.E.; Krishna, S.S.; Rife, C.L.; Weekes, D.; Brittain, S.M.; Abdubek, P.; Ambing, E.; et al. Structure of a tryptophanyl-tRNA synthetase containing an iron-sulfur cluster. *Acta Crystallogr. Sect. F Struct. Biol. Cryst. Commun.* **2010**, *66*, 1326–1334. [CrossRef] [PubMed]
39. Zetsche, B.; Gootenberg, J.S.; Abudayyeh, O.O.; Slaymaker, I.M.; Makarova, K.S.; Essletzbichler, P.; Volz, S.E.; Joung, J.; van der Oost, J.; Regev, A.; et al. Cpf1 Is a Single RNA-Guided Endonuclease of a Class 2 CRISPR-Cas System. *Cell* **2015**, *163*, 759–771. [CrossRef] [PubMed]
40. Roy, H.; Ibba, M. Phenylalanyl-tRNA synthetase contains a dispensable RNA-binding domain that contributes to the editing of noncognate aminoacyl-tRNA. *Biochemistry* **2006**, *45*, 9156–9162. [CrossRef] [PubMed]
41. Kapps, D.; Cela, M.; Théobald-Dietrich, A.; Hendrickson, T.; Frugier, M. OB or Not OB: Idiosyncratic utilization of the tRNA-binding OB-fold domain in unicellular, pathogenic eukaryotes. *FEBS Lett.* **2016**, *590*, 4180–4191. [CrossRef] [PubMed]
42. Rao, S.T.; Rossmann, M.G. Comparison of super-secondary structures in proteins. *J. Mol. Biol.* **1973**, *76*, 241–256. [CrossRef]
43. Moutiez, M.; Belin, P.; Gondry, M. Aminoacyl-tRNA-Utilizing Enzymes in Natural Product Biosynthesis. *Chem. Rev.* **2017**. [CrossRef] [PubMed]
44. Wang, C.; Sobral, B.W.; Williams, K.P. Loss of a universal tRNA feature. *J. Bacterial.* **2007**, *189*, 1954–1962. [CrossRef] [PubMed]
45. Yuan, J.; Gogakos, T.; Babina, A.M.; Söll, D.; Randau, L. Change of tRNA identity leads to a divergent orthogonal histidyl-tRNA synthetase/tRNA^{His} pair. *Nucleic Acids Res.* **2011**, *39*, 2286–2293. [CrossRef] [PubMed]
46. Bonnefond, L.; Frugier, M.; Giegé, R.; Rudinger-Thirion, J. Human mitochondrial TyrRS disobeys the tyrosine identity rules. *RNA* **2005**, *11*, 558–562. [CrossRef] [PubMed]
47. Bonnefond, L.; Frugier, M.; Touzé, E.; Lorber, B.; Florentz, C.; Giegé, R.; Sauter, C.; Rudinger-Thirion, J. Crystal Structure of Human Mitochondrial Tyrosyl-tRNA Synthetase Reveals Common and Idiosyncratic Features. *Structure* **2007**, *15*, 1505–1516. [CrossRef] [PubMed]
48. Buddha, M.R.; Crane, B.R. Structure and activity of an aminoacyl-tRNA synthetase that charges tRNA with nitro-tryptophan. *Nat. Struct. Mol. Biol.* **2005**, *12*, 274–275. [CrossRef] [PubMed]
49. Kitabatake, M.; Ali, K.; Demain, A.; Sakamoto, K.; Yokoyama, S.; Söll, D. Indolmycin resistance of *Streptomyces coelicolor* A3(2) by induced expression of one of its two tryptophanyl-tRNA synthetases. *J. Biol. Chem.* **2002**, *277*, 23882–23887. [CrossRef] [PubMed]
50. Rubio, M.Á.; Napolitano, M.; Ochoa de Alda, J.A.; Santamaría-Gómez, J.; Patterson, C.J.; Foster, A.W.; Bru-Martínez, R.; Robinson, N.J.; Luque, I. Trans-oligomerization of duplicated aminoacyl-tRNA synthetases maintains genetic code fidelity under stress. *Nucleic Acids Res.* **2015**, *43*, 9905–9917. [CrossRef] [PubMed]
51. Buddha, M.R.; Keery, K.M.; Crane, B.R. An unusual tryptophanyl tRNA synthetase interacts with nitric oxide synthase in *Deinococcus radiodurans*. *Proc. Natl. Acad. Sci. USA* **2004**, *101*, 15881–15886. [CrossRef] [PubMed]
52. Larson, E.T.; Kim, J.E.; Castaneda, L.J.; Napuli, A.J.; Zhang, Z.; Fan, E.; Zucker, F.H.; Verlinde, C.L.; Buckner, F.S.; Van Voorhis, W.C.; et al. The double-length tyrosyl-tRNA synthetase from the eukaryote *Leishmania major* forms an intrinsically asymmetric pseudo-dimer. *J. Mol. Biol.* **2011**, *409*, 159–176. [CrossRef] [PubMed]

Review

Future of the Genetic Code

Hong Xue and J. Tze-Fei Wong *

Division of Life Science and Applied Genomics Center, Hong Kong University of Science & Technology, Clear Water Bay, Hong Kong, China; hxue@ust.hk
* Correspondence: bcjtw@ust.hk; Tel.: +852-2358-7288; Fax: +852-2358-1552

Academic Editor: Koji Tamura
Received: 6 January 2017; Accepted: 23 February 2017; Published: 28 February 2017

Abstract: The methods for establishing synthetic lifeforms with rewritten genetic codes comprising non-canonical amino acids (NCAA) in addition to canonical amino acids (CAA) include proteome-wide replacement of CAA, insertion through suppression of nonsense codon, and insertion via the pyrrolysine and selenocysteine pathways. Proteome-wide reassignments of nonsense codons and sense codons are also under development. These methods enable the application of NCAAs to enrich both fundamental and applied aspects of protein chemistry and biology. Sense codon reassignment to NCAA could incur problems arising from the usage of anticodons as identity elements on tRNA, and possible misreading of NNY codons by UNN anticodons. Evidence suggests that the problem of anticodons as identity elements can be diminished or resolved through removal from the tRNA of all identity elements besides the anticodon, and the problem of misreading of NNY codons by UNN anticodon can be resolved by the retirement of both the UNN anticodon and its complementary NNA codon from the proteome in the event that a restrictive post-transcriptional modification of the UNN anticodon by host enzymes to prevent the misreading cannot be obtained.

Keywords: synthetic life; non-canonical amino acid; rewritten genetic code; restrictive post-transcriptional modification; anticodon identity element

1. Introduction

The 100th anniversary of the birth of Francis Crick is an occasion that calls for celebration of the double helix, triplet genetic code, tRNA, wobble rules, and aspects of molecular biology and neurobiology advanced by his gifted insights.

The cooperation between RNA and protein founded on the triplet genetic code is pivotal to life, and the development of the standard genetic code is the centerpiece in life's emergence. The sharing of the same protein alphabet by all living species suggests that the alphabet determined by the standard code predated the earliest divergence of organisms. Yet evidence suggests that the early code began with prebiotically-derived amino acids and expanded to include biosynthetically derived amino acids, thus predicting that the code is intrinsically a mutable code. To put this prediction to the test, experiments were performed in 1983 to determine the code's mutability. These experiments led to the isolation of genetic code mutants of the Trp-auxotroph *Bacillus subtilis* QB928 in which Trp has been either replaced in the code by its normally toxic fluoro analogues 4FTrp (4-fluoroTrp), 5FTrp, and 6FTrp; or even displaced entirely by 4FTrp to become an inhibitory analogue [1–5]. This proof of the mutability of the code opens up the code to revision and expansion, encoding NCAAs alongside CAAs in the protein alphabet.

2. Synthetic Lifeform Production

Since the genetic code is the most basic attribute of living systems, genetic code mutants represent the ultimate test-tube evolution [6]. Accordingly, organisms such as the *B. subtilis* strains that have rejected Trp from their genetic codes may be designated as synthetic lifeforms, distinct from synthetic biological constructs that contain novel genes and gene ensembles, but adhere strictly to the universal protein and nucleic acid alphabets [7]. The synthetic lifeforms with altered protein alphabets can be either optional or mandatory in their utilization of NCAAs, and the NCAAs can be incorporated proteome-wide or localized to specific protein sites. Synthetic lifeforms using a revised DNA alphabet where thymine is replaced by 5-chlorouracil on an optional or mandatory basis [8–10], or an extra unnatural base pair exemplified by dNaM-d5SICS has been added to the A-T and G-C pairs [11–13], have since been isolated as well.

Since the isolation of the first synthetic lifeforms from *Bacillus subtilis*, a number of different methods have been developed for the production of synthetic lifeforms equipped with mutated genetic codes and altered protein alphabets:

Proteome-wide replacement of CAA by NCAA: NCAAs that have acquired encoding by the genetic code include fluoroTrps in *B. subtilis* [1,4], *E. coli* and coliphages [14–16], and L-β-(thieno[3,2-b]-pyrrolyl)alanine in *E. coli* [17]. With respect to 4FTrp, genome sequencing of the successive mutant strains leading from wild-type *B. subtilis* to the HR23 strain which rejects Trp from its genetic code, and on to revertant TR7 strains where Trp rejoins the code, revealed how 4FTrp and Trp could be admitted into or excluded from the code as the result of a limited number of mutations in oligogenic-barrier genes that have preserved the protein alphabet against change through the ages [2,4,5].

NCAA insertion through suppression of a nonsense codon: To safeguard the fidelity of translation, aminoacyl-tRNA synthetases (ARSs) react with tRNAs cognate to the amino acid substrate, but not tRNAs cognate to other amino acids. Surprisingly, however, when *E. coli* ARSs were presented with tRNAs from other species, they reacted well with most tRNAs sourced from other bacteria, but poorly with numerous tRNAs from another biological domain: with *Halobacterium cutirubrum* tRNAs, they displayed only 1%–3% reactivity with tRNAs for Phe, Asp, Lys, and Pro, 0.4%–0.5% with tRNAs for Tyr, Leu, and Arg, and 0% with tRNAs for Ser [18,19]. Such strikingly low inter-domain reactivities made possible the design of 'orthogonal' ARS-tRNA pairs that do not interact productively with tRNAs and ARSs of the host cell [20], notably based on the archaeal *Methanocaldococcus* TyrRS, *Methanosarcina* PylRS, and SepRS from methanogens [21]. Using this approach, a wide range of site-specific NCAA incorporations in bacterial and eukaryotic hosts including *C. elegans* and *Xenopus* oocytes have been achieved based on suppression of nonsense codon [22–26]. Limitations of NCAA insertion through nonsense codon suppression include low efficiency of NCAA translation at levels close to those displayed by near-cognate CAAs, and competition from release factor for the nonsense codon [21]. These limitations could be overcome by directed evolution, with up to three-fold reduction of promiscuous aminoacylation of orthogonal tRNA by endogenous ARS through a combination of positive selection for amber suppression activity and negative selection toward ARS [27,28]. Efficiency also may be enhanced by the use of quadruplet-decoding ribosomes [29].

NCAA encoding through pyrrolysine and selenocysteine pathways: Since the nonsense codon UAG can perform double duty as a termination signal and a codon for pyrrolysine (Pyl) depending on the sequence context, PylRS and its cognate tRNA(Pyl) represent an endogenous orthogonal ARS-tRNA pair [30], and NCAA encoding has been achieved through their adaptation to NCAAs [31–33]. Although PylRS is comparable in catalytic efficiency to other ARSs with regard to Pyl-tRNA(Pyl) formation, engineered PylRS displayed 1000-fold reduction in catalytic efficiency toward NCAA [21], again indicating the need for enhancement. The nonsense codon UGA, likewise, can perform double duty as a termination signal and a codon for selenocysteine (Sec). The adaptation of the Sec pathway to NCAA encoding is rendered difficult by the requirement for a selenocysteine insertion sequence (SECIS) motif, and the non-binding of Sec-tRNA(Sec) to EF-Tu. However, both of these hurdles have been removed through the development of an effective tRNA(Sec) that incorporated Sec into proteins via EF-Tu binding, thus converting the highly restricted occurrence of Sec only at SECIS-directed mRNA contexts to unconstrained insertion as an NCAA anywhere in the proteome [34,35]. This liberated Sec pathway therefore can be adapted to the incorporation of NCAAs as in the case of the Pyl pathway.

Proteome-wide reassignment of nonsense codon: To overcome the restriction of site-specific incorporation of NCAA, nonsense codon encoding of NCAAs is being extended to the proteome-wide scale by replacing all 314 UAG nonsense codons in *E. coli* by the synonymous UAA nonsense codon using multiplex automated genome engineering (MAGE) followed by hierarchical conjugative assembly genome engineering (CAGE). The retired UAG codon is thus ready for proteome-wide reassignment to NCAA [36].

Proteome-wide reassignment of sense codon: The genetic code contains only three nonsense codons but 61 sense codons that might be reassignable to NCAAs. Reserving two codons for each CAA other than Met and Trp still leaves many sense codons for reassignment to NCAAs. Reassignments of the rare AGG and CGG codons of Arg to NCAAs have been achieved for homoArg and N6-(1-iminoethyl)Lys via the pyrrolysine pathway [37–39]. As well, genome-wide replacement of 13 rare sense codons from a panel of 42 highly-expressed essential genes of *E. coli* by their synonymous codons was functionally tolerable [40], and an orthogonal IleRS-tRNA(Ile) pair from *Mycoplasma mobile* was found that could decipher the AUA sense codon of *E. coli*, indicating the feasibility of reassigning this codon to NCAA [41]. General reassignment of non-rare codons on a proteome-wide basis will entail large scale genome reengineering employing either genome editing procedures or synthetic genomes exemplified by the *Mycoplasma mycoides* JCV1-syn1.0 genome [42]. To cope with the possibly severe adverse effects of such reassignment on the overall performance of the proteome, it might be necessary in some cases to increase gradually the NCAA/CAA incorporation ratio at the codon positions involved instead of an abrupt switch from CAA to NCAA. In addition, important issues that need to be resolved include the usage of anticodon as an identity element on tRNA, and the inherent ability of the UNN anticodon to read all four codons in a family codon box.

3. Anticodons as Identity Elements

The selection of tRNA substrate by ARS for aminoacylation is guided by identity elements on tRNA, e.g., the identity elements on *B. subtilis* tRNA(Trp) include G73 and its anticodon as major, and A1-U72, G5-C68, and A9 as minor, identity elements [43]. The discriminator base N73 and the anticodon are the most commonly encountered identity elements [44]. To reassign the CGA and CGG codons of Arg to an NCAA, for instance, these codons need to be replaced by synonymous Arg codons, and the tRNA(Arg)s that read CGA and CGG have to be deleted from the genome, followed by the introduction of orthogonal NcaaARS and tRNA(Ncaa)s bearing UCG anticodon, or UCG and CCG anticodons into the host cell. However, where ArgRS recognizes the anticodon as an identity element, as in *E. coli* and yeast [44], the endogenous ArgRS in the host cell could misread the anticodon on the newly-introduced tRNA(Ncaa)s as its cognate, and proceed to arginylate these tRNA(Ncaa)s [21,38,45]. If the anticodon represents the sole tRNA identity element recognized by ArgRS, the misreading error may be difficult to avoid. On the other hand, if ArgRS recognizes the anticodon along with other co-identity elements, the magnitude of the misreading error would depend on whether the various identity elements act additively or synergistically.

Earlier, when cloned bovine tRNA(Trp) bearing the A73 and G1-C72 identity elements was reacted with human TrpRS, highly efficient Trp-tRNA(Trp) formation was observed. When either A73 or G1-C72 was altered, efficiency was reduced to ∼25%. When both A73 and G1-C72 were altered, or when the 1-72 base pair was disrupted, efficiency was reduced to <2.5% (Figure 1). These findings showed that the identity elements A73 and G1-C72 acted synergistically, rather than additively in attracting Trp-tRNA formation, such that alteration of either A73 or G1-C72 diminished the Trp-tRNA formation by more than half. The physical basis of the synergism between A73 and G1-C72 was demonstrated by NMR spectroscopy: a change in the chemical shift for the G1-C72 base pair became manifest upon replacement of A73 by G73, with the G1 resonance moved in the ^{1}H dimension from 11.75 to 12.15 ppm [46].

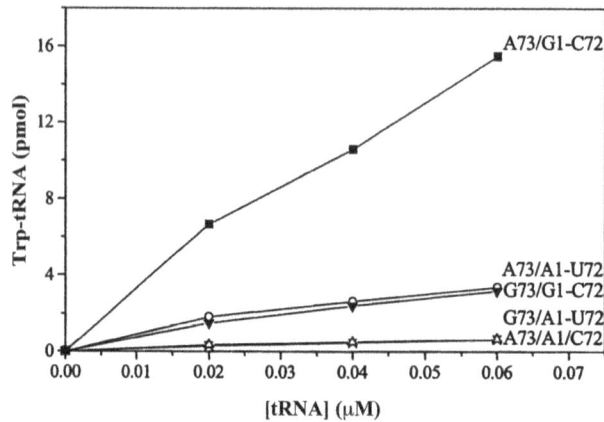

Figure 1. Human TrpRS tryptophanylation of bovine tRNA(Trp) with varied N73 and 1-72 configurations (after [46]).

Moreover, the relative rate of Trp-tRNA formation by *B. subtilis* TrpRS was 100% for *B. subtilis* tRNA, 0.5% for bovine tRNA, and 1.0% for *Archaeglobus fulgidus* tRNA. On the other hand, the relative rate of Trp-tRNA formation by *A. fulgidus* TrpRS was 100% for *A. fulgidus* tRNA, 4% for bovine tRNA and 5% for *B. subtilis* tRNA. Therefore, although the anticodon was a major identity element for *B. subtilis* TrpRS, and both *A. fulgidus* tRNA(Trp) and bovine tRNA(Trp) carried the same CCA anticodon as *B. subtilis* tRNA(Trp), *B. subtilis* TrpRS reacted poorly with bovine and *A. fulgidus* tRNAs, indicating that the presence of a correct anticodon identity element was insufficient to attract substantial Trp-tRNA formation by *B. subtilis* TrpRS [46]. Altogether the findings suggest that the problem of anticodon as identity element could be overcome or diminished in at least some instances by removing from the orthogonal tRNA(Ncaa) all identity elements outside of the anticodon that could attract misacylation by the ARS(Caa) for the CAA donating the codons to the NCAA.

4. Misreading by the UNN Anticodon

Since codon boxes are, as a rule, read by less than four anticodons, reading of multiple codons by an anticodon is commonplace. This suggests that reassignment of just a single sense codon to an NCAA can easily trigger misreading of a codon by a non-cognate anticodon belonging to the same box except for specialized instances illustrated by the AUA codon [41]. On the other hand, because Ser is allocated six codons in the code and the anticodon on tRNA(Ser) is also not an identity element in host species such as *E. coli* and yeast [44], reassignment of the entire UCN Ser codon box to an NCAA could be free of both the anticodon identity element problem and the UNN anticodon misreading problem. However, this would leave only the two codons AGU and AGC for Ser. Since there are comparable numbers of Ser and Glu residues in the *E. coli* proteome, and Glu functions well with only two codons, Ser might manage with two as well. In contrast, splitting up an existing 1aa codon box may incur misreading of NNU and NNC codons by the UNN anticodon.

Readings of codons by anticodons are governed by Crick's wobble rules [47]. Evolution of tRNA sequences indicates that there have been four major stages of wobble development [48]. In Stage 1 wobble, a single tRNA with a UNN anticodon reads all four codons in a codon box. As a result, no codon box could accommodate more than a single amino acid. Stage 2 wobble allocates a GNN-UNN anticodon duo to each box, thus allowing both 1aa and 2aa codon boxes. All eight standard 1aa boxes and five standard 2aa boxes of *Methanopyrus kandleri* (Mka), which is most closely related to the Last Universal Common Ancestor (LUCA), employ Stage 2 wobble. Some archaeons employ in their standard boxes both Stage 2 wobble and Stage 3 wobble which allocates a GNN-UNN-CNN anticodon trio to each box. The majority of *Archaea* species use a Stage 3 wobble for all of their standard boxes. Stage 4 wobble, mainly used by eukaryotes, adds yet another anticodon viz. A(I)NN to its anticodon ensemble.

In Stage 1, the U-1st anticodon reads all four codons of a box in a two-out-of-three reading mode [49,50]. In Stage 2, a G-1st anticodon reads the Y-3rd codons in a box; the U-1st anticodon reads mainly or only the R-3rd codons in a 1aa box, but only the R-3rd codons in a 2aa box. Since the UNN anticodon is inherently capable of two-out-of-three reading, some solution has to be developed to disallow misreading of Y-3rd codons by UNN anticodon in a 2aa box [51,52]. In the course of genetic code evolution, primitive organisms about to acquire Stage 2 wobble had to choose between two solutions to this challenge [53,54]:

Solution I: Prohibit the reading of Y-3rd codons by UNN anticodon through its post-transcriptional modification to form a restricted *UNN anticodon capable of reading the R-3rd but not the Y-3rd codons in a 2aa box; or

Solution II: Abandon the use of the A-3rd codon and U-1st anticodon in the 2aa box altogether, leaving only a C-1st anticodon to read the G-3rd codon and a G-1st anticodon to read the Y-3rd codons.

Since UNN anticodons were already in use in all the codon boxes of Stage 1 organisms, it was expeditious for the nascent Stage 2 organisms to adopt Solution I, using post-transcriptional modification enzymes to convert UNN anticodons into *UNN. Such conversion proceeded readily in the CAN, AAN, GAN, and AGN boxes in the right half of the code that employs R-2nd codons, but not in the codon boxes in the left half of the code that employs Y-2nd codons owing to the different allowable codon-anticodon configurations in the two halves [55–57]. Among extant organisms, modifications of U34 to form xm^5U34, s^2U34, and Um34 represent the major routes to a restricted *UNN [58]. The recognition signals for these modifications are largely non-elucidated for any organism, except that the signal for ribose methylation on N34 appears to reside outside the anticodon [59].

In splitting the four codons in a 1aa codon box such as the CCN box for Pro into two pairs, one may leave the CCY codons to Pro and give the CCR codons to an NCAA, or vice versa. Giving the CCR codons to NCAA could be preferable unless the existing UGG anticodon for Pro is already in the *UGG form. This requires a search for suitable orthogonal tRNA(Ncaa)s bearing UGG and CGG anticodons. Since a UGG anticodon might misread the CCU and CCC codons, the UGG anticodon on the sought-for tRNA(Ncaa) should be conducive to being modified into a restricted *UGG by host enzymes. There is a paucity of information on recognition signals on tRNA that direct the enzymic conversion of a UNN anticodon to *UNN. More broadly, we earlier cloned the genes for *B. subtilis*, *A. fulgidus*, and bovine tRNA(Trp)s into *E. coli* host, and examined the post-transcriptional modifications on their gene products [46]. The results showed that (Table 1):

(a) All the cloned tRNA(Trp)s displayed s^4U in resemblance to native *E. coli* tRNA(Trp) even though native *B. subtilis* and native bovine tRNA(Trp)s were devoid of this modification. Both cloned *B. subtilis* and native *E. coli* tRNA(Trp)s contained Cm, although native *B. subtilis* tRNA(Trp) lacked this modification. Native bovine tRNA(Trp) contained m^1A and m^2G, whereas cloned bovine and native *E. coli* tRNA(Trp)s were devoid of these modifications. These findings showed the decisive influence of host enzymes regarding some modifications on exogenous tRNAs.

(b) On the other hand, native and cloned *B. subtilus* tRNA(Trp)s both contained i^6A, but native *E. coli* tRNA(Trp) did not. Also, native and cloned bovine tRNA(Trp)s both contained Gm, but native *E. coli* tRNA(Trp) did not. These findings showed that the sequence of an exogenous tRNA could be a more important determinant than host enzymes for other modifications.

Accordingly, when a UNN anticodon-bearing orthogonal tRNA(Ncaa) is introduced into a host cell, both the sequence of the tRNA and the specificity of host enzymes could impact on the post-transcriptional modification profile of the tRNA. As a result, there is a possibility, but no certainty, that any orthogonal tRNA(Ncaa) with a UNN anticodon would be converted to a restricted tRNA(Ncaa) with a *UNN anticodon. Therefore multiple candidate tRNA(Ncaa)s may have to be screened to find one that yields a *UNN. Insofar that the UNN anticodons for all the 2aa codon boxes of any host cell are prone to conversion to *UNN by host enzymes, the chances of a tRNA(Ncaa) attracting a similar conversion by host enzymes might be increased if its anticodon loop sequence resembles those of the UNN anticodon-bearing tRNAs for 2aa boxes in the host cell.

Table 1. Post-transcriptional modifications of tRNA(Trp)s (after [46]).

tRNA(Trp)	s^4U	Gm	Cm	m^1A	i^6A	m^2G
Cloned *B. subtilis*	0.2	0.2	1.0	–	0.9	–
Cloned bovine	0.3	0.1	0.9	–	–	–
Cloned *A. fulgidus*	0.2	0.1	–	–	–	–
Native *E. coli*	0.5	–	1.1	–	–	–
Native *B. subtilis*	–	–	–	–	+	–
Native bovine	–	+	+	+	–	+
Native *H. volcanii*	–	–	+	–	–	–

In the event that repeated trials fail to produce a *UGG-bearing tRNA(Ncaa) to read a CCA codon to be reassigned to an NCAA, Solution II can be adopted by retiring all CCA codons from the proteome, and deleting from the genome all tRNA(Pro)s that read the CCA codon. Although the presence of unassigned codons that are not assigned to any amino acid or termination signal could be detrimental to an organism, the extent of the detriment might not always be severe [53], and unassigned codons are also known to exist in organisms [60,61]. Moreover, when the CCA codon is entirely retired from the proteome, it becomes an absent codon rather than an unassigned codon and unable to cause damage except where it reappears through random point mutations. Such random reappearances would be tolerated by the cells as in the case of a great majority of random point mutations if they do not cause significant damage, and eliminated by natural selection if they do. Thus, either way, their perturbation of the reassignment process could be limited.

5. Discussion

By means of electric discharge, high-energy particle bombardment, and conveyance on meteorites, etc., prebiotic Earth was endowed with the presence of a range of different amino acids [62]. After primitive RNA genes accumulated through selection of aptamers and ribozymes over non-functional RNAs to initiate the RNA world, attachment of amino acids to these functional RNAs (fRNAs) furnished much needed extra sidechains to enhance catalytic and transporter activities. The fRNAs developed mRNA domains to encode peptide prosthetic groups composed of Phase 1 prebiotically-derived amino acids [48,63]. The sidechain imperative continued to propel the coevolution of genetic code and amino acid biosynthesis, adding biosynthetically derived Phase 2 amino acids to the code [64,65]. Even after establishment of the standard genetic code, more Phase 3 amino acid residues are added to proteins via post-translational modifications (PTM) [66]. There are more than 87,000 experimentally-identified PTMs on 530,264 proteins including phosphorylation, acetylation, glycosylation, amidation, hydroxylation, and D-amino acids, etc., such that the rate of detection of PTM sites is outpacing biological knowledge of the function of those modifications [67,68]. This flourish of PTMs underlines the sustained strength of the sidechain imperative as a factor in protein evolution. In just over three decades since the isolation of the *B. subtilis* genetic code mutants [1], a wide range of mandatory and optional synthetic lifeforms have been developed incorporating a comparable number of code mutation-induced Phase 4 NCAAs as PTM-induced Phase 3 NCAAs into proteins [1,4,14–17,20–33,37], attesting to the continuing need, now scientifically-perceived instead of evolution-driven, for more amino acid sidechains.

The methods that have been developed to produce Phase 4 protein alphabets include proteome-wide replacement of CAA by NCAA, NCAA insertion through suppression of nonsense codons, and NCAA encoding through the pyrrolysine and selenocysteine pathways. Efforts to encode NCAAs through proteome-wide reassignment of nonsense or sense codons are also ongoing. Reassignment of non-rare codons is a large-scale operation, but genome editing and synthetic genomes are powerful tools well suited to the task. Such reassignments may meet with hurdles posed by anticodons that double as tRNA identity elements and, in the event of codon box splitting, misreading of NNY codons by UNN anticodon. However, the results shown in Figure 1 and Table 1, together with the option of Solution II abandoning both the NNA codon and UNN anticodon for a split box suggest that these hurdles could be surmountable in at least some instances. Deepened insight into the structure-function relations of ARSs and tRNAs, especially archaeal ones as a rich source of orthogonal components, together with an understanding of the recognition signals on tRNA for the conversion of UNN anticodons to *UNN in different host systems would be most valuable. In addition to NCAAs that are unknown to living cells, Phase 3 NCAA residues produced by PTM likewise merit encoding so they can become Phase 4 NCAAs, freed from their hitherto strict dependence on PTM enzymes and insertable anywhere in the proteome. These PTM-generated NCAA residues have long served important functions in vivo, and their catabolic products are non-toxic at modest levels.

In the triplet genetic code, the UCN Ser, CUN Leu, and CGN Arg codon boxes can be reassigned to three NCAAs without codon box splitting. Allowing box splitting and leaving two codons to each CAA outside of Met and Trp can accommodate up to 12 NCAAs. Switching from triplet to quadruplet codons [29] gives rise to 256 codons, and addition of an extra X-Y base pair to the current A-T and G-C base pairs [11–13] gives rise to 216 triplet codons, both vastly increasing the number of NCAAs that might be packaged into the same code. Since different synthetic lifeforms may be allocated different sets of NCAAs, there will be ample scope to construct revised triplet codes using just the A-T and G-C base pairs, each encoding 20 CAAs, Sec, Pyl plus one or more NCAAs, for the exploration of wide ranging applications of NCAA-enhanced protein alphabets including the following:

5.1. Protein Structure-Activity Relationships

Active sites of enzymes are often strongly conserved in localized sequence space owing to an optimally evolved protein configuration. However, the strong conservation renders it difficult to assess the uniqueness of the evolved configuration. In this regard, when the genomes of TR7 revertants of synthetic lifeform HR23, in which Trp has regained the ability to support cell propagation, were sequenced, their RNA polymerases were found to harbor the β-Glu433Lys, β'-Ile280Thr or β'-Pro277His mutations even though the β-Glu433, β'-Ile280 and β'-Pro277 residues at the outer claw-like region of bacterial RNAP where sigma factor binds are strongly conserved [5]. Without the immense stress placed on the cells by the proteome-wide displacement of Trp by 4FTrp, it would be highly unlikely to encounter the presence of β-Lys433, β'-Thr280, or β'-His277 in an alternative active configuration of the RNAP claw. Consequently, when NCAAs are inserted into diverse positions throughout the proteome, they can reveal unforeseen insights into the structure-activity relationships of many proteins.

5.2. Peptide and Protein Drugs

Novel peptide/protein drugs, antibodies and industrial proteins are applications that can benefit from NCAAs. Eight NCAA residues, including D- and DL-amino acids, have been incorporated into icatibant, a bradykinin B2 receptor antagonist decapeptide [69], and D-amino acid containing peptides furnish candidate drugs to inhibit the aggregation of amyloid peptides in Alzheimer's disease [70]. Conversion of cystine to diselenide in alpha-selenoconotoxins increases drug efficacy through prevention of reduction by glutathione or serum albumin [71]. NCAA encoding would accelerate development of such peptide or protein drugs where the presence of NCAA expands the physicochemical properties of their sidechains. Since the $\Delta\Delta G$ for transfer from octanol to water of 4FTrp, 6FTrp, or 5FTrp compared to Trp amounts to 0.42, 0.80, or 0.90 kcal/mole, respectively, replacement of Trp by 4FTrp, 6FTrp or 5FTrp can bring about a graded increase in hydrophobicity [72];

gamma-carboxyGlu can enhance charge intensity [73]; and irreversible attachment of the drugs to their target sites can be achieved using NCAAs capable of click chemistry [21,22,74].

MicroRNAs (miRNAs), small interfering RNAs (siRNAs), and peptide motifs such as zinc fingers, play important roles in gene regulation and represent potential therapeutic agents [75–78]. Nuclease-resistant nucleobase-peptides and nucleobase-proteins comprising both regular peptide segments and polyamide nucleic acid (PNA) [79] segments can be synthesized through the encoded incorporation of NCAAs bearing U(T), C, A, and G nucleobase sidechains. Given the capability of PNA for strong complementary base-pairing, nucleobase-peptides can provide therapeutic agents that interact with DNA advantageously with cooperation between the nucleobase sidechains and other amino acid sidechains.

5.3. Enhancement of Biological Fitness

The evolution of the genetic code was guided by the increased biological fitness made possible by extra amino acid sidechains in the protein alphabet, and this can be continued through expansion of the code to include NCAAs as exemplified by the finding that placement of 3-iodoTyr into the Tyr39 position of holing enhanced the competitive fitness of T7 phage [80]. Asn and Gln are highly prone to thermal deamidation [81], and the more thermostable albizzine (α-amino-β-ureidopropionic acid) has been proposed as a possible Gln replacement [82]. Since the deamidation of Asn and Gln in proteins could contribute to senescence, as well as amyloid formation in dementia, type 2 diabetes, cataracts and Parkinson's disease [83–85], it would be instructive to replace Asn and Gln residues in proteins with competent yet more stable analogues, if such can be found, to assess the roles of Asn and Gln as factors of senescence and human diseases.

5.4. Metabolic and Biomimetic Engineering

The opening up of selenoprotein biochemistry via unconstrained proteome-wide Sec encoding enables the replacement of disulfide bridges by diselenide bridges in proteins [34,35]. As a result, metabolic engineering can be carried out replacing glutathione (GSSG) as a cellular redox buffer by selenoglutathione (GSeSeG): otherwise selenoglutathione would destabilize disulfide bridges in proteins, for the $E^{\circ\prime}$ of -407 mV for GSeSeG is much more negative than that of -256 mV for GSSG [86]. The finding of 10^2 to 10^4-fold faster reduction of thioredoxin by synthetic seleno-glutaredoxin 3 compared to glutaredoxin 3 [87] suggests that replacement of glutathione by selenoglutathione could facilitate reductive processes such as nitrogen fixation [88,89].

The enzyme farnesyltransferase adds the 15-carbon farnesyl group to Cys in a CaaX motif at the carboxyl terminus of a protein, thereby favoring attachment of the protein to membranes. Ras proteins are activated by farnesylation and association with the inner surface of the plasma membrane [90]. Encoded incorporation of farnesylated NCAAs into various soluble proteins (if necessary using a modified EF-Tu that accommodates a bulky amino acid sidechain, or through sidechain add-on by means of click chemistry) can therefore relocate them on to membranes, thereby allowing a shift of protein distribution from cytosol to membranes. Such studies will contribute to delineation of the relationships between cellular architecture, regulation, and carcinogenesis.

Lipoproteins play important roles in metabolism and biological structures, but they are typically maintained by non-covalent rather than covalent binding between lipid and protein. However, akin to covalent nucleobase-proteins, covalent lipoproteins can be formed by incorporating lipoidal NCAAs with for example sphingosine, sphingomyelin, fatty acid, isoprenoid, or sterol type sidechains into proteins. The covalent lipoproteins can potentially be employed to produce artificial myelin sheaths and skin grafts, or circulating amphiphiles for coating the inner wall of blood vessels to prevent vascular plaques. They can also be used in a new generation of liposomes embedded with protein receptors, transporters, and ion channels as integral parts of the membrane.

An important goal in artificial photosynthesis for hydrogen fuel production is to find responsive matrices that mimic the protein matrices of natural photosynthesis, which through concerted motion can reduce entropy production and maximize the performance of essential components of the process [91–95]. Like natural photosynthesis, which proceeds in photosynthetic membranes where chlorophylls are held in place by specific proteins, covalent lipoprotein membranes containing ordered arrays of light harvesters, reaction centers, hydrogenase catalytic sites, etc. may provide suitable materials for engineering the necessary responsive matrices.

In conclusion, synthetic lifeforms employing rewritten genetic codes can be produced by a number of different methods. These methods will widen the scope of synthetic life research, bringing unique insights into protein chemistry and biology as well as a wide range of applications. Building the rewritten genetic codes and the novel protein alphabets ushered in by them, optimizing their uses and preventing all possibilities of misuse, will represent a momentous development that advances science, medicine and biotechnology.

Acknowledgments: The study was supported by grant FSGRF12SC30 from the Hong Kong University of Science and Technology.

Author Contributions: H.X. and J.T.-F.W. jointly conceived the study and wrote the paper.

Conflicts of Interest: The authors declare no conflict of interest.

References

1. Wong, J.T. Membership mutation of the genetic code: Loss of fitness by tryptophan. *Proc. Natl. Acad. Sci. USA* **1983**, *80*, 6303–6306. [CrossRef] [PubMed]
2. Wong, J.T. Evolution and mutation of the amino acid code. In *Dynamics of Biochemical Systems*; Ricard, J., Cornish-Bowden, A., Eds.; Plenum Press: New York, NY, USA, 1984; pp. 247–258.
3. Bronskill, P.M.; Wong, J.T. Suppression of fluorescence of tryptophan residues in proteins by replacement with 4-fluorotryptophan. *Biochem. J.* **1988**, *249*, 305–308. [CrossRef] [PubMed]
4. Mat, W.K.; Xue, H.; Wong, J.T. Genetic code mutations: The breaking of a three billion year invariance. *PLoS ONE* **2010**, *5*, e12206. [CrossRef] [PubMed]
5. Yu, A.C.; Yim, A.K.; Mat, W.K.; Tong, A.H.; Lok, S.; Xue, H.; Tsui, S.K.; Wong, J.T.; Chan, T.F. Mutations enabling displacement of tryptophan by 4-fluorotryptophan as a canonical amino acid of the genetic code. *Genome Biol. Evol.* **2014**, *6*, 629–641. [CrossRef] [PubMed]
6. Hesman, T. Code breakers: Scientists are altering bacteria in a most fundamental way. *Sci. News* **2000**, *157*, 360–362. [CrossRef]
7. Wong, J.T.; Xue, H. Synthetic genetic codes as the basis of synthetic life. In *Chemical Synthetic Biology*; Luisi, P.L., Chiarabelli, C., Eds.; Wiley: New York, NY, USA, 2010; pp. 178–199.
8. Marliere, P.; Patrouix, J.; Doring, V.; Herdewijn, P.; Tricot, S.; Cruveiller, S.; Bouzon, M.; Mutzel, R. Chemical evolution of a bacterium's genome. *Angew Chem. Int. Ed. Engl.* **2011**, *50*, 7109–7114. [CrossRef] [PubMed]
9. Marliere, P. Charting the xenobiotic continent. In Proceedings of the First Conference on Xenobiology, Genoa, Italy, 6–8 May 2014.
10. Acevedo-Rocha, C.G.; Budisa, N. On the road towards chemically modified organisms endowed with a genetic firewall. *Angew Chem. Int. Ed. Engl.* **2011**, *50*, 6960–6962. [CrossRef] [PubMed]
11. Benner, S.A.; Chen, F.; Yang, Z. Synthetic biology, tinkering biology and artificial biology: A perspective from chemistry. In *Chemical Synthetic Biology*; Luisi, P.L., Chiarabelli, C., Eds.; Wiley: New York, NY, USA, 2010; pp. 69–106.
12. Li, L.; Degardin, M.; Lavergne, T.; Malyshev, D.A.; Kirandeep, D.; Ordoukhanian, P.; Romesberg, F.E. Natural-like replication of an unnatural base pair for the expansion of the genetic alphabet and biotechnology applications. *J. Am. Chem. Soc.* **2014**, *136*, 826–829. [CrossRef] [PubMed]
13. Zhang, Y.; Lamb, B.M.; Feldman, A.W.; Zhou, A.X.; Lavergne, T.; Li, L.; Romesberg, F.E. A semisynthetic organism engineered for the stable expansion of the genetic alphabet. *Proc. Nat. Acad. Sci. USA* **2017**, *114*, 1317–1322. [CrossRef] [PubMed]
14. Bacher, J.M.; Ellington, A.D. Selection and characterization of *Escherichia coli* variants capable of growth on an otherwise toxic tryptophan analogue. *J. Bacteriol.* **2001**, *183*, 5414–5425. [CrossRef] [PubMed]
15. Bacher, J.M.; Bull, J.J.; Ellington, A.D. Evolution of phage with chemically ambiguous proteomes. *BMC Evol. Biol.* **2003**, *3*. [CrossRef] [PubMed]

16. Bacher, J.M.; Hughes, R.A.; Wong, J.T.; Ellington, A.D. Evolving new genetic codes. *Trends Ecol. Evol.* **2004**, *19*, 69–75. [CrossRef] [PubMed]

17. Hoesl, M.G.; Oehm, S.; Durkin, P.; Darmon, E.; Peil, L.; Aerni, H.R.; Rappsilber, J.; Rinehart, J.; Leach, D.; Soll, D.; et al. Chemical evolution of a bacterial proteome. *Angew Chem. Int. Ed. Engl.* **2015**, *54*, 10030–10034. [CrossRef] [PubMed]

18. Kwok, Y.; Wong, J.T. Evolutionary relationship between Halobacterium cutirubrum and eukaryotes determined by use of aminoacyl-tRNA synthetases as phylogenetic probes. *Can. J. Biochem.* **1980**, *58*, 213–218. [CrossRef] [PubMed]

19. Wong, J.T. Emergence of life: From functional RNA selection to natural selection and beyond. *Front Biosci.* **2014**, *19*, 1117–1150. [CrossRef]

20. Santoro, S.W.; Anderson, J.C.; Lakshman, V.; Schultz, P.G. An archaebacteria-derived glutamyl-tRNA synthetase and tRNA pair for unnatural amino acid mutagenesis of proteins in *Escherichia coli*. *Nucleic Acids Res.* **2003**, *31*, 6700–6709. [CrossRef] [PubMed]

21. O'Donoghue, P.; Ling, J.; Wang, Y.S.; Soll, D. Upgrading protein synthesis for synthetic biology. *Nat. Chem. Biol.* **2013**, *9*, 594–598. [CrossRef] [PubMed]

22. Liu, C.C.; Schultz, P.G. Adding new chemistries to the genetic code. *Annu. Rev. Biochem.* **2010**, *79*, 413–444. [CrossRef] [PubMed]

23. Hoesl, M.G.; Budisa, N. Recent advances in genetic code engineering in *Escherichia coli*. *Curr. Opin. Biotechnol.* **2012**, *23*, 751–757. [CrossRef] [PubMed]

24. Greiss, S.; Chin, J.W. Expanding the genetic code of an animal. *J. Am. Chem. Soc.* **2011**, *133*, 14196–14199. [CrossRef] [PubMed]

25. Parrish, A.R.; She, X.; Xiang, Z.; Coin, I.; Shen, Z.; Briggs, S.P.; Dillin, A.; Wang, L. Expanding the genetic code of Caenorhabditis elegans using bacterial aminoacyl-tRNA synthetase/tRNA pairs. *ACS Chem. Biol.* **2012**, *7*, 1292–1302. [CrossRef] [PubMed]

26. Ye, S.; Riou, M.; Carvalho, S.; Paoletti, P. Expanding the genetic code in Xenopus laevis oocytes. *ChemBioChem* **2013**, *14*, 230–235. [CrossRef] [PubMed]

27. Ellefson, J.W.; Meyer, A.J.; Hughes, R.A.; Cannon, J.R.; Brodbelt, J.S.; Ellington, A.D. Directed evolution of genetic parts and circuits by compartmentalized partnered replication. *Nat. Biotechnol.* **2014**, *32*, 97–101. [CrossRef] [PubMed]

28. Maranhao, A.C.; Ellington, A.D. Evolving orthogonal suppressor tRNAs to incorporate modified amino acids. *ACS Synth. Biol.* **2016**, 27600875. [CrossRef] [PubMed]

29. Neumann, H.; Wang, K.; Davis, L.; Garcia-Alai, M.; Chin, J.W. Encoding multiple unnatural amino acids via evolution of a quadruplet-decoding ribosomes. *Nature* **2010**, *464*, 441–444. [CrossRef] [PubMed]

30. Nozawa, K.; O'Donoghue, P.; Gundllapalli, S.; Araiso, Y.; Ishitani, R.; Umehara, T.; Soll, D.; Nureki, O. Pyrrolysyl-tRNA synthetase-tRNA(Pyl) structure reveals the molecular basis of orthogonality. *Nature* **2009**, *457*, 1163–1167. [CrossRef] [PubMed]

31. Mukai, T.; Kobayashi, T.; Hino, N.; Yanagishawa, T.; Sakamoto, K.; Yokoyama, S. Adding l-lysine derivatives to the genetic code of mammalian cells with engineered pyrrolysyl-tRNA synthetases. *Biochem. Biophys. Res. Commun.* **2008**, *371*, 818–822. [CrossRef] [PubMed]

32. Hancock, S.M.; Uprety, R.; Deiters, A.; Chin, J.W. Expanding the genetic code of yeast for incorporation of diverse unnatural amino acids via a pyrrolysyl-tRNA synthetase/tRNA pair. *J. Am. Chem. Soc.* **2010**, *132*, 14819–14824. [CrossRef] [PubMed]

33. Fekner, T.; Chan, M.K. The pyrrolysine translational machinery as a genetic code-expansion tool. *Curr. Opin. Chem. Biol.* **2011**, *15*, 387–391. [CrossRef] [PubMed]

34. Miller, C.; Brocker, M.J.; Prat, L.; Ip, K.; Chirathivat, N.; Felock, A.; Veszpremi, M.; Soll, D. A synthetic tRNA for EF-Tu mediated selenocysteine incorporation in vivo and in vitro. *FEBS Lett.* **2015**, *589*, 2194–2199. [CrossRef] [PubMed]

35. Thyer, R.; Robotham, S.A.; Brodbelt, J.S.; Ellington, A.D. Evolving tRNA[Sec] for efficient incorporation of selenocysteine. *J. Am. Chem. Soc.* **2015**, *137*, 46–49. [CrossRef] [PubMed]

36. Isaacs, F.J.; Carr, P.A.; Wang, H.H.; Lajoie, M.J.; Sterling, B.; Kraal, L.; Tolonen, A.C.; Gianoulis, T.A.; Goodman, D.B.; Reppas, N.B.; et al. Precise manipulation of chromosomes in vivo enables genome-wide codon replacement. *Science* **2011**, *333*, 348–353. [CrossRef] [PubMed]

37. Mukai, T.; Yamaguchi, A.; Ohtake, K.; Takahashi, M.; Hayashi, A.; Iraha, F.; Kira, S.; Yanagisawa, T.; Yokoyama, S.; Hoshi, H.; et al. Reassignment of a rare sense codon to a non-canonical amino acid in *Escherichia coli. Nucleic Acid Res.* **2015**, *43*, 8111–8122. [CrossRef] [PubMed]

38. Krishnakumar, R.; Prat, L.; Aerni, H.; Ling, J.; Merryman, C.; Glass, J.I.; Rinehart, J.; Söll, D. Transfer RNA misidentification scrambles sense codon recoding. *ChemBioChem* **2013**, *14*, 1967–1972. [CrossRef] [PubMed]

39. Zeng, Y.; Wang, W.; Liu, W.R. Toward reassigning the rare AGG codon in *Escherichia coli. ChemBioChem* **2014**, *15*, 1750–1754. [CrossRef] [PubMed]

40. Lajoie, M.J.; Kosuri, S.; Mosberg, J.A.; Gregg, C.J.; Zhang, D.; Church, G.M. Probing the limits of genetic recoding in essential genes. *Science* **2013**, *342*, 361–363. [CrossRef] [PubMed]

41. Bohlke, N.; Budisa, N. Sense codon emancipation for proteome-wide incorporation of noncanonical amino acids; rare isoleucine codon AUA as a target for genetic code expansion. *FEMS Mcrobiol. Lett.* **2014**, *351*, 133–144. [CrossRef] [PubMed]

42. Gibson, D.G.; Glass, J.I.; Lartigue, C.; Noskov, V.N.; Chuang, R.Y.; Algire, M.A.; Benders, G.A.; Montague, M.G.; Ma, L.; Moodie, M.M.; et al. Creation of a bacterial cell controlled by a chemically synthesized genome. *Science* **2010**, *329*, 52–56. [CrossRef] [PubMed]

43. Xue, H.; Shen, W.; Giege, R.; Wong, J.T. Identity elements of tRNA(Trp). Identification and evolutionary conservation. *J. Biol. Chem.* **1993**, *268*, 9316–9322. [PubMed]

44. Giege, R.; Sissler, M.; Florentz, C. Universal rules and idiosyncratic features in tRNA identity. *Nucleic Acid Res.* **1998**, *26*, 5017–5035. [CrossRef] [PubMed]

45. Krishnakumar, R.; Ling, J. Experimental challenges of sense codon reassignment: An innovative approach to genetic code expansion. *FEBS Lett.* **2014**, *588*, 383–388. [CrossRef] [PubMed]

46. Guo, Q.; Gong, Q.; Grosjean, H.; Zhu, G.; Wong, J.T.; Xue, H. Recognition by tryptophanyl-tRNA synthetases of discriminator base on the tRNA^{Trp} from three biological domains. *J. Biol. Chem.* **2002**, *277*, 14343–14349. [CrossRef] [PubMed]

47. Crick, F.H. Codon—Anticodon pairing: The wobble hypothesis. *J. Mol. Biol.* **1966**, *19*, 548–555. [CrossRef]

48. Wong, J.T.; Ng, S.K.; Mat, W.K.; Hu, T.; Xue, H. Coevolution theory of the genetic code at age forty: Pathway to translation and synthetic life. *Life* **2016**, *6*, 12. [CrossRef] [PubMed]

49. Lagerkvist, U. Two-out of three: An alternative method for codon reading. *Proc. Natl. Acad. Sci. USA* **1978**, *75*, 1759–1762. [CrossRef] [PubMed]

50. Rogalski, M.; Karcher, D.; Bock, R. Superwobbling facilitates translation with reduced tRNA sets. *Nat. Struct. Mol. Biol.* **2008**, *15*, 192–198. [CrossRef] [PubMed]

51. Grosjean, H.J.; de Henau, S.; Crothers, D.M. On the physical basis for ambiguity in genetic coding interaction. *Proc. Natl. Acad. Sci. USA* **1978**, *75*, 610–614. [CrossRef] [PubMed]

52. Yokoyama, S.; Nishimura, S. Modified nucleosides and codon recognition. In *tRNA Structure, Biosynthesis and Function*; Soll, D., RajBhandary, U.L., Eds.; American Society for Microbiology Press: Washington, DC, USA, 1995; pp. 207–223.

53. Van der Gulik, P.T.S.; Hoff, W.D. Unassigned codons, nonsense suppression, and anticodon modifications in the evolution of the genetic code. *J. Mol. Evol.* **2011**, *73*, 59–69. [CrossRef] [PubMed]

54. Van der Gulik, P.T.S.; Hoff, W.D. Anticodon modifications in the tRNA set of LUCA and the fundamental regularity in the standard genetic code. *PLoS ONE* **2016**, *11*, e0158342. [CrossRef] [PubMed]

55. Crick, F.H.C. The origin of the genetic code. *J. Mol. Biol.* **1968**, *38*, 367–379. [CrossRef]

56. Rumer, I.B. On codon systemization in the genetic code. *Dokl Akad Nauk SSSR* **1966**, *167*, 1393–1394. [PubMed]

57. Lehmann, J.; Libchaber, A. Degeneracy of the genetic code and stability of the base pair at the second position of the anticodon. *RNA* **2008**, *14*, 1264–1269. [CrossRef] [PubMed]

58. Curran, J.F. Modified nucleosides in translation. In *Modification and Editing of RNA*; Grosjean, H., Benne, R., Eds.; American Society for Microbiology Press: Washington, DC, USA, 1998; pp. 493–516.

59. Bjork, G.R. Biosynthesis and function of modified nucleosides. In *tRNA Structure, Biosynthesis and Function*; Soll, D., RajBhandary, U.L., Eds.; American Society for Microbiology Press: Washington, DC, USA, 1995; pp. 165–205.

60. Oba, T.; Andachi, Y.; Muto, A.; Osawa, S. CGG: An unassigned or nonsense codon in Mycoplasma capricolum. *Proc. Natl. Acad. Sci. USA* **1991**, *88*, 921–925. [CrossRef] [PubMed]

61. Kano, A.; Ohama, T.; Abe, R.; Osawa, S. Unassigned or nonsense codons in Micrococcus luteus. *J. Mol. Biol.* **1993**, *230*, 51–56. [CrossRef] [PubMed]

62. Miller, S.L. The formation of organic compounds on the primitive Earth. *Ann. N. Y. Acad. Sci.* **1957**, *69*, 260–275. [CrossRef] [PubMed]

63. Wong, J.T. Origin of genetically encoded protein synthesis: A model based on selection for RNA peptidation. *Ori. Life Evol. Biosph.* **1991**, *21*, 165–176. [CrossRef]

64. Wong, J.T. A co-evolution theory of the genetic code. *Proc. Natl. Acad. Sci. USA* **1975**, *72*, 1909–1912. [CrossRef] [PubMed]

65. Wong, J.T. Coevolution theory of the genetic code: A proven theory. *Ori. Life Evol. Biosph.* **2007**, *37*, 403–408. [CrossRef] [PubMed]

66. Wong, J.T. Coevolution of the genetic code and amino acid biosynthesis. *Trends Biochem. Sci.* **1981**, *16*, 33–35. [CrossRef]

67. Khoury, G.A.; Baliban, R.C.; Floudas, C.A. Proteome-wide post-translational modification statistics: Frequency analysis and curation of the swiss-prot database. *Sci. Rep.* **2011**, *1*, 90. [CrossRef] [PubMed]

68. Naegle, K.M.; Gymrek, M.; Joughin, B.A.; Wagner, J.P.; Welsch, R.E.; Yaffe, M.B.; Lauffenburger, D.A.; White, F.M. PTMScout, a Web resource for analysis of high throughput post-translational proteomics studies. *Mol. Cell. Proteom.* **2010**, *9*, 2558–2570. [CrossRef] [PubMed]

69. Cicardi, M.; Banerji, A.; Bracho, F.; Malbran, A.; Rosenkranz, B.; Riedl, M.; Bork, K.; Lumry, W.; Aberer, W.; Bier, H.; et al. Icatibant, a new bradykinin-receptor antagonist, in hereditary angioedema. *N. Engl. J. Med.* **2010**, *363*, 532–541. [CrossRef] [PubMed]

70. Kumar, J.; Sim, V. D-amino acid-based peptide inhibitors as early or preventive therapy in Alzheimer disease. *Prion* **2014**, *8*. [CrossRef]

71. Armishaw, C.J.; Daly, N.L.; Nevis, S.T.; Adams, D.J.; Craik, D.J.; Alewood, P.F. Alpha-selenoconotoxins, a new class of potent alpha7 neuronal nicotinic receptor antagonists. *J. Biol. Chem.* **2006**, *281*, 14136–14143. [CrossRef] [PubMed]

72. Xu, Z.J.; Love, M.L.; Ma, L.Y.Y.; Blum, M.; Bronskill, P.M.; Bernstein, J.; Grey, A.A.; Hofmann, T.; Camerman, N.; Wong, J.T. Tryptophanyl-tRNA synthetase from Bacillus subtilis: Characterization and role of hydrophobicity in substrate recognition. *J. Biol. Chem.* **1989**, *264*, 4304–4311. [PubMed]

73. Stenflo, J. Contributions of Gla and EGF-like domains to the function of vitamin K-dependent coagulation factors. *Crit. Rev. Eukaryot. Gene Expr.* **1999**, *9*, 59–88. [PubMed]

74. Wang, J.; Zhang, W.; Song, W.; Wang, Y.; Yu, Z.; Li, J.; Wu, M.; Wang, L.; Zang, J.; Lin, Q. A biosynthetic route to photoclick chemistry on proteins. *J. Am. Chem. Soc.* **2010**, *132*, 14812–14818. [CrossRef] [PubMed]

75. Wang, V.; Wu, W. MicroRNA-based therapeutics for cancer. *BioDrugs* **2009**, *23*, 15–23. [CrossRef] [PubMed]

76. Broderick, J.A.; Zamore, P.D. MicroRNA therapeutics. *Gene Ther.* **2011**, *18*, 1104–1110. [CrossRef] [PubMed]

77. Xu, G.L.; Bestor, T.H. Cytosine methylation targeted to pre-determined sequences. *Nat. Genet.* **1997**, *17*, 376–378. [CrossRef] [PubMed]

78. Gommans, W.M.; McLaughlin, P.M.; Lindhout, B.I.; Segal, D.J.; Wiegman, D.J.; Haisma, H.J.; van der Zaal, B.J.; Rots, M.G. Engineering zinc finger protein transcription factors to downregulate the epithelial glycoprotein-2 promotor as a novel anti-cancer treatment. *Mol. Carcinog.* **2007**, *46*, 391–401. [CrossRef] [PubMed]

79. Nielsen, P.E.; Egholm, M.; Berg, R.H.; Buchardt, O. Sequence-selective recognition of DNA by strand displacement with a thymine-substituted polyamide. *Science* **1991**, *254*, 1497–1500. [CrossRef] [PubMed]

80. Hammerling, M.J.; Ellefson, J.W.; Boutz, D.R.; Marcotte, E.M.; Ellington, A.D.; Barrick, J.E. Bacteriophages use an expanded genetic code on evolutionary paths to higher fitness. *Nat. Chem. Biol.* **2014**, *10*, 178–180. [CrossRef] [PubMed]

81. Wong, J.T.; Bronskill, P.M. Inadequacy of prebiotic synthesis as origin of proteinous amino acids. *J. Mol. Evol.* **1979**, *13*, 115–125. [CrossRef] [PubMed]

82. Weber, A.L.; Miller, S.L. Reasons for the occurrence of the twenty coded protein amino acids. *J. Mol. Evol.* **1981**, *17*, 273–284. [CrossRef] [PubMed]

83. Chavous, D.A.; Jackson, F.R.; O'Connor, C.M. Extension of the Drosophila lifespan by overexpression of a protein repair methyltransferase. *Proc. Natl. Acad. Sci. USA* **2001**, *98*, 14814–14818. [CrossRef] [PubMed]

84. Adav, S.S.; Gallart-Palau, X.; Tan, K.H.; Lim, S.K.; Tam, J.P.; Sze, S.K. Dementia-linked amyloidosis is associated with brain protein deamidation as revealed by proteomic profiling of human brain tissues. *Mol. Brain* **2016**, *9*, 20. [CrossRef] [PubMed]

85. Dunkelberger, E.B.; Buchanan, L.E.; Marek, P.; Cao, P.; Raleigh, D.P.; Zanni, M.T. Deamidation accelerates amyloid formation and alters amylain fiber structure. *J. Am. Chem. Soc.* **2012**, *134*, 12658–12667. [CrossRef] [PubMed]
86. Beld, J.; Woycechowsky, J.; Hilvert, D. Selenoglutathione: Efficient oxidative protein folding by a diselenide. *Biochemistry* **2007**, *46*, 5382–5390. [CrossRef] [PubMed]
87. Metanis, N.; Keinan, E.; Davison, P.E. Synthetic seleno-glutaredoxin 3 analogs are highly reducing oxidoreductases with enhanced catalytic efficiency. *J. Am. Chem. Soc.* **2006**, *128*, 16684–16691. [CrossRef] [PubMed]
88. Commans, S.; Bock, A. Selenocysteine inserting tRNAs: An overview. *FEMS Microbiol. Rev.* **1999**, *23*, 335–351. [CrossRef] [PubMed]
89. Fay, P. Oxygen relations of nitrogen fixation in cyanobacteria. *Microbiol. Rev.* **1992**, *56*, 340–373. [PubMed]
90. Rowinsky, E.K.; Windle, J.J.; Von Hoff, D.D. Ras protein farnesyltransferase: A strategic target for anticancer therapeutic development. *J. Clin. Oncol.* **1999**, *17*, 3631–3652. [CrossRef] [PubMed]
91. Purchase, R.L.; de Groot, H.J.M. Biosolar cells: Global artificial photosynthesis needs responsive matrices with quantum coherent kinetic control for high yield. *Interface Focus* **2015**, *5*. [CrossRef] [PubMed]
92. Magnuson, A.; Anderlund, M.; Johansson, O.; Lindblad, P.; Lomoth, R.; Polivka, T.; Ott, S.; Stensjö, K.; Styring, S.; Sundström, V.; et al. Biomimetic and microbial approaches to solar fuel generation. *Acc. Chem. Res.* **2009**, *42*, 1899–1909. [CrossRef] [PubMed]
93. Stripp, S.T.; Happe, T. How algae produce hydrogen-news from the photosynthetic hydrogenase. *Dalton Trans.* **2009**. [CrossRef] [PubMed]
94. English, C.M.; Eckert, C.; Brown, K.; Seibert, M.; King, P.W. Recombinant and in vitro expression systems for hydrogenases: New frontiers in basic and applied studies for biological and synthetic H_2 production. *Dalton Trans.* **2009**. [CrossRef] [PubMed]
95. Tard, C.; Pickett, C.J. Structural and functional analogues of the active sites of the [Fe]-, [NiFe]- and [FeFe]-hydrogenases. *Chem. Rev.* **2009**, *109*, 2245–2274. [CrossRef] [PubMed]

Review

Efforts and Challenges in Engineering the Genetic Code

Xiao Lin, Allen Chi Shing Yu and Ting Fung Chan *

School of Life Sciences, The Chinese University of Hong Kong, Sha Tin, NT, Hong Kong, China;
xlin@link.cuhk.edu.hk (X.L.); allenyu@cuhk.edu.hk (A.C.S.Y.)
* Correspondence: tf.chan@cuhk.edu.hk; Tel.: +852-3943-6876

Academic Editor: Koji Tamura
Received: 26 January 2017; Accepted: 10 March 2017; Published: 14 March 2017

Abstract: This year marks the 48th anniversary of Francis Crick's seminal work on the origin of the genetic code, in which he first proposed the "frozen accident" hypothesis to describe evolutionary selection against changes to the genetic code that cause devastating global proteome modification. However, numerous efforts have demonstrated the viability of both natural and artificial genetic code variations. Recent advances in genetic engineering allow the creation of synthetic organisms that incorporate noncanonical, or even unnatural, amino acids into the proteome. Currently, successful genetic code engineering is mainly achieved by creating orthogonal aminoacyl-tRNA/synthetase pairs to repurpose stop and rare codons or to induce quadruplet codons. In this review, we summarize the current progress in genetic code engineering and discuss the challenges, current understanding, and future perspectives regarding genetic code modification.

Keywords: frozen accident; genetic code; genetic engineering; evolution; synthetic biology

---

1. Introduction

In 1968, Francis Crick first proposed the frozen accident theory of the genetic code [1]. The 20 canonical amino acids were once believed to be immutable elements of the code. The genetic code appears to be universal, from simple unicellular organisms to complex vertebrates. Yet in contrast to studies of the natural selection of lifeforms wherein the gradual evolution of species can be observed in a myriad of taxa, relatively few examples of natural genetic code variations (e.g., selenocysteine [2], pyrrolysine [3,4], and stop codon read through [5,6]) have been observed. Different explanations have been proposed to address these variations, such as the codon capture hypothesis [7], the ambiguous intermediate hypothesis [8], and the genome streamlining hypothesis [9]. These hypotheses have been reviewed elsewhere [10,11]. Although some existing noncanonical amino acids (NCAAs) are known to be compatible with enzymatic aminoacylation [12–20], the 20 canonical amino acids in the standard genetic code have been stringently selected over the course of biological evolution. Organisms that require peptides with modified side chains will often resort to pre-translational or post-translational modifications to incorporate NCAAs [21–26]. Some organisms require alternative genetic codes to survive in harsh living conditions [27].

Genetic code engineering refers to the modification, or the directed evolution of cellular machineries, in order to incorporate NCAAs into the proteome of an organism. In general, NCAAs can be artificially incorporated in a site-specific or proteome-wide manner. In the former, scientists have attempted to artificially engineer organisms for compatibility with various NCAAs by employing orthogonal tRNA/aminoacyl-tRNA synthetase pairs [28–32]. In the latter, an organism is forced to take up specific NCAAs, followed by isolating mutants in media containing NCAAs [33–35]. Currently, researchers are recording cellular responses and genetic changes in engineered organisms to understand the mechanisms behind the use of alternative genetic codes. Such efforts enable an understanding of the evolutionary course of the genetic code and provide a foundation for the derivation of additional alternative codes, a particularly important feature in the era of synthetic biology given the increased focus on engineering synthetic organisms with modified genetic codes [36,37]. Engineered genetic code holds tremendous potential in the field of protein engineering and xenobiology, which was extensively reviewed by Budisa et al. in 2017 [38].

In this review, we first give a brief introduction of current studies on both site-specific and proteome-wide incorporation of NCAAs. Next, we will focus on the challenges of engineering organisms to use modified genetic codes and their implications, such as inhibitory effects caused by NCAAs. Finally, we will discuss current trends in this research area.

2. Genetic Code Engineering

2.1. Incorporation of NCAAs into Specific Sites

Currently, three major approaches are used to engineer the genetic code in a site-specific manner: (1) amber codon suppression; (2) rare sense codon reassignment; and (3) quadruplet codon. Figure 1 provides a schematic illustration of each method. Because organisms such as *E. coli* BL21 rarely use the amber stop codon (UAG) (only 275 of 4160 stop codons in BL21 are amber codons), which minimizes disturbances to existing protein termination signals, this codon has been preferably selected for NCAA encoding [30,39,40]. To enhance the efficiency of amber codon recognition by the orthogonal $tRNA_{CUA}$, Release Factor 1 [41,42] is usually mutated or knocked out [43], thus enabling orthogonal $tRNA_{CUA}$ to recognize and increase its competitive binding to the amber codon [30] (Figure 1a). The role of different artificial tRNA/tRNA synthetase pairs, as well as their structural relationship with different NCAAs, were extensively reviewed by Anaëlle et al. [44].

Rare sense codon assignment [45–47], which is based on a similar principle, repurposes rare sense codons, particularly rare codons including AGG [45,46] and AUA [47], using newly designed tRNA/aminoacyl-tRNA synthetase pairs. In this method, the introduction of a NCAA during protein synthesis requires either competition between a genetically modified tRNA and the corresponding wild-type tRNA [45] or the inhibition of wild-type tRNA via the deletion of its tRNA synthetase [47] (Figure 1b).

To circumvent the limitations of reprogramming existing codons, some researchers have explored NCAA encoding via expansion of the genetic code using quadruplet codons [48–53]. In brief, a single-base (e.g., "U") is inserted after a canonical triplet codon (e.g., a "CUC" triplet codon) to form a frameshift mutation at the specific position (Figure 1c). The additional base also creates a new quadruplet codon (e.g., "CUCU") at this position, which can be recognized by an engineered quadruplet tRNA (e.g., $tRNA_{AGAG}$). Early versions of the quadruplet in vivo coding system were initially tested in *E. coli* [48,51,52], followed by *Xenopus* oocytes [49] and mammalian cells [50,53]. It is also worth mentioning that noncanonical RNA translations, such as the use of tetra- and penta-codon, were observed in mitochondria; however, the 4th and 5th nucleotides were found to be silent during translation [54]. More mechanistic studies would be required to establish their roles in genetic code engineering.

Although the site-specific incorporation approach is arguably the most widely used to produce artificial proteins with NCAAs, some challenges can limit the stability of the engineered code. The efficiency of an engineered tRNA/aminoacyl-tRNA synthetase pair must be high enough to minimize the generation of truncated proteins [55]. Methods such as orthogonal ribosome use can lead to a threefold improvement in the efficiency of unnatural amino acid incorporation [55]. Endogenous tRNA/aminoacyl-tRNA synthetase pairs can also be engineered to incorporate unnatural amino acids. For example, by changing the phenylalanyl-tRNA synthetase amino acid recognition site, phenylalanine analogs such as p-Cl-phenylalanine or p-Br-phenylalanine can be successfully charged to $tRNA^{Phe}$ [14,56]. Advances in genome editing techniques, such as multiplex automated genomic engineering [31] and CRISPR/Cas [51], may further increase the efficiency and accuracy of NCAA incorporation in specific sites of the proteome.

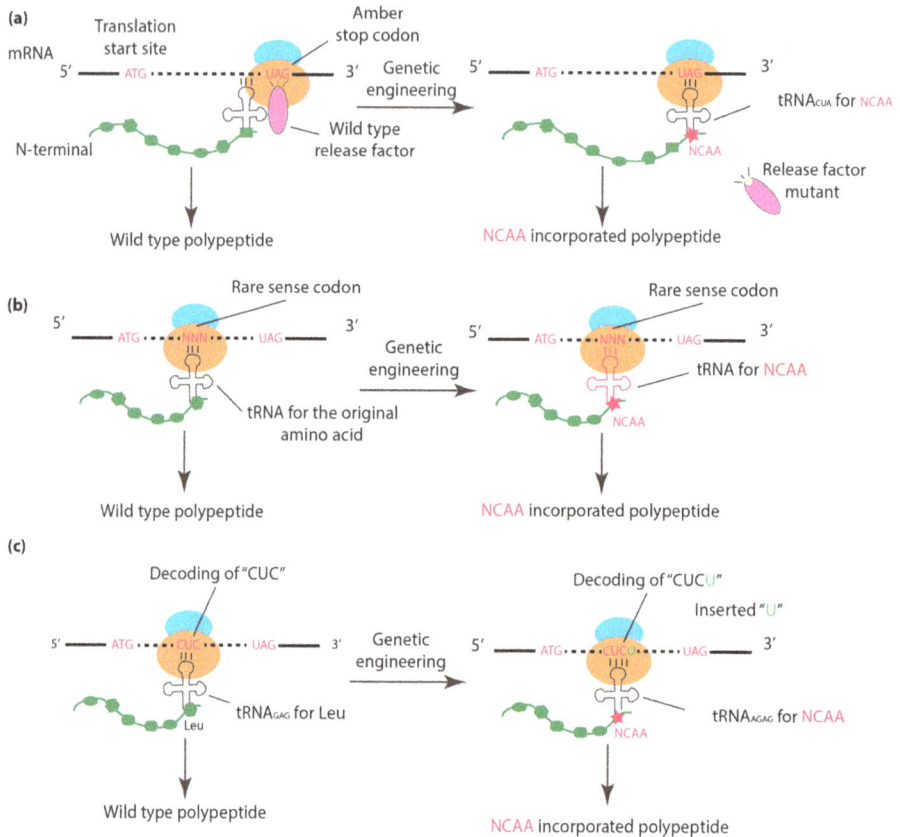

Figure 1. An overview of approaches to incorporate NCAAs into specific sites. (**a**) The wild-type release factor is mutated or knocked out, allowing the newly introduced tRNA$_{CUA}$ to read through the stop codon, followed by NCAA incorporation with assistance from the compatible aminoacyl-tRNA synthetase. (**b**) The tRNA and corresponding tRNA synthetase for a rare sense codon are genetically engineered to confer the ability to encode NCAA. (**c**) A single-base is inserted after the canonical codon (e.g. "CUC" for Leu). The newly introduced quadruplet tRNA (e.g., tRNA$_{AGAG}$) can encode NCAA by targeting the quadruplet codon "CUCU."

2.2. Proteome-Wide Incorporation of NCAAs

Proteome-wide incorporation offers an alternative approach toward unnatural amino acid incorporation. In the most common approach, amino acid uptake is artificially controlled by feeding auxotrophs with NCAAs [33–35] (Figure 2a). Attempts to control NCAA synthesis have involved supplying organisms with NCAA precursors [57–62] (Figure 2b), which is also known as metabolic engineering [63,64]. In one example, the precursor L-β-thieno [3,2-b]pyrrolyl ([3,2]Trp) was fed to a tryptophan (Trp)-auxotrophic *E. coli* capable of synthesizing [3,2]Tpa (a Trp analog) to generate mutants that could propagate on L-β-(thieno [3,2-b]pyrrolyl)alanine ([3,2]Tpa) [57] (Figure 2b). Although directly feeding auxotrophs with NCAAs is a simpler approach, metabolic engineering could reduce the unwanted effects of impure commercial NCAAs [57].

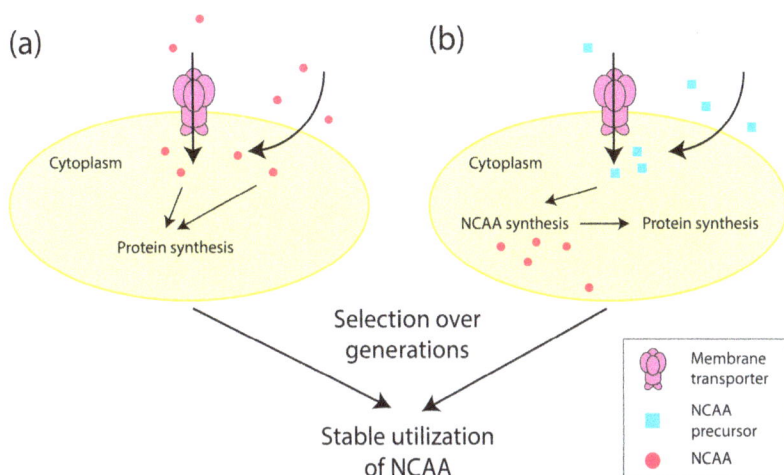

Figure 2. An overview of proteome-wide approaches to incorporate NCAAs. (**a**) The NCAA enters a cell via membrane transporters or diffusion across the membrane. (**b**) The NCAA precursor similarly enters a cell in which it will be used to synthesize NCAAs. Following several generations of propagation with either the NCAA or its precursor, cells that can stably utilize the NCAA are selected.

Regardless of approach, the incorporation of NCAAs in the proteome may negatively affect the growth of an organism. The inherent toxicities of many unnatural amino acids could suppress propagation of the wild-type strain and select mutants that respond favorably to the NCAA, ultimately causing rejection of the expanded genetic code [34,35]. In the following section, we will focus on the challenges in genetic code modification.

3. Challenges of Genetic Code Engineering

3.1. Inhibitory Effects of Engineered Genetic Codes

The growth inhibitory effects caused by NCAAs, which have been demonstrated in different species including bacteria [65–67], yeasts [68], insects [69,70], and mammals [71], comprise one major challenge encountered during genetic code modification. The inhibitory effects of NCAAs are mainly attributable to two aspects. First, minor structural and chemical differences between NCAAs and their canonical counterparts can drastically affect enzymatic activities [72–75]. Second, these structural and chemical differences may also negatively affect protein synthesis, as some NCAAs cannot be efficiently charged to tRNAs by aminoacyl-tRNA synthetases [15,76]. A better understanding of the key genes and cellular responses associated with these modified genetic codes is of paramount importance to alleviating these inhibitory effects.

3.2. Discovering the Key Genes Controlling the Genetic Code

The growth inhibitory potentials of NCAAs create negative selective pressure, while the organism adapts to the modified genetic code. One effective strategy for overcoming this evolutionary barrier comprises an increase in the mutation rate via mutagenesis with the expectation of generating beneficial mutations that would favor the NCAA. Wong and colleagues isolated mutants from a Trp auxotroph (*Bacillus subtilis* str. QB928) via sequential mutagenesis in an early attempt to modify the genetic code. The resultant HR23 strain could propagate indefinitely on 4-fluoro-tryptophan (4FTrp) but became inviable on canonical Trp [34,35]. As Trp is encoded by a single codon (UGG), the research by Wong and colleagues provided the first evidence of codon membership malleability under external selection pressure. Subsequently, Yu et al. traced mutations in intermediate mutants, as well as the HR23 strain [77]. A nonsense mutation in the Trp operon RNA-binding attenuation protein (TRAP), which controls transcriptional attenuation of the Trp operon [78,79] and translational repression of Trp transporters [80–83], was shared by all mutants. This lack of TRAP would increase 4-FTrp uptake to compensate for the relatively low charge rate of 4-FTrp to tRNA^{Trp} [15].

In a separate attempt, Bacher et al. isolated *E. coli* mutants that could propagate in medium wherein 4-FTrp comprised ~99% of available Trp. However, the mutant strains could not grow indefinitely under these conditions and required minimal canonical Trp [33]. *E. coli* mutants were found to harbor several mutations affecting genes such as *aroP*, which encodes an aromatic amino acid transporter [84], and *tyrR*, which encodes the associated regulator [85]. Mutated *aroP* and *tyrR* might cooperatively increase 4-FTrp uptake, similar to the effect of TRAP knockout in *B. subtilis*. Taken together, these findings suggest that an efficient NCAA uptake system is essential to accommodation of the modified genetic codes.

RNA polymerase might also play a key role in controlling the genetic code. The above-mentioned *B. subtilis* mutant HR23 was found to harbor a nonsynonymous mutation in the RNA polymerase subunit gene (*rpoB*) that was absent from all other intermediate strains that could still propagate on Trp, suggesting a potential role for this mutation in switching membership of the UGG codon from Trp to 4-FTrp [77]. In an independent study of amber codon-directed 3-iodotyrosine (3-iodoTyr) incorporation in *E. coli*, a *rpoB* mutation was found to confer rifampicin resistance via amber suppression at Gln513 [86], and the same research group also engineered a bacteriophage, T7, that could incorporate 3-iodoTyr at amber codons [29]. In that study, Hammerling et al. observed high mutation frequencies in genes encoding RNA polymerase and the lysis timing regulator type II holin. The authors suggested that these two genes played important roles in the evolution of the expanded genetic code [29]. These studies have shed light on the previously unexplored roles of key genes in genetic code identity.

3.3. Lack of Transcriptomic and Proteomic Studies Related to Engineered Genetic Codes

In addition to mutations, gene and protein expression profiles might also reveal key factors needed to fine-tune the use of modified genetic codes. Technologies such as RNA-seq and mass spectrometry can be used to investigate the cellular responses of organisms in high resolution. RNA-seq was used to compare the cellular responses between mutant (grown on 4-FTrp) and wild-type strains of the above-mentioned *B. subtilis* HR23 mutant (unpublished data). Here, a gene ontology analysis of the gene expression profiles of these strains demonstrated enrichment of genes related to reactive oxygen species responses and branched-chain amino acid biosynthetic processes among upregulated genes, and enrichment of genes related to siderophore biosynthetic processes among downregulated genes (unpublished data). Unsurprisingly, stress response genes were modulated in response to the new genetic code, and the downregulation of siderophore biosynthetic process related genes was consistent with a previous observation of the reduced growth rate of HR23 cells grown on 4-FTrp [77] because iron homeostasis is closely related to bacterial growth [87]. This unique set of data was the first to demonstrate the adaptation of an organism to a new genetic code at the transcriptomic level.

Methanosarcina acetivorans is a methanogenic archaea strain that uses the alternative genetic codon UAG to encode pyrrolysine (Pyl) [88]. O'Donoghue et al. attempted to reduce the genetic code of this strain by deleting tRNA^{Pyl}, thus blocking the incorporation of Pyl in the proteome. A comparison of the proteomes of mutant and wild-type *M. acetivorans* strains revealed that most upregulated peptides were related to methanogenesis, protein synthesis, and the stress response [89], suggesting that, in this organism, various stress response genes must be fine-tuned before a reduced genetic code can be used.

Very few transcriptomic and proteomic studies of organisms with modified genetic codes have been conducted, and we have only glimpsed the potential factors involved in adaptation to modified genetic codes. Additional genes that contribute to this adaptation might remain to be discovered. In the future, studies of gene and protein expression in organisms with modified genetic codes will be necessary.

3.4. Environmental Factors Affecting Adaptation to Engineered Genetic Codes

Environmental factors, such as the growth medium and selection method, are important when optimizing the use of a modified genetic code. The amino acid source is the first and most obvious factor, as an organism can either take up NCAAs directly from the environment or synthesize them using environmentally available molecules. If the source of NCAAs is from the environment, mutations in amino acid transporters are often needed to facilitate NCAA uptake [33,77].

Positive selection pressure is also needed to maintain stability of the modified genetic code. In a previous study, incorporation of the methionine analog azidohomoalanine (Aha) into the coat protein of a human adenovirus and the subsequent addition of a folate group to Aha facilitated adenoviral infection in mouse hosts [90]. In other words, adenovirus strains that can use modified Aha have a selective survival advantage over other strains. In a more recent study of different *E. coli* strains, the site-specific incorporation of two tyrosine analogs in β-lactamase was selected, and enzymatic function was found to depend on the presence of these analogs [91]. As described above regarding adenovirus, *E. coli* mutants that could utilize NCAAs enjoyed a selective advantage under growth mediums containing certain classes of antibiotics [91]. In one interesting example, even the carbon source may affect the selection of genetic codes by the Pyl-utilizing bacteria *Acetohalobium arabaticum* [92]. *A. arabaticum* used the standard genetic code when grown on pyruvate, but gained the ability to use an expanded genetic code that included Pyl in the presence of the alternative carbon source trimethylamine [92].

4. Future Directions

Current efforts in genetic code engineering have reshaped our ideas regarding genetic code evolution and have paved the way for expanding the genetic alphabet. Based on these studies, we have outlined the key steps by which an organism accommodates a modified genetic code (Figure 3). During adaptation, mutations in amino acid transporters and/or their key regulators allow more efficient NCAA uptake, possibly by increasing the number of amino acid transporters [33,77]. Mutations in the key genes might also favor the use of a modified genetic code [29,33,77,86]. Additionally, environmental positive selection forces contribute to stability of the modified genetic code [77,91,92]. Currently, the genomic changes in organisms with modified genetic codes have been well explored [29,33,77,86] relative to transcriptomic (unpublished data) and proteomic [89] changes. Future trends in elucidation of the biological mechanisms underlying genetic code modifications include the integration of genomic, transcriptomic, and proteomic data and the refining of functional study targets.

From the viewpoint of synthetic biology and xenobiology, genetic code engineering increases the repertoire of building blocks available for protein engineering, thus enabling the development of novel proteins that would be impossible with canonical amino acids [38,93]. Xenobiology is an emerging field that involves synthesizing xenonucleic acids other than the canonical nucleic acids with adenine (A), thymine (T), cytosine (C), and guanine (G) as bases, with alternative pairing rules for protein engineering [94]. It has been demonstrated experimentally that two such xenonucleic acids can be integrated into the current DNA backbone [95–97], and more have been tested for their potentials as novel building blocks of DNA [98]. With the addition of xenonucleic acids, the number of encoded amino acids is likely to be increased to far beyond 20 [94].

High-throughput genome editing technologies, such as MAGE [36] and the emerging CRISPR/Cas technology [99], allow an organism's genetic code to be directly rewritten [37,100] and facilitate the creation of synthetic life [101]. Although the first synthetic minimal bacterial genome still uses the standard genetic code [101], it is now possible to synthesize genomes based on alternative genetic codes. A full exploration of the possibilities enabled by genetic code engineering requires an understanding of the key molecular biological and biochemical mechanisms underlying the modifications. Gradual efforts to address this main question may improve our understanding of the process of genetic code evolution and lay a better foundation for future synthetic biology research.

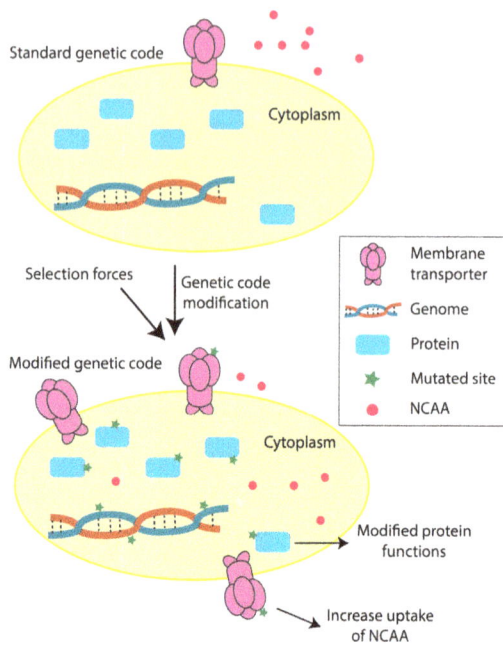

Figure 3. Key steps in the accommodation of a modified genetic code.

Acknowledgments: This work is partially supported by a CUHK Direct Grant 4053041, the Lo Kwee-Seong Biomedical Research Fund, Lee Hysan Foundation, and the Partner State Key Laboratory of Agrobiotechnology, The Chinese University of Hong Kong to T.F.C.

Author Contributions: X.L., A.C.S.Y., and T.F.C. wrote the paper.

Conflicts of Interest: The authors declare no conflicts of interest.

References

1. Crick, F.H.C. The origin of the genetic code. *J. Mol. Biol.* **1968**, *38*, 367–379. [CrossRef]
2. Cone, J.E.; Del Río, R.M.; Davis, J.N.; Stadtman, T.C. Chemical characterization of the selenoprotein component of clostridial glycine reductase: Identification of selenocysteine as the organoselenium moiety. *Proc. Natl. Acad. Sci. USA* **1976**, *73*, 2659–2663. [CrossRef] [PubMed]
3. Hao, B.; Gong, W.; Ferguson, T.K.; James, C.M.; Krzycki, J.A.; Chan, M.K. A new UAG-encoded residue in the structure of a methanogen methyltransferase. *Science* **2002**, *296*, 1462–1466. [CrossRef] [PubMed]
4. Srinivasan, G.; James, C.M.; Krzycki, J.A. Pyrrolysine encoded by UAG in archaea: Charging of a UAG-decoding specialized tRNA. *Science* **2002**, *296*, 1459–1462. [CrossRef] [PubMed]
5. Adachi, M.; Cavalcanti, A.R.O. Tandem stop codons in ciliates that reassign stop codons. *J. Mol. Evol.* **2009**, *68*, 424–431. [CrossRef] [PubMed]
6. Beznosková, P.; Gunišová, S.; Valášek, L.S. Rules of UGA-N decoding by near-cognate tRNAs and analysis of readthrough on short uORFs in yeast. *RNA* **2016**, *22*, 456–466. [CrossRef] [PubMed]
7. Osawa, S.; Jukes, T.H. Codon reassignment (codon capture) in evolution. *J. Mol. Evol.* **1989**, *28*, 271–278. [CrossRef] [PubMed]
8. Schultz, D.W.; Yarus, M. Transfer RNA mutation and the malleability of the genetic code. *J. Mol. Biol.* **1994**, *235*, 1377–1380. [CrossRef] [PubMed]
9. Andersson, S.G.; Kurland, C.G. Genomic evolution drives the evolution of the translation system. *Biochem. Cell Biol.* **1995**, *73*, 775–787. [CrossRef] [PubMed]

10. Söll, D.; RajBhandary, U.L. The genetic code—Thawing the "frozen accident". *J. Biosci.* **2006**, *31*, 459–463. [CrossRef] [PubMed]

11. Koonin, E.V.; Novozhilov, A.S. Origin and evolution of the genetic code: The universal enigma. *IUBMB Life* **2009**, *61*, 99–111. [CrossRef] [PubMed]

12. Hartman, M.C.T.; Josephson, K.; Lin, C.-W.; Szostak, J.W. An expanded set of amino acid analogs for the ribosomal translation of unnatural peptides. *PLoS ONE* **2007**, *2*, e972. [CrossRef] [PubMed]

13. Wang, L.; Xie, J.; Schultz, P.G. Expanding the genetic code. *Annu. Rev. Biophys. Biomol. Struct.* **2006**, *35*, 225–249. [CrossRef] [PubMed]

14. Ibba, M.; Hennecke, H. Relaxing the substrate specificity of an aminoacyl-tRNA synthetase allows in vitro and in vivo synthesis of proteins containing unnatural amino acids. *FEBS Lett.* **1995**, *364*, 272–275. [CrossRef]

15. Xu, Z.J.; Love, M.L.; Ma, L.Y.; Blum, M.; Bronskill, P.M.; Bernstein, J.; Grey, A.A.; Hofmann, T.; Camerman, N.; Wong, J.T. Tryptophanyl-tRNA synthetase from Bacillus subtilis. Characterization and role of hydrophobicity in substrate recognition. *J. Biol. Chem.* **1989**, *264*, 4304–4311. [PubMed]

16. Lovett, P.S.; Ambulos, N.P.; Mulbry, W.; Noguchi, N.; Rogers, E.J.; Rogers, E.J. UGA can be decoded as tryptophan at low efficiency in *Bacillus subtilis*. *J. Bacteriol.* **1991**, *173*, 1810–1812. [CrossRef] [PubMed]

17. Cataldo, F.; Iglesias-Groth, S.; Angelini, G.; Hafez, Y. Stability toward high energy radiation of non-proteinogenic amino acids: Implications for the origins of life. *Life* **2013**, *3*, 449–473. [CrossRef] [PubMed]

18. Richmond, M.H. The effect of amino acid analogues on growth and protein synthesis in microorganisms. *Bacteriol. Rev.* **1962**, *26*, 398–420. [PubMed]

19. Lea, P.J.; Norris, R.D. The use of amino acid analogues in studies on plant metabolism. *Phytochemistry* **1976**, *15*, 585–595. [CrossRef]

20. Rodgers, K.J.; Hume, P.M.; Dunlop, R.A.; Dean, R.T. Biosynthesis and turnover of DOPA-containing proteins by human cells. *Free Radic. Biol. Med.* **2004**, *37*, 1756–1764. [CrossRef] [PubMed]

21. Commans, S.; Böck, A. Selenocysteine inserting tRNAs: An overview. *FEMS Microbiol. Rev.* **1999**, *23*, 335–351. [CrossRef] [PubMed]

22. Berry, M.J.; Banu, L.; Chen, Y.; Mandel, S.J.; Kieffer, J.D.; Harney, J.W.; Larsen, P.R. Recognition of UGA as a selenocysteine codon in Type I deiodinase requires sequences in the 3′ untranslated region. *Nature* **1991**, *353*, 273–276. [CrossRef] [PubMed]

23. Silva, R.M.; Paredes, J.A.; Moura, G.R.; Manadas, B.; Lima-Costa, T.; Rocha, R.; Miranda, I.; Gomes, A.C.; Koerkamp, M.J.G.; Perrot, M.; et al. Critical roles for a genetic code alteration in the evolution of the genus Candida. *EMBO J.* **2007**, *26*, 4555–4565. [CrossRef] [PubMed]

24. Krzycki, J.A. Function of genetically encoded pyrrolysine in corrinoid-dependent methylamine methyltransferases. *Curr. Opin. Chem. Biol.* **2004**, *8*, 484–491. [CrossRef] [PubMed]

25. Polycarpo, C.; Ambrogelly, A.; Bérubé, A.; Winbush, S.M.; McCloskey, J.A.; Crain, P.F.; Wood, J.L.; Söll, D. An aminoacyl-tRNA synthetase that specifically activates pyrrolysine. *Proc. Natl. Acad. Sci. USA* **2004**, *101*, 12450–12454. [CrossRef] [PubMed]

26. Mann, M.; Jensen, O.N. Proteomic analysis of post-translational modifications. *Nat. Biotechnol.* **2003**, *21*, 255–261. [CrossRef] [PubMed]

27. Rother, M.; Krzycki, J.A. Selenocysteine, pyrrolysine, and the unique energy metabolism of methanogenic archaea. *Archaea* **2010**, *2010*, 453642. [CrossRef] [PubMed]

28. Bacher, J.M.; Hughes, R.A.; Tze-Fei Wong, J.; Ellington, A.D. Evolving new genetic codes. *Trends Ecol. Evol.* **2004**, *19*, 69–75. [CrossRef] [PubMed]

29. Hammerling, M.J.; Ellefson, J.W.; Boutz, D.R.; Marcotte, E.M.; Ellington, A.D.; Barrick, J.E. Bacteriophages use an expanded genetic code on evolutionary paths to higher fitness. *Nat. Chem. Biol.* **2014**, *10*, 178–180. [CrossRef] [PubMed]

30. Wang, L.; Brock, A.; Herberich, B.; Schultz, P.G. Expanding the genetic code of *Escherichia coli*. *Science* **2001**, *292*, 498–500. [CrossRef] [PubMed]

31. Park, H.-S.; Hohn, M.J.; Umehara, T.; Guo, L.-T.; Osborne, E.M.; Benner, J.; Noren, C.J.; Rinehart, J.; Söll, D. Expanding the genetic code of *Escherichia coli* with phosphoserine. *Science* **2011**, *333*, 1151–1154. [CrossRef] [PubMed]

32. Mukai, T.; Yanagisawa, T.; Ohtake, K.; Wakamori, M.; Adachi, J.; Hino, N.; Sato, A.; Kobayashi, T.; Hayashi, A.; Shirouzu, M.; et al. Genetic-code evolution for protein synthesis with non-natural amino acids. *Biochem. Biophys. Res. Commun.* **2011**, *411*, 757–761. [CrossRef] [PubMed]

33. Bacher, J.M.; Ellington, A.D. Selection and characterization of *Escherichia coli* variants capable of growth on an otherwise toxic tryptophan analogue. *J. Bacteriol.* **2001**, *183*, 5414–5425. [CrossRef] [PubMed]

34. Wong, J.T. Membership mutation of the genetic code: Loss of fitness by tryptophan. *Proc. Natl. Acad. Sci. USA* **1983**, *80*, 6303–6306. [CrossRef] [PubMed]

35. Mat, W.K.; Xue, H.; Wong, J.T.F. Genetic code mutations: The breaking of a three billion year invariance. *PLoS ONE* **2010**, *5*, e12206. [CrossRef] [PubMed]

36. Gallagher, R.R.; Li, Z.; Lewis, A.O.; Isaacs, F.J. Rapid editing and evolution of bacterial genomes using libraries of synthetic DNA. *Nat. Protoc.* **2014**, *9*, 2301–2316. [CrossRef] [PubMed]

37. Ostrov, N.; Landon, M.; Guell, M.; Kuznetsov, G.; Teramoto, J.; Cervantes, N.; Zhou, M.; Singh, K.; Napolitano, M.G.; Moosburner, M.; et al. Design, synthesis, and testing toward a 57-codon genome. *Science* **2016**, *353*, 819–822. [CrossRef] [PubMed]

38. Budisa, N.; Völler, J.-S.; Koksch, B.; Acevedo-Rocha, C.G.; Kubyshkin, V.; Agostini, F. Xenobiology meets enzymology: Exploring the potential of unnatural building blocks in biocatalysis. *Angew. Chem. Int. Ed.* **2017**. [CrossRef]

39. Wang, L.; Brock, A.; Schultz, P.G. Adding L-3-(2-Naphthyl)alanine to the genetic code of *E. coli*. *J. Am. Chem. Soc.* **2002**, *124*, 1836–1837. [CrossRef] [PubMed]

40. Chin, J.W.; Martin, A.B.; King, D.S.; Wang, L.; Schultz, P.G. Addition of a photocrosslinking amino acid to the genetic code of *Escherichia coli*. *Proc. Natl. Acad. Sci. USA* **2002**, *99*, 11020–11024. [CrossRef] [PubMed]

41. Scolnick, E.; Tompkins, R.; Caskey, T.; Nirenberg, M. Release factors differing in specificity for terminator codons. *Proc. Natl. Acad. Sci. USA* **1968**, *61*, 768–774. [CrossRef] [PubMed]

42. Petry, S.; Brodersen, D.E.; Murphy, F.V.; Dunham, C.M.; Selmer, M.; Tarry, M.J.; Kelley, A.C.; Ramakrishnan, V. Crystal structures of the ribosome in complex with release factors RF1 and RF2 bound to a cognate stop codon. *Cell* **2005**, *123*, 1255–1266. [CrossRef] [PubMed]

43. Johnson, D.B.F.; Xu, J.; Shen, Z.; Takimoto, J.K.; Schultz, M.D.; Schmitz, R.J.; Xiang, Z.; Ecker, J.R.; Briggs, S.P.; Wang, L. RF1 knockout allows ribosomal incorporation of unnatural amino acids at multiple sites. *Nat. Chem. Biol.* **2011**, *7*, 779–786. [CrossRef] [PubMed]

44. Dumas, A.; Lercher, L.; Spicer, C.D.; Davis, B.G. Designing logical codon reassignment—Expanding the chemistry in biology. *Chem. Sci.* **2015**, *6*, 50–69. [CrossRef]

45. Zeng, Y.; Wang, W.; Liu, W.R. Towards reassigning the rare AGG codon in *Escherichia coli*. *ChemBioChem* **2014**, *15*, 1750–1754. [CrossRef] [PubMed]

46. Mukai, T.; Yamaguchi, A.; Ohtake, K.; Takahashi, M.; Hayashi, A.; Iraha, F.; Kira, S.; Yanagisawa, T.; Yokoyama, S.; Hoshi, H.; et al. Reassignment of a rare sense codon to a non-canonical amino acid in *Escherichia coli*. *Nucleic Acids Res.* **2015**, *43*, 8111–8122. [CrossRef] [PubMed]

47. Bohlke, N.; Budisa, N. Sense codon emancipation for proteome-wide incorporation of noncanonical amino acids: Rare isoleucine codon AUA as a target for genetic code expansion. *FEMS Microbiol. Lett.* **2014**, *351*, 133–144. [CrossRef] [PubMed]

48. Anderson, J.C.; Wu, N.; Santoro, S.W.; Lakshman, V.; King, D.S.; Schultz, P.G. An expanded genetic code with a functional quadruplet codon. *Proc. Natl. Acad. Sci. USA* **2004**, *101*, 7566–7571. [CrossRef] [PubMed]

49. Rodriguez, E.A.; Lester, H.A.; Dougherty, D.A. In vivo incorporation of multiple unnatural amino acids through nonsense and frameshift suppression. *Proc. Natl. Acad. Sci. USA* **2006**, *103*, 8650–8655. [CrossRef] [PubMed]

50. Niu, W.; Schultz, P.G.; Guo, J. An expanded genetic code in mammalian cells with a functional quadruplet codon. *ACS Chem. Biol.* **2013**, *8*, 1640–1645. [CrossRef] [PubMed]

51. Magliery, T.J.; Anderson, J.C.; Schultz, P.G. Expanding the genetic code: Selection of efficient suppressors of four-base codons and identification of "shifty" four-base codons with a library approach in *Escherichia coli*. *J. Mol. Biol.* **2001**, *307*, 755–769. [CrossRef] [PubMed]

52. Anderson, J.C.; Magliery, T.J.; Schultz, P.G. Exploring the limits of codon and anticodon size. *Chem. Biol.* **2002**, *9*, 237–244. [CrossRef]

53. Taki, M.; Matsushita, J.; Sisido, M. Expanding the genetic code in a mammalian cell line by the introduction of four-base codon/anticodon pairs. *ChemBioChem* **2006**, *7*, 425–428. [CrossRef] [PubMed]

54. Seligmann, H. Natural chymotrypsin-like-cleaved human mitochondrial peptides confirm tetra-, pentacodon, non-canonical RNA translations. *Biosystems* **2016**, *147*, 78–93. [CrossRef] [PubMed]

55. Wang, K.; Neumann, H.; Peak-Chew, S.Y.; Chin, J.W. Evolved orthogonal ribosomes enhance the efficiency of synthetic genetic code expansion. *Nat. Biotechnol.* **2007**, *25*, 770–777. [CrossRef] [PubMed]

56. Kast, P.; Hennecke, H. Amino acid substrate specificity of *Escherichia coli* phenylalanyl-tRNA synthetase altered by distinct mutations. *J. Mol. Biol.* **1991**, *222*, 99–124. [CrossRef]

57. Hoesl, M.G.; Oehm, S.; Durkin, P.; Darmon, E.; Peil, L.; Aerni, H.-R.; Rappsilber, J.; Rinehart, J.; Leach, D.; Söll, D.; et al. Chemical evolution of a bacterial proteome. *Angew. Chem. Int. Ed.* **2015**, *54*, 10030–10034. [CrossRef] [PubMed]

58. Ehrlich, M.; Gattner, M.J.; Viverge, B.; Bretzler, J.; Eisen, D.; Stadlmeier, M.; Vrabel, M.; Carell, T. Orchestrating the biosynthesis of an unnatural pyrrolysine amino acid for its direct incorporation into proteins inside living cells. *Chem. Eur. J.* **2015**, *21*, 7701–7704. [CrossRef] [PubMed]

59. Ma, Y.; Biava, H.; Contestabile, R.; Budisa, N.; di Salvo, M. Coupling bioorthogonal chemistries with artificial metabolism: Intracellular biosynthesis of azidohomoalanine and its incorporation into recombinant proteins. *Molecules* **2014**, *19*, 1004–1022. [CrossRef] [PubMed]

60. Exner, M.P.; Kuenzl, T.; To, T.M.T.; Ouyang, Z.; Schwagerus, S.; Hoesl, M.G.; Hackenberger, C.P.R.; Lensen, M.C.; Panke, S.; Budisa, N. Design of S-allylcysteine in situ production and incorporation based on a novel pyrrolysyl-tRNA synthetase variant. *Chembiochem* **2017**, *18*, 85–90. [CrossRef] [PubMed]

61. Ou, W.; Uno, T.; Chiu, H.-P.; Grunewald, J.; Cellitti, S.E.; Crossgrove, T.; Hao, X.; Fan, Q.; Quinn, L.L.; Patterson, P.; et al. Site-specific protein modifications through pyrroline-carboxy-lysine residues. *Proc. Natl. Acad. Sci. USA* **2011**, *108*, 10437–10442. [CrossRef] [PubMed]

62. Mehl, R.A.; Anderson, J.C.; Santoro, S.W.; Wang, L.; Martin, A.B.; King, D.S.; Horn, D.M.; Schultz, P.G. Generation of a bacterium with a 21 amino acid genetic code. *J. Am. Chem. Soc.* **2003**, *125*, 935–939. [CrossRef] [PubMed]

63. Acevedo-Rocha, C.G.; Budisa, N. Xenomicrobiology: A roadmap for genetic code engineering. *Microb. Biotechnol.* **2016**, *9*, 666–676. [CrossRef] [PubMed]

64. Völler, J.-S.; Budisa, N. Coupling genetic code expansion and metabolic engineering for synthetic cells. *Curr. Opin. Biotechnol.* **2017**, *48*, 1–7. [CrossRef] [PubMed]

65. Grant, M.M.; Brown, A.S.; Corwin, L.M.; Troxler, R.F.; Franzblau, C. Effect of L-azetidine 2-carboxylic acid on growth and proline metabolism in *Escherichia coli*. *Biochim. Biophys. Acta* **1975**, *404*, 180–187. [CrossRef]

66. Unger, L.; DeMoss, R.D. Action of a proline analogue, L-thiazolidine-4-carboxylic acid, in *Escherichia coli*. *J. Bacteriol.* **1966**, *91*, 1556–1563. [PubMed]

67. Moran, S.; Rai, D.K.; Clark, B.R.; Murphy, C.D. Precursor-directed biosynthesis of fluorinated iturin A in *Bacillus* spp. *Org. Biomol. Chem.* **2009**, *7*, 644. [CrossRef] [PubMed]

68. Téllez, R.; Jacob, G.; Basilio, C.; George-Nascimento, C. Effect of ethionine on the in vitro synthesis and degradation of mitochondrial translation products in yeast. *FEBS Lett.* **1985**, *192*, 88–94. [CrossRef]

69. Rosenthal, G.; Lambert, J.; Hoffmann, D. Canavanine incorporation into the antibacterial proteins of the fly, *Phormia terranovae* (Diptera), and its effect on biological activity. *J. Biol. Chem.* **1989**, *26417*, 9768–9771.

70. Teramoto, H.; Kojima, K. Incorporation of methionine analogues into bombyx mori silk fibroin for click modifications. *Macromol. Biosci.* **2015**, *15*, 719–727. [CrossRef] [PubMed]

71. Poirson-Bichat, F.; Lopez, R.; Bras Gonçalves, R.A.; Miccoli, L.; Bourgeois, Y.; Demerseman, P.; Poisson, M.; Dutrillaux, B.; Poupon, M.F. Methionine deprivation and methionine analogs inhibit cell proliferation and growth of human xenografted gliomas. *Life Sci.* **1997**, *60*, 919–931. [CrossRef]

72. Merkel, L.; Budisa, N. Organic fluorine as a polypeptide building element: In vivo expression of fluorinated peptides, proteins and proteomes. *Org. Biomol. Chem.* **2012**, *10*, 7241. [CrossRef] [PubMed]

73. Schlesinger, S. The effect of amino acid analogues on alkaline phosphatase. Formation in *Escherichia coli* K-II. Replacement of tryptophan by azatryptophan and by tryptazan. *J. Biol. Chem.* **1968**, *243*, 3877–3883. [PubMed]

74. Pine, M.J. Comparative physiological effects of incorporated amino acid analogs in *Escherichia coli*. *Antimicrob. Agents Chemother.* **1978**, *13*, 676–685. [CrossRef] [PubMed]

75. Wong, H.; Kwon, I. Effects of non-natural amino acid incorporation into the enzyme core region on enzyme structure and function. *Int. J. Mol. Sci.* **2015**, *16*, 22735–22753. [CrossRef] [PubMed]

76. Kwon, I.; Tirrell, D.A. Site-specific incorporation of tryptophan analogues into recombinant proteins in bacterial cells. **2007**, *129*, 10431–10437. [CrossRef] [PubMed]

77. Yu, A.C.-S.; Yim, A.K.-Y.; Mat, W.-K.; Tong, A.H.-Y.; Lok, S.; Xue, H.; Tsui, S.K.-W.; Wong, J.T.-F.; Chan, T.-F. Mutations enabling displacement of tryptophan by 4-fluorotryptophan as a canonical amino acid of the genetic code. *Genome Biol. Evol.* **2014**, *6*, 629–641. [CrossRef] [PubMed]

78. Gollnick, P.; Ishino, S.; Kuroda, M.I.; Henner, D.J.; Yanofsky, C. The mtr locus is a two-gene operon required for transcription attenuation in the trp operon of *Bacillus subtilis*. *Proc. Natl. Acad. Sci. USA* **1990**, *87*, 8726–8730. [CrossRef] [PubMed]

79. Yanofsky, C. RNA-based regulation of genes of tryptophan synthesis and degradation, in bacteria. *RNA* **2007**, *13*, 1141–1154. [CrossRef] [PubMed]

80. Sarsero, J.P.; Merino, E.; Yanofsky, C. A Bacillus subtilis gene of previously unknown function, *yhaG*, is translationally regulated by tryptophan-activated TRAP and appears to be involved in tryptophan transport. *J. Bacteriol.* **2000**, *182*, 2329–2331. [CrossRef] [PubMed]

81. Yakhnin, H.; Yakhnin, A.V.; Babitzke, P. The trp RNA-binding attenuation protein (TRAP) of Bacillus subtilis regulates translation initiation of *ycbK*, a gene encoding a putative efflux protein, by blocking ribosome binding. *Mol. Microbiol.* **2006**, *61*, 1252–1266. [CrossRef] [PubMed]

82. Yakhnin, H.; Zhang, H.; Yakhnin, A.V.; Babitzke, P. The trp RNA-binding attenuation protein of Bacillus subtilis regulates translation of the tryptophan transport gene trpP (*yhaG*) by blocking ribosome binding. *J. Bacteriol.* **2004**, *186*, 278–286. [CrossRef] [PubMed]

83. Du, H.; Tarpey, R.; Babitzke, P. The trp RNA-binding attenuation protein regulates TrpG synthesis by binding to the trpG ribosome binding site of *Bacillus subtilis*. *J. Bacteriol.* **1997**, *179*, 2582–2586. [CrossRef] [PubMed]

84. Honoré, N.; Cole, S.T. Nucleotide sequence of the *aroP* gene encoding the general aromatic amino acid transport protein of *Escherichia coli* K-12: Homology with yeast transport proteins. *Nucleic Acids Res.* **1990**, *18*, 653. [CrossRef] [PubMed]

85. Wang, J.G.; Fan, C.S.; Wu, Y.Q.; Jin, R.L.; Liu, D.X.; Shang, L.; Jiang, P.H. Regulation of *aroP* expression by *tyrR* gene in *Escherichia coli*. *Acta Biochim. Biophys. Sin.* **2003**, *35*, 993–997. [PubMed]

86. Hammerling, M.J.; Gollihar, J.; Mortensen, C.; Alnahhas, R.N.; Ellington, A.D.; Barrick, J.E. Expanded genetic codes create new mutational routes to rifampicin resistance in *Escherichia coli*. *Mol. Biol. Evol.* **2016**, *33*, 2054–2063. [CrossRef] [PubMed]

87. Andrews, S.C.; Robinson, A.K.; Rodríguez-Quiñones, F. Bacterial iron homeostasis. *FEMS Microbiol. Rev.* **2003**, *27*, 215–237. [CrossRef]

88. Borrel, G.; Gaci, N.; Peyret, P.; O'Toole, P.W.; Gribaldo, S.; Brugère, J.F. Unique characteristics of the pyrrolysine system in the 7th order of methanogens: Implications for the evolution of a genetic code expansion cassette. *Archaea* **2014**, *2014*, 374146. [CrossRef] [PubMed]

89. O'Donoghue, P.; Prat, L.; Kucklick, M.; Schäfer, J.G.; Riedel, K.; Rinehart, J.; Söll, D.; Heinemann, I.U. Reducing the genetic code induces massive rearrangement of the proteome. *Proc. Natl. Acad. Sci. USA* **2014**, *111*, 17206–17211. [CrossRef] [PubMed]

90. Banerjee, P.S.; Ostapchuk, P.; Hearing, P.; Carrico, I.S. Unnatural amino acid incorporation onto adenoviral (Ad) coat proteins facilitates chemoselective modification and retargeting of Ad type 5 vectors. *J. Virol.* **2011**, *85*, 7546–7554. [CrossRef] [PubMed]

91. Tack, D.S.; Ellefson, J.W.; Thyer, R.; Wang, B.; Gollihar, J.; Forster, M.T.; Ellington, A.D. Addicting diverse bacteria to a noncanonical amino acid. *Nat. Chem. Biol.* **2016**, *12*, 138–140. [CrossRef] [PubMed]

92. Prat, L.; Heinemann, I.U.; Aerni, H.R.; Rinehart, J.; O'Donoghue, P.; Söll, D. Carbon source-dependent expansion of the genetic code in bacteria. *Proc. Natl. Acad. Sci. USA* **2012**, *109*, 21070–21075. [CrossRef] [PubMed]

93. Zhang, W.H.; Otting, G.; Jackson, C.J. Protein engineering with unnatural amino acids. *Curr. Opin. Struct. Biol.* **2013**, *23*, 581–587. [CrossRef] [PubMed]

94. Schmidt, M. Xenobiology: A new form of life as the ultimate biosafety tool. *Bioessays* **2010**, *32*, 322–331. [CrossRef] [PubMed]

95. Yang, Z.; Sismour, A.M.; Sheng, P.; Puskar, N.L.; Benner, S.A. Enzymatic incorporation of a third nucleobase pair. *Nucleic Acids Res.* **2007**, *35*, 4238–4249. [CrossRef] [PubMed]

96. Sismour, A.M.; Lutz, S.; Park, J.-H.; Lutz, M.J.; Boyer, P.L.; Hughes, S.H.; Benner, S.A. PCR amplification of DNA containing non-standard base pairs by variants of reverse transcriptase from Human Immunodeficiency Virus-1. *Nucleic Acids Res.* **2004**, *32*, 728–735. [CrossRef] [PubMed]

97. Yang, Z.; Hutter, D.; Sheng, P.; Sismour, A.M.; Benner, S.A. Artificially expanded genetic information system: A new base pair with an alternative hydrogen bonding pattern. *Nucleic Acids Res.* **2006**, *34*, 6095–6101. [CrossRef] [PubMed]

98. Leconte, A.M.; Hwang, G.T.; Matsuda, S.; Capek, P.; Hari, Y.; Romesberg, F.E. Discovery, characterization, and optimization of an unnatural base pair for expansion of the genetic alphabet. *J. Am. Chem. Soc.* **2008**, *130*, 2336–2343. [CrossRef] [PubMed]

99. Cong, L.; Ran, F.A.; Cox, D.; Lin, S.; Barretto, R.; Habib, N.; Hsu, P.D.; Wu, X.; Jiang, W.; Marraffini, L.A.; et al. Multiplex genome engineering using CRISPR/Cas systems. *Science* **2013**, *339*, 819–823. [CrossRef] [PubMed]

100. Lajoie, M.J.; Rovner, A.J.; Goodman, D.B.; Aerni, H.-R.; Haimovich, A.D.; Kuznetsov, G.; Mercer, J.A.; Wang, H.H.; Carr, P.A.; Mosberg, J.A.; et al. Genomically recoded organisms expand biological functions. *Science* **2013**, *342*, 357–360. [CrossRef] [PubMed]

101. Hutchison, C.A.; Chuang, R.-Y.; Noskov, V.N.; Assad-Garcia, N.; Deerinck, T.J.; Ellisman, M.H.; Gill, J.; Kannan, K.; Karas, B.J.; Ma, L.; et al. Design and synthesis of a minimal bacterial genome. *Science* **2016**, *351*, aad6253. [CrossRef] [PubMed]

Review

The Genetic Code and RNA-Amino Acid Affinities

Michael Yarus

Department of Molecular, Cellular and Developmental Biology, University of Colorado, Boulder, CO 80309-0347, USA; yarus@stripe.colorado.edu; Tel.: +1-303-817-6018; Fax: +1-303-492-7744

Academic Editor: Koji Tamura
Received: 17 February 2017; Accepted: 17 March 2017; Published: 23 March 2017

Abstract: A significant part of the genetic code likely originated via a chemical interaction, which should be experimentally verifiable. One possible verification relates bound amino acids (or perhaps their activated congeners) and ribonucleotide sequences within cognate RNA binding sites. To introduce this interaction, I first summarize how amino acids function as targets for RNA binding. Then the experimental method for selecting relevant RNA binding sites is characterized. The selection method's characteristics are related to the investigation of the RNA binding site model treated at the outset. Finally, real binding sites from selection and also from extant natural RNAs (for example, the *Sulfobacillus* guanidinium riboswitch) are connected to the genetic code, and by extension, to the evolutionary progression that produced the code. During this process, peptides may have been produced directly on an instructive amino acid binding RNA (a DRT; Direct RNA Template). Combination of observed stereochemical selectivity with adaptation and co-evolutionary refinement is logically required, and also potentially sufficient, to create the striking order conserved throughout the present coding table.

Keywords: binding; triplet; codon; anticodon; DRT

1. Introduction

1.1. The Argument

In all likelihood, construction of the genetic code required specific interactions between amino acids and RNAs, acting alone, before peptides could be encoded. Close study of this molecular interaction, therefore, is one of the most promising routes we possess to the origin of the code and translation itself. Here we test for unexpectedly frequent cognate coding triplets within, taking an essential role in, a specific set of RNA-amino acid binding sites.

1.2. Amino Acids as RNA Ligands

Amino acids, though they are much smaller (MW $\approx$ 110) than nucleotides (MW $\approx$ 340), present two faces for interaction by nucleotides in RNA. As judged from crystal structures of riboswitches [1], RNA sites necessarily allow conserved, highly polar α-carbon groups (like carboxyl and amino) to be fixed in space by a convergence of highly directional polar interactions, such as hydrogen bonds [2]. With such a fixed, common foundation, an RNA binding site can also make bonds to a side chain group (Figure 1). This double-ended "polar profile" [1], of course, only applies to amino acids with two polar centers to offer. Further, even with possible bi-directional interactions in hand, other constraints (such as the selection for small site size) will favor interactions with one locus or the other.

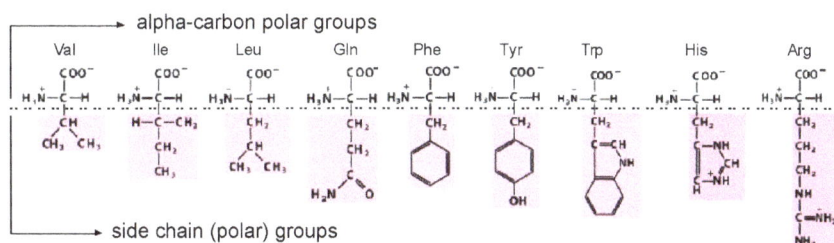

Figure 1. Amino acids studied by selection. The drawing divides the amino acids into two sites of possible interaction, divided by a horizontal dashed line. Firstly, α-carbon groups, which can make favorable polar interactions with RNA in every case. Then side chains, which can make stabilizing RNA contacts when their varied polar character allows. Val = valine; Ile = isoleucine; Leu = leucine; Gln = glutamine; Phe = phenylalanine; Tyr = tyrosine; Trp = tryptophan; His = histidine; Arg = arginine. His imidazole is drawn protonated, because the major His site [3] and the His-Phe peptide site as well [4] prefer protonated His imidazole.

For the purposes of biological structure and coding, we will be interested only in sites that include interactions with a side chain. Thus, relevant RNAs will bind both α-carbon/side chain or side chain only. Such sites are amino acid specific, and thus allow encoding of the amino acid. These distinctions are crucial to the function of RNA binding sites, because double-ended sites yield greater energies of interaction ($\Delta G < 0$) than single ended sites. For similar reasons, double-ended sites have greater stereoselectivities because they localize the sidechain (which transits the carbon tetrahedron when an L-amino acid becomes a D-amino acid).

We can make these descriptions quantitative [1]. Based on 337 independently-derived binding sites for nine of the standard protein amino acids, potentially single-ended RNA sites have K_D from 10^{-2} to 10^{-3} M/ΔG_{bind} −2.8 to −4 kcal/mol. The more intimately engaged amino acids, presenting two sites of interaction, have K_D from 10^{-4} to 10^{-6} M/ΔG_{bind} from −5.5 to −8 kcal/mol at 25 °C. The stronger affinities are clearly consistent with several intermolecular bonds, to two sites.

Switching to stereoselectivity, apparently single-ended sites range from 1-fold (no distinction) to ≈30-fold (0–2 kcal/mol), and 10- to several thousand-fold (1–5 kcal/mol) in double-ended amino acid binding sites.

1.3. A Substantially Single-Ended Example Site, Isoleucine (Ile)

Notably, a less polar side chain does not rule out all amino acid selectivity. Hydrophobic sidechains like Val (valine) and Ile (isoleucine) (Figure 1) are of interest because they are observed in spark tube experiments [5], and are therefore thought of as primitive [6]. Despite these distinctions, they do not offer polar sidechain interactions. Nevertheless, an RNA site selected for L-Val [7] prefers it by 1.6 kcal/mol to L-α-amino-butyrate (one methylene group smaller). A site selected for l-Ile [8] prefers it by 0.82 kcal/mol to L-valine (one methylene group smaller). These findings raised the possibility of specific RNA bonds to aliphatic sidechains. However, these specificities are now instead believed to result from use of the size of the sidechain [1] as an essential site structural element, because further decreases in side chain size after removal of the first methylene have little effect.

1.4. A Frequently Double-Ended Example Site, Arginine (Arg)

The arginine (Arg) side chain features a terminal guanidinium ion. The ion is planar, aromatic, positively charged, and offers a pattern of hydrogen bonding that matches the edge of nucleobases extremely well. This makes Arg sites very frequent in RNA; for example, such sites are smaller than sites for other amino acids [9]. In tallies of the content of RNA-protein interfaces, arginine provides the most numerous contacts [10]. This significance extends to regulatory interactions, where Arg contacts with RNA mediate regulatory modulation, for example, in the TAR peptide of HIV [11,12]. Further, Arg is unique in having unusual general interactions with folding RNAs, where it uniquely destabilizes tertiary folding, both slowing formation and speeding the breaking of a tertiary RNA contact [13]. For parallel reasons, it is no surprise that the first-detected specific amino acid binding site on RNA was for Arg [14]; the amino acid competes with G nucleotides for interaction with the splicing cosubstrate site on *Tetrahymena* self-splicing rRNA [15].

The ability of specific RNA folds to bind one or both amino acid domains will be a crucial point of discussion below.

2. The General Study of Amino Acid-RNA Binding by RNA

In order to generalize about amino acid-ribonucleotide interfaces, it is productive to study a number of them, involving bound amino acids of different types (Figure 1). The selection and cloning of RNAs [16–18] specifically eluted by cognate free amino acids [19] from carboxyl-immobilized amino acid columns provided this opportunity. Immediately above, some properties of binding sites obtained in this way have been listed. Below, I characterize the way the selection method produces its results to provide context for interpretation of now-numerous, newly-selected binding site sequences.

2.1. The Affinity Method

Amino acids are immobilized at concentrations of several mM to tens of mM, usually by coupling their carboxyls to make amides, using amines linked to a chromatographic support via a neutral connecting arm. Large populations of randomized RNA sequences will contain some active amino acid sites that bind to such an immobilized amino acid. These bound RNAs can be eluted, after washing away unbound molecules, with solutions of amino acid. In effect, a small minority of RNA amino acid sites declare themselves by first becoming immobile on the fixed amino acids of the column, then being mobilized by the minor solution change produced by addition of a low concentration of, say, dissolved mM L-histidine (His) in column buffer. Because initial randomized sequences are usually flanked by constant sequences complementary to primers, RNAs that bind pure free L-amino acid, D-amino acid, or derivatives can be saved as DNAs, which are amplified, then transcribed from a promoter in a constant region to later use them.

Such affinity chromatographic procedures purify L-Ile-binding RNAs by ≈100-fold when first applied [20], typically declining to 1-fold (no purification) after five or six chromatography-amplification cycles, at which time ≈20% of transcripts are eluted by isoleucine.

2.2. Simple, Abundant Sites

It is vital to appreciate the target(s) detected by affinity selection. Experiments on the origin of the code do not seek sites with optimized performance, but instead, the simplest sites. That is, shorter RNAs that may exhibit less impressive affinities and selectivity [21]. This is because a primitive environment is likely to be restrictive to RNA synthesis and survival. Accordingly, the molecules most easily accessed, least sensitive to physical or chemical attack, seem the appropriate targets.

That is—it certainly is possible to do selections that optimize a function. Selecting RNAs that slowly release a ligand selects most stable binding, for example, by L-Arg [22]. Alternatively, if a selection allows RNAs to compete for reaction at a limited number of sites, selection of the fastest reacting can be the result [23,24]. However, in the absence of such functional pressures, the most numerous RNAs, or most probable, or the simplest, are the ones readily isolated.

The latter case describes affinity selection. A 1 ml affinity column containing 1 mM ligand has 6×10^{17} potential RNA binding sites. Roughly 10^{15} total RNAs are added to initiate a selection, and a small fraction of these fold to produce amino acid binding sites. If 10^{-10} of random sequences have active sites [25], 10^5 molecules of RNAs assort themselves among 6×10^{17} loci. Competition is vanishingly rare, even after selection has greatly increased the active RNA fraction.

Using equations for affinity chromatography at equilibrium [26], it can be shown [19] that a 'typical' column affinity selection recovers RNAs with $K_D \leq$ approximately half the eluant concentration; $K_D \leq 2.5$ mM for free ligand when RNA is eluted with 5 mM ligand. This ability to examine simple RNAs with affinities into the mM range is another of the qualities that specifically suit affinity chromatography to coding studies.

2.2.1. Number of Essential Nucleotides

Usually, one can define nucleotides essential to RNA site functions using straightforward biochemical criteria. Such nucleotides are conserved in independent isolates; protected or sensitized to chemical probes by interaction with specific RNA ligands; or alter RNA activities if they are previously altered chemically or by mutation (e.g., [27]). The biochemically defined active site is the sum of such functional nucleotides, the number of "Implicated Site Nucleotides" (ISN). Implicated Site Nucleotides differ from the constellation of atoms also called nucleotides by a structural biologist, and sometimes the distinction is essential.

Though usually obvious, site nucleotides can occasionally be elusive. In the simplest L-tryptophan (Trp) site [28], a G flanking the amino acid binding loop is absolutely required for function, but so variable in position and in surrounding structure that it was not evidently conserved, and so was not initially detected [29]. Nevertheless, such cryptic requirements still affect the frequency of Trp-binding activity. For purposes of thought, the simplest L-His RNA site contained a mean of 20.1 ISN [4], the sufficient L-Trp site about 18 ISN [29], and the simplest L-phenylalanine (Phe) site 17.5 ISN [4].

In 1 A_{260} of the above partially randomized RNA (with flanking constant sequences), all contiguous 24-mer sequences will likely be present [19]. Shorter chains of essential nucleotides will be multiply present, and are more likely to be recovered. Thus, a simple summary is: the shortest contiguous sequences, usually having ≤24 essential nucleotides, should be the most likely to be isolated. These 24 "essential nucleotides" are defined by statistics. If only purines occur at a given position, this is twice as likely to occur as one specific nucleotide. In this case, selection can isolate twice as many such "essential nucleotides".

To put these ideas in another useful way, increasing the scale of an experiment by using 10-fold more RNA usually provides access to 1.66 additional essential nucleotides [19]. Thus, there are two kinds of selection experiments. One can do large experiments to seek large active motifs, but this usually implies looking among sparsely sampled molecules, because not all sequences of long lengths are present. Alternatively, one can look for smaller motifs, using RNA populations that contain many copies of them. Such an experiment tests every possible sequence of shorter length for the selected activity, which is often desirable. Increasing the amount of RNA moves the size boundary between these two experimental goals, 1.66 nucleotides for every 10-fold in RNA. This quantitative argument therefore also bears on the scope of small experiments. Because, typically, essential nucleotides ≤ ISN, selection experiments of practical laboratory size, even small ones, easily recover RNAs with enough ISN to fold functional amino acid binding sites.

2.2.2. Modularity

However, real RNA active sites are not usually made of the contiguous essential nucleotides discussed above. An active internal loop, for example, may be composed of two active 'single-stranded' loop modules which combine to yield an active two-sided loop surrounded by helices—with little regard to the initial spacing between the conserved loop modules. This is very important to real tertiary structures because the more modules, and the more even their sizes, the more ways there are to place them—thus the more frequently they occur within a randomized sequence [30]. Therefore, being composed of many pieces, in the best case pieces of similar size, can also determine whether an RNA structure can be isolated. Selections tend to isolate the most modular structures, as well as the ones containing the fewest essential nucleotides [31].

The reasoning that makes modules helpful also suggests that space is similarly good. Longer RNAs for selection should have more ways of, and be more capable of, manifesting a structure. To an extent, this is true experimentally; up to ca. 60 randomized nucleotides, the Ile RNA binding site becomes more frequent [20]. However, then in violating theory, it is less frequent in longer molecules. Perhaps long RNAs go to Uhlenbeck's alternative conformer hell [32].

2.2.3. Partially Conserved Nucleotides

Even nucleotides not usually defined as conserved must be recruited to form an active site, like those that form variable paired regions around a more conserved internal loop. These requirements also reduce the frequency of sites, and can be subtle.

For example, complementary primer sequences reduce the frequency of the prevalent Ile-binding site ca. 7.5-fold [20]. This effect can be traced to a displacement of site-bounding helices. The preexisting constant helical structure favors one permutation of the Ile site, because one bounding helix is easier to form from random sequences, thus also decreasing the accessible sequence space and total frequency of the Ile motif.

Adding site-defining stable helices to flank active Ile loop modules decreases active site occurrence by orders of magnitude [33]. The result is that about 4.1×10^9 100-mers or 0.2 nanograms or 7 femtomol of RNA chains must be searched to find the folded Ile-binding RNA. Judging from folding calculations, inhibitory folding effects appear to be a much smaller impediment than effects of the rarity of these bounding helical structures themselves. Nevertheless, these populations are orders smaller than the usual laboratory selection experiment. They therefore suggest that an RNA world with amino acid binding RNAs is more accessible than intuition at first suggests.

2.2.4. Constant Promoter/Primers

The above Ile effect introduces the effects of flanking sequences, which can become directly or indirectly involved in the active sites. Such direct effects of constant sequences are easily found. Flanking sequences can be incorporated as ISN, thereby changing the most likely site. The incorporation of an AAA run from constant sequences completely changed the outcome of a selection for Ile-binding sites [34], reducing the most frequent motif in any other selection to a minority. Two-nucleotide constant tag sequences introduced for another reason led to isolation of a previously unseen motif for D-His binding [35]. When the unique tags were not supplied, the novel site did not appear at all in later selections.

As might be expected, the effects of constant sequences fade as the random region is lengthened, and the selected site (for Ile; [20]), on average, moves away from constant influence. However, the goal of coding experiments is to persuasively eliminate outside effects on the selection. This kind of spurious effect can be eliminated by re-isolation of the same site in the context of different constant sequences. For example, this has been done for the simplest Ile, His, and Trp-binding RNA motifs. A more specific strategy, for coding studies, is to bar an amino acid's codons and anticodons from fixed sequences (and thus bar them from inducing complements in selected sequences), as was done for L-His [3,35].

Thus, a selection experiment also selects the constant sequences in the RNA transcript. Usually, this is of no concern. However, in a rare case the end(s) of the RNA are crucial to activity, and the RNA selected can change dramatically when a bounding sequence is changed or eliminated [36].

2.3. Sequentially Squeezed Selection

As this discussion shows, many factors alter the occurrence of a selected RNA sequence. To simplify selection outcomes, and make them more easily interpretable, amino acid binding selections have been conducted in random regions of decreasing size. For example, L-Ile binding was sought within 26, 22, and 16 contiguous randomized nucleotides [25]. This size range is narrow enough to avoid size selection based on slower replication of longer molecules [37]. Moreover, the experimental design accentuates two well-known benefits. First, short RNA populations contain sequences, like the Ile binding site, at frequencies close to calculated from probability, whereas long RNAs are deficient [20]. Second, as pointed out above, short sequences can be fully represented in initial selection populations, so that RNAs derived are plausibly the only functional ones existing at that size.

As we hoped, one Ile site sequence was prominent at larger lengths, the majority sequence with selected activity at a shorter length, and then disappeared, leaving no *bona fide* L-Ile-binding RNAs at the shortest length. Thus, there is a predominant active structure, which persists as space for it is shortened. Squeezing appears to establish a reliable limit—when selection requires more nucleotides than randomized tracts provide, no shorter site is selected.

Related experiments apparently yield the simplest amino acid site for L-Ile [20,25], L-His [35], L-Trp [28,29], and L-Arg [9]. This is not trivial in any case, but arginine is especially interesting. L-Arg-RNA interactions are unusually strong and versatile (see above). Thus, numerous L-Arg sites had been isolated. However, despite repeated selection, no L-Arg binding site had been observed more than once. Nevertheless, under sequentially squeezed selection, a simplest L-Arg site emerged. Note particularly that the shortest, simplest site in these experiments is required to be sidechain-specific (otherwise an amino acid cannot be meaningfully encoded). Thus, a squeezed specific selection probably focuses the site profile toward sidechain features.

Moreover, study of two activities side-by-side allows investigation of which is the simpler RNA function (takes place in the smaller site). Simultaneous mixed squeezed selection of affinity for D-His and L-His attached to a non-chiral glass support suggests that D-ribose RNA has an intrinsic chiral preference. It folds the simplest site for L-His using about one less essential nucleotide than required for the simplest D-His site [35]. The simplest L-His site was the same one [3] previously isolated for L-His alone using a different column matrix, different fixed sequences and solution conditions, strengthening the argument for selection of simple sites. This same chiral L-His RNA site has been taken through the looking glass, by synthesizing a Spiegelmer containing L-ribose rather than D-ribose. Ruta, et al [38] confirm that an enantiomeric switch in ribose also switches the RNA binding site to favor D-His.

2.4. Reproducible Selections

I emphasize a general conclusion about amino acid affinity selections. Within appropriate limits, for example, attributable to the need for fixed flanking sequences that do not intrude, selections have a predictable outcome. There are, reproducibly, simplest sites. The simplest L-Ile site has been independently isolated 267 times [20]. Even for a versatile amino acid like L-Arg, which binds quite variable, small ribonucleotide sequences—nonetheless a properly constrained search repeatedly finds particular simple, recurring binding sites [9]. By extension, given predictable selection, evolution at the amino acid-RNA level of complexity can be productively interrogated by experiments, and reliable relations between amino acids and RNA sequences can be derived.

3. Amino Acid Binding Sites and Coding Triplets

We now consider one of those "reliable relations" in selected RNA-amino acid binding sites. What follows (and what came before) is based on data for eight amino acids of varied chemical classification (Figure 1): charged polar (Arg+, His+), uncharged polar (Tyr, Gln), aliphatic hydrophobes (Ile, Leu) and aromatics (Phe, Trp). The survey is partial, but quite broad (Figure 2). There are 464 independently derived sites in the characterized populations, Implicated Site Nucleotides number 7137, and total nucleotides, inside and outside amino acid sites, are 21,938. Tested amino acids emerged from the evolution of the code with six, three, two, and one triplet(s). Site sequences have been examined for 44 coding triplets altogether, 22 cognate codons and 22 cognate anticodons. The results surveyed are those referenced earlier [1], updated for the sequentially squeezed selection for L-Arg [9].

P_{codon} and $P_{anticodon}$ are probabilities that the associated coding triplets are equally frequent outside each site and inside (within the ISN of) each site. That is, Figure 2 tabulates the probability that frequencies outside and inside are equal, by the G test—related to Chi-squared, but more versatile [39]. Equality is not the rule, as shown by probabilities with triple-digit negative exponents observed in Figure 2. Instead, seven cognate anticodons and two codons are very significantly elevated (marked by shaded backgrounds for probabilities) in the ISN that are most closely connected to a bound amino acid. The control is initially randomized nucleotides also in the selected RNAs, also selected using the same procedures, but outside the ISN of the active binding site. Further, coding triplets in boldly outlined white boxes in Figure 2 are the one codon and four anticodons concentrated in sequentially squeezed selections for RNAs binding Trp [29], His [35], Ile [25], and Arg [9]. No amino acid site concentrates codons alone; real cases either present both codons and anticodons (Arg, Ile) or anticodons alone (His, Phe, Trp, Tyr). Notably, the positive results can be called sparse: only two of 12 Arg triplets are significantly implicated by selected sites, or two of six triplets for Ile. Other cases have found only one triplet concentrated in RNA binding sites. Sparseness is a crucial finding, whose implications reappear below.

Indep sites	Site nt	Total nt	Amino acid	Codons	P_{codon}	Anticodons	$P_{anticodon}$
				CGU	0.666	ACG	0.0043
				CGC	0.011	GCG	0.0058
161	2653	4830	Arg	CGA	0.901	UCG	3.9×10^{-5}
				CGG	0.0031	CCG	0.974
				AGA	0.957	UCU	0.953
				AGG	5.6×10^{-19}	CCU	4.0×10^{-26}
2	42	156	Gln	CAA	0.042	UUG	0.970
				CAG	·	CUG	0.950
54	969	3644	His	CAU	0.870	AUG	0.010
				CAC	0.120	GUG	1.6×10^{-8}
185	2508	9915	Ile	AUU	8.0×10^{-110}	AAU	1.00
				AUC	1.00	GAU	1.00
				AUA	1.00	UAU	3.2×10^{-131}
				UUA	0.980	UAA	·
				UUG	0.029	CAA	0.710
1	37	73	Leu	CUU	·	AAG	0.950
				CUC	0.990	GAG	0.250
				CUA	0.300	UAG	0.0060
				CUG	0.300	CAG	·
2	35	160	Phe	UUU	0.980	AAA	0.012
				UUC	0.980	GAA	5.5×10^{-5}
56	763	2889	Trp	UGG	1.00	CCA	2.7×10^{-13}
3	130	271	Tyr	UAU	0.026	AUA	6.0×10^{-6}
				UAC	0.0041	GUA	0.0020
Totals	464	7137	21938				

Figure 2. Probability of uniform distribution of codon and anticodon triplets. Here the Implicated Site Nucleotides (ISN) are compared with other initially randomized positions in individual selected amino acid binding RNAs. Fixed sequences, of course, are not considered. Under the null hypothesis that cognate triplets (listed in columns) are equally frequent inside and outside the ISN, the probabilities of equal triplet distributions for eight kinds of amino acid sites are tabulated. Probabilities come from a two-tailed G test with Williams correction [39]. Probability boxes containing dashes are triplets that did not exist in the experimental sample. Among probabilities, shaded boxes with italicized numbers are significant. To evaluate significance in a conservative way, I compute $P_{sig} = 1 - (1 - P_{err})^{1/n}$ where P_{sig} is the maximum acceptable probability and P_{err} is the target error for each of the n trials in the Figure. To limit the probability of error to $P_{err} = 0.01$ in 44 individual trials, the maximum probability regarded as significant is $P_{sig} = 2.3 \times 10^{-4}$. Among triplets, italic triplets on white backgrounds are those concentrated by sequential squeezed selections for the cognate amino acid. Binding and sequence data can be found in: Ile [8,20,25], leucine (Leu) (I. Majerfeld, M. Illangasekare, M. Yarus, unpublished; see [1], Gln (C. Scerch and G. Tocchini-Valentini, pers. comm; see [1]), Phe [40], Tyr [41], Trp [28,29], His [3,35], Arg [9,22,26,42].

As an example, an arginine site is shown in Figure 3, where one of the most prevalent L-Arg binding motifs is drawn. Gray circles mark Implicated Site Nucleotides. RNAs closely related to this one, which bind L-Arg near the junction of a short helix and a highly-conserved 8-membered hairpin loop (Figure 3), comprised 62% of all isolated RNAs. Related small sites conserve the L-Arg anticodon marked at the entry to the hairpin loop (Figure 3) in 94% of all sequences. These motifs are well-represented even when given only 17 initially randomized nucleotides to fold.

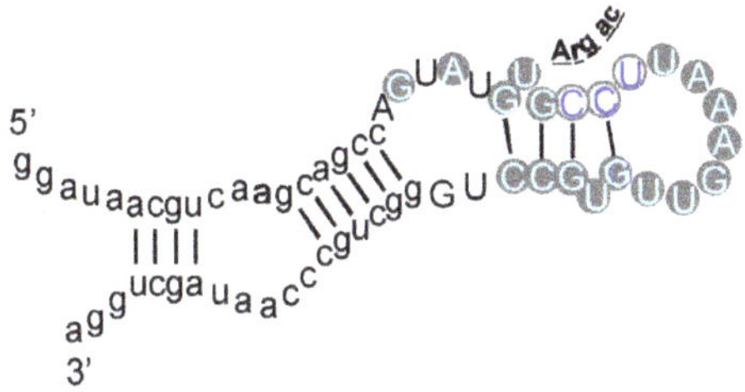

Figure 3. An example: the most prevalent Arg-binding RNA. Arg-606 [9], derived from a 25-nucleotide randomized region, is shown. Lower-case letters are fixed sequences, capital letters represent originally randomized nucleotide positions. The nucleotide sequence is threaded through the probable secondary structure for all related isolates, deduced by BayesFold [43]. Gray circles mark Implicated Site Nucleotides, and the three open gray circles are a very highly conserved arginine anticodon (cognate to codon AGG). Arg-606 had K_D = 0.5 mM, and D/L ≈ 35, consistent with the idea that the smallest sidechain-specific sites are predominantly single-ended. Comparable simplest His, Ile, and Trp sites from separate sequentially squeezed selections have been reviewed [27].

4. Tiny Probabilities

Below, I argue that minute probabilities in Figure 2 are reliable guides—cognate coding triplets are improbably elevated within RNA binding sites. These particular minute magnitudes are produced by the experimental context. Sequentially squeezed selections generate many new, independently derived binding sites. If a conserved cognate triplet appears in the simplest site, more sites with this non-random outcome force the probability of an unbiased distribution progressively down. This is evident in Figure 2, where the tiniest P_{codon} and $P_{anticodon}$ are in white boxes associated with squeezed selections. However, that being said, what of it? This behavior characterizes any true hypothesis. The more experimental evidence, the less probable that we will contradict a true finding. Moreover, Figure 2 contains cases like Phe and Tyr, where characterization of a few motifs from a normal selection turn up an improbably concentrated cognate triplet. This was true for sequentially squeezed selections also, before they were squeezed. Therefore, association of cognate triplets with RNA binding sites does not depend on a special experiment—it was evident, in all cases, among initial examples isolated.

4.1. Observed Triplet Concentration Is Not Attributable to the Statistical Test

The test used in Figure 2 (G test for goodness of fit with the Williams correction [39]) is related to one universally used to test ratios in genetic crosses, and is therefore employed widely in Biology. However, no test, nor any assumption whatever about the natural distribution of triplets within RNAs is needed to reach the conclusion that the null hypothesis (triplets equivalent everywhere) is very improbable. For L-Arg [9], nucleotide sequences of isolated RNAs were randomized 10^6 times, and the resulting "binding sites" at previous positions were retested. The concentration of the Arg CCU anticodon in real binding sites, for example (Figures 2 and 3), was not observed in a million such tries.

4.2. Triplet Concentrations Have the Logic of Real Coding: Reversed Triplets

5′ to 3′ reversed codons (e.g., UUC Phe > CUU) and anticodons have the same compositions and the same predicted random frequencies as true triplets. Such reversals would be concentrated in binding sites by any accidental process. Moreover, if binding sites (or nonbinding sites) express an underlying preference for certain nucleotides or triplet compositions, reversed triplets would succumb. Thus, it is striking that, tested for multiple RNAs binding each of six amino acids, multiple observed excesses of cognate triplets of both kinds vanish when tested triplets are reversed [44]. Because binding sites contain several triplets (compare Figure 3), one might argue that at a significant frequency, cognate triplets will recur by chance. Evidently, this is rare, since reversed codons and anticodons do not observably do so, given 42 triplets evaluated in 22 site sequences of six specificities.

4.3. Triplet Concentrations Have the Logic of Real Coding: Variation of the Code

Fifty million randomized codes have also been tested for triplet localization in experimental binding sites [44]. Notably, 10^6 new codes were derived in five ways: with codons placed randomly in the Coding Table, amino acids assigned randomly among real coding blocks, amino acid identities assorted to blocks of the same size, randomization of triplet position 1 and 2, and reassignment of initial codon doublets. In short, 99.2% to 99.5% of these randomized codes give less association with observed binding sites than the real code, and those that do yield association tend to be those retaining fragments of initial code structure. There are important positive and negative implications. Positively, triplet excesses in experimental RNA binding sites are strongly associated with assignments made during evolution of the bona fide coding table. Negatively, these data are further strong evidence against accidental links between triplets and cognate amino acids as a result of these procedures (Figure 2).

4.4. Relation to Natural Cases

Remarkably, pooled experimental results in Figure 2 overlap evidence from natural RNA sequences. The *Tetrahymena* self-splicing group I intron binds arginine [14], and guanidinium ion as an analogue of the Arg side chain, using the G of a conserved Arg codon AGA/CGA/AGG [45]. Arg guanidinium (terminus of the Arg side chain in Figure 1) emulates the base-pairing of G, so it can bind at the same site [15,46]. Because this is within the active site for the co-splicing substrate, a guanosine nucleotide, Arg and guanidinium inhibit splicing [14]. This behavior overlaps the concentration of AGG triplets within newly selected RNA structures that bind Arg (Figure 2). Thus, the initial evidence which initiated studies of amino acid-RNA binding is echoed in present selection results.

An even more surprising case appears in riboswitches regulated by guanidinium ion, in bacteria that need to control its modification and export [47]. Riboswitches regulate linked messages by changing structure on binding metabolites. There are, for example, RNA riboswitch domains that bind Gly [48], Lys [49], and Gln [50]. Such RNAs usually have complex structures and functions, and so are not plausibly related to selected simplest amino acid sites.

However, guanidinium ion may be an exception. This small-molecule analogue of the Arg side chain terminus (Figure 1) is bound within the conjunction of three conserved Arg codons, AGA/CGG/CGG (Figure 4). Nucleotides of the three Arg codons are not only in close contact with the ligand, but completely fill the space around guanidinium and engage all of the polar groups of the ion [51]. Using an adjacent G surface, this three-Arg-triplet site also includes close contact with the top and bottom of the Arg side chain analog. The *Tetrahymena* site binds Arg [14], though for lack of space, the *Sulfobacillus* site does not admit the complete amino acid [47]. However, both natural examples display extreme concentration on the distal amino acid side chain of Arg, accompanied by cognate coding triplets. Thus, sites in *Tetrahymena* rRNA and *Sulfobacillus* riboswitch aptamers suggest that for Arg, the chemical connection between arginine/guanidinium affinity and coding triplets has found biological uses which persist into modern organisms. Such contemporary interactions may be much more frequent—anticodons in rRNA appear appreciably concentrated close to cognate amino acid sidechains in four crystallographically defined ribosomes [52].

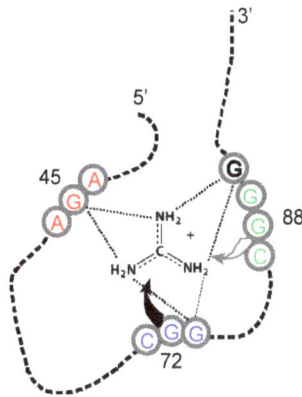

Figure 4. Schematic structure of a *Sulfobacillus* guanidinium riboswitch. Gray numbered circles are nucleotides of the crystallographic structure for guanidinium ion bound to the sensor of the guanidinium-I riboswitch. Dotted lines are hydrogen bonds; the gray and black curved arrows indicate that G72 covers the top, and G88 forms the bottom of the guanidinium binding site, respectively. Arg triplet nucleotides are colored; the ones centered at G45 and G72 are almost completely conserved; at G88, ≈75% conserved. G90 (black) is a non-triplet site nucleotide. Drawn from [51] and Protein Data Base (PDB) structure 5T83.

It is unexpected that an amino acid affinity purification isolates RNA sequences that repeatedly show a specific formal relation to the genetic code. Moreover, similar interactions for Arg appear in natural RNAs. These data (Figures 2 and 3) are particularly interesting because squeezed sites, and natural RNAs that emulate squeezed sites by concentrating on the terminus of a side chain, also elevate the probability of essential coding triplets. In light of these repeated findings, there does not seem to be a plausible alternative to the conclusion that RNA binding sites recapitulate an essential event during the evolution of the amino acid code—but what event? Association of triplets and cognate sites is itself objectively demonstrable. However, to reason about the foundational events of the genetic code, a bit of speculation is required.

5. Direct RNA Templates (DRT)

The apparent simplest way to use these findings in primordial translation uses an RNA template that directly binds (activated) amino acids side by side, so they subsequently react to form ordered, encoded peptides. This emulates the mechanism of the ribosomal peptidyl transferase itself—it accelerates its reaction principally by apposing reactants [53]. Cognate RNA triplets within amino acid binding sites subsequently evolve to act as anticodons in tRNAs and codons in mRNAs [1]. In fact, the potential co-occurrence of amino acid specificity, anticodons, and codons together in one RNA binding site is an intrinsically striking property. RNAs studded with multiple aminoacyl-RNA synthesis centers at a potential mean spacing of only a few nucleotides are also well known [54], and similar aminoacyl transfer centers can be supplied with activated amino acids by a ribozyme [55,56]. These data together make possible RNA encoded peptide synthesis resident in one small RNA complex. The advantages of DRT simplicity have been argued before [1], though there are other possibilities [57].

Because we were interested in the molecular constraints on a DRT, we selected RNAs that bind [4] two amino acids in peptide linkage, NH$_2$-His-Phe-COOH, retaining specificity for both side chains. His and Phe were used because their binding as free amino acids was already understood (see references,

Figure 2). This experiment required counterselection against affinity for His and Phe individually, because singly-directed sites require fewer nucleotides. Thus, affinity for a single side chain (usually protonated His, Figure 1) is selected preferentially. When the census of ISN is taken on these sequenced and characterized His-Phe RNAs, His sites required 20.1 ISN, Phe 17.5 ISN, and His-Phe 24.4 ISN (averaging all RNAs in the two prevalent motifs for the latter). As an example, RNA 16 has $K_D = 90$ µM for His-Phe, 13 mM for L-His, and 100 mM for L-Phe. Thus, a peptide-binding RNA, even one that contacts both side chains, is not the sum of two amino acid affinities. Instead, the peptide site is only ≈35% larger than a site for one amino acid. Consistent with these counts, neither the previously known His site, nor the known Phe site, appear in these selected His-Phe RNAs. A new dual, smaller site is selected instead. An example of the most frequent His-Phe site is shown in Figure 5.

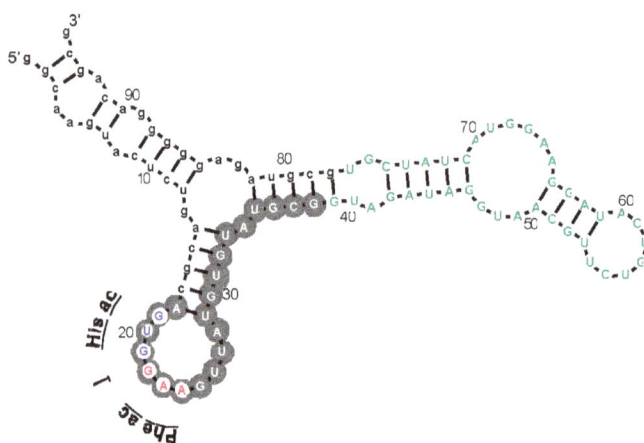

Figure 5. RNA 8, which has affinity for His and Phe in His-Phe. Lower case letters are fixed sequences, capital letters are initially randomized positions [4]. The RNA is threaded through the most probable secondary structure computed for its independent isolates by BayesFold [43]. Gray circles mark Implicated Site Nucleotides; those with white centers are potential coding triplets labeled "Phe ac" (ac = anticodon) and "His ac". Green nucleotides are non-site, but initially randomized nucleotides.

A ready rationale exists for smaller individual amino acid sites, still side chain specific. These can be extreme single-ended sites (see above), forced to be small because of the crowding of two sites produced by the short single covalent peptide bond between His and Phe. To be consistent, this kind of L-His site was not produced by sequential squeezed selection [35], so its structure must depend on the adjacent Phe residue or site. The existence of this kind of molecule supports the DRT, because it shows that RNA that binds DRT substrates (which are like free amino acids) can also bind the peptide product (His-Phe). Thus, for catalysis, only the binding of the transition state for peptide bond formation has not been shown, and this predicted activity can now be subjected to experimental search.

However, support for a DRT from this work has another, more surprising dimension. The sequence of the ISN for His-Phe RNA (Figure 5) contains adjacent His and Phe anticodons (white centers, Figure 5). Further, these are the same triplets over-represented in newly selected separate His and Phe binding sites (Figure 2). In this experiment, we do not have the statistical power (Figure 2) or structural resolution (Figure 4) of the general investigation of amino acid sites, whose interpretation presently relies on almost 100-fold more sites than for His-Phe peptide. Thus, caution is appropriate. Nevertheless, anticodon triplets (Figure 5) are noticeably conserved. There are seven independent parental molecules (12 isolates) of the His-Phe RNA shown. Three of seven have the Phe anticodon shown, two of those also have the adjacent His anticodon [4].

It would be unexpected to discover a new series of amino acid sites connected to the genetic code, in peptide binding sites. So, it is probably not a new set of sites, but simply a more radically squeezed structure. In other words, a partial site, not stable without the adjacent amino acid, but containing the same cognate anticodon as in the free amino acid site (Figure 2). This idea merits further investigation. Meanwhile, specific His-Phe peptide affinity, accompanied by individual sidechain contacts and cognate anticodons, are remarkably consistent with a primordial DRT.

6. The Origin of the Genetic Code Is a Puzzle Whose Pieces Fit Together

Two other major accounts of the code's history, co-evolution [58], and adaptation [59], also have major roles to play. These roles are, in fact, now explicitly defined by data in Figure 2, in the following sense.

Co-evolution is the idea that an early code ceded codons to later amino acids or acquired unused codons, as biochemical pathways extended the amino acid repertoire. This idea can be analyzed by comparing the coding table to biosynthetic pathways [60]. Adaptation theories propose that the code was created by optimization, most explicitly by reducing errors created by mistranslation [61]. Adaptation can be supported by showing resemblance between the genetic code's order and an optimized arrangement on the basis of similar amino acid chemical properties [62].

Both co-evolution and adaptation require a pre-existing code. There must be coding to be extended as biosynthesis advances. There must be coding to be optimized by adaptation. Therefore, both hypotheses require something like the stereochemically-defined core suggested by RNA binding data (Figure 2). In one sense, this pre-existing stereochemical core is likely to be substantial. Six of eight arbitrarily characterized amino acids (Figure 4) concentrate their anticodons in the ISN of binding sites selected from random RNA sequences. Thus, excepting Gln and Leu, traces of a canonical core are observed for 75% of amino acids surveyed.

6.1. The Nature of the Stereochemical Basis

However, 75% overstates the results in an important way. As pointed out above, coverage of the 48 possible triplets in binding sites is sparse. Arg is the high extreme: one of its six codons, and two of six anticodons are implicated by selection results (Figures 2 and 3). If one adds the group I self-splicing RNA [45], the count rises to three Arg codons and two anticodons. Provisionally adding the guanidinium specific riboswitch site yields one new codon [51]. Thus, this extensive dataset yields associations with six of 12 possible Arg triplets. Moreover, in the more complete survey (Figure 2) of eight amino acids, 12 of 48 possible associations have been detected. As Arg surely illustrates, we can be surprised by new data. However, it is more plausible that exacting chemical requirements for participation in a specific RNA binding site's tertiary structure can only be satisfied by a few cognate triplets, of all those available. The final result might be estimated close to the current average for eight amino acids, 25%, and less than the maximum 50% of triplets for Arg, the most RNA-accessible amino acid. That is: given present accounting (Figures 2–5), the majority of triplets may have entered the code another way, rather than via RNA-amino acid specificity.

6.2. Co-Evolution Is Needed to Reach Barren Areas

This reasoning implies a role for co-evolution and adaptation. How might one extend coding to triplets not touched by amino acid sites, like those for Gln (Figure 2)? A clear possibility is: one can co-evolve to adopt them. In fact, it has been suggested [63] that the existence of Glu-tRNA^{Gln}, a modern metabolite and possible co-evolutionary intermediate in the incorporation of the Gln codons, is strong support for co-evolution to Gln coding. This Glu-tRNA^{Gln} argument also complements negative RNA binding evidence for Gln triplets from selection (Figure 2).

6.3. Adaptation Is Needed to Fill Boxes

As for adaptation: how might one fill in the six kinds of partially occupied coding boxes sparsely created by RNA affinities (Figure 2)? The logic of RNA binding sites has no apparent reason to respect the neat groups of six or four or three or two triplets so characteristic of the code. However, this is easily rationalized as the result of a process which minimized the effect of translational ambiguity by evolving to use sets of related triplets. In fact, it can be shown that even levels of pre-existing stereochemical assignment we have found still allow resulting codes to be optimized [64]. There is no logical inconsistency in believing both stereochemistry and adaptation were influential in code history.

7. Conclusions

We decisively confirm the hypothesis in this review's first paragraph. The RNA-amino acid interface does contain the logic of (some of) the genetic code, relating triplets to amino acid side chains. Cognate triplets, though their functions may vary, are unexpectedly close to their amino acids. The conclusion is unequivocal—the probability that the contrary is true hovers in negative exponential triple digits (Figures 2–5). These data together strongly confirm intuitions of Crick [65], Orgel [66], and Woese [62], who thought that such a connection would exist.

It is presently less clear how to incorporate this finding into the code's history, but early data on a Direct RNA Template are very positive (Figure 5). Among the most probable His-Phe RNAs are frequent molecules contacting both amino acid side chains, held at a spacing appropriate to peptide synthesis, and containing both cognate His and Phe anticodons.

Accordingly, events attending the birth of the genetic code are still remarkably evident in modern RNAs and amino acids. This implies that modern molecules are very similar to their ancestors. This is consistent with the tree of life on Earth [67], which shows that the code and translation are virtually universal, so their molecules trace back at least to the Last Common Ancestor [68]. In the experiments above, we show that these agents are older yet, likely surviving from the first encoded ancestral peptides. This is crucial data; modern biochemicals are tacitly assumed relevant in many studies of molecular evolution.

Finally, though study of the route to the full code is just beginning, several strong constraints have empirical support (Figure 2). Despite persuasive evidence for cognate triplets in RNA binding sites, neither the resulting stereochemistry, nor adaptation, nor co-evolution are plausibly sufficient to create the entire code, acting alone. Stereochemical affinities are uniquely capable of initiating coding, but extension of such initial assignments via co-evolution and adaptation are probably essential to complete the modern coding table.

Acknowledgments: My continuing gratitude goes to the scientists whose names appear alongside mine below. As all who have lived in a laboratory know, this essay would not exist without their minds and hands, freely lent. Early work was supported by NIH research grant GM48080 and the NASA Astrobiology Center NCC2-1052.

Conflicts of Interest: The author declares no conflict of interest.

References

1. Yarus, M.; Widmann, J.J.; Knight, R. RNA-amino acid binding: A stereochemical era for the Genetic Code. *J. Mol. Evol.* **2009**, *69*, 406–429. [CrossRef] [PubMed]
2. Montange, R.K.; Batey, R.T. Structure of the *S*-adenosylmethionine riboswitch regulatory mRNA element. *Nature* **2006**, *441*, 1172–1175. [CrossRef] [PubMed]
3. Majerfeld, I.; Puthenvedu, D.; Yarus, M. RNA affinity for molecular L-histidine; genetic code origins. *J. Mol. Evol.* **2005**, *61*, 226–235. [CrossRef] [PubMed]
4. Turk-Macleod, R.M.; Puthenvedu, D.; Majerfeld, I.; Yarus, M. The plausibility of RNA-templated peptides: Simultaneous RNA affinity for adjacent peptide side chains. *J. Mol. Evol.* **2012**, *74*, 217–225. [CrossRef] [PubMed]

5. Ring, D.; Wolman, Y.; Friedmann, N.; Miller, S.L. Prebiotic Synthesis of Hydrophobic and Protein Amino Acids. *Proc. Natl. Acad. Sci. USA* **1972**, *69*, 765–768. [CrossRef] [PubMed]

6. Weber, A.L.; Miller, S.L. Reasons for the Occurrence of the Twenty Coded Protein Amino Acids. *J. Mol. Evol.* **1981**, *17*, 273–284. [CrossRef] [PubMed]

7. Majerfeld, I.; Yarus, M. An RNA pocket for an aliphatic hydrophobe. *Nat. Struct. Biol.* **1994**, *1*, 287–292. [CrossRef] [PubMed]

8. Majerfeld, I.; Yarus, M. Isoleucine: RNA sites with essential coding sequences. *RNA* **1998**, *4*, 471–478. [PubMed]

9. Janas, T.; Widmann, J.J.; Knight, R.; Yarus, M. Simple, recurring RNA binding sites for L-arginine. *RNA* **2010**, *16*, 805–816. [CrossRef] [PubMed]

10. Ellis, J.J.; Broom, M.; Jones, S. Protein-RNA interactions: Structural analysis and functional classes. *Proteins* **2007**, *66*, 903–911. [CrossRef] [PubMed]

11. Pugilisi, J.D.; Tan, R.; Calnan, B.J.; Frankel, A.D.; Williamson, J.R. Conformation of the TAR RNA-Arginine Complex by NMR Spectroscopy. *Science* **1992**, *257*, 76–80. [CrossRef]

12. Tao, J.; Frankel, A.D. Specific binding of arginine to TAR RNA. *Proc. Natl. Acad. Sci. USA* **1992**, *89*, 2723–2726. [CrossRef] [PubMed]

13. Sengupta, A.; Sung, H.L.; Nesbitt, D.J. Amino Acid Specific Effects on RNA Tertiary Interactions: Single-Molecule Kinetic and Thermodynamic Studies. *J. Phys. Chem. B* **2016**, *120*, 10615–10627. [CrossRef] [PubMed]

14. Yarus, M. A specific amino acid binding site composed of RNA. *Science* **1988**, *240*, 1751–1758. [CrossRef] [PubMed]

15. Michel, F.; Hanna, M.; Green, R.; Bartel, D.P.; Szostak, J.W. The guanosine binding site of the Tetrahymena ribozyme. *Nature* **1989**, *342*, 391–395. [CrossRef] [PubMed]

16. Ellington, A.D.; Szostak, J.W. In vitro selection of RNA molecules that bind specific ligands. *Nature* **1990**, *346*, 818–822. [CrossRef] [PubMed]

17. Robertson, D.L.; Joyce, G.F. Selection in vitro of an RNA enzyme that specifically cleaves single-stranded DNA. *Nature* **1990**, *344*, 467–468. [CrossRef] [PubMed]

18. Tuerk, C.; Gold, L. Systematic Evolution of Ligands by Exponential Enrichment: RNA Ligands to Bacteriophage T4 DNA Polymerase. *Science* **1990**, *249*, 505–510. [CrossRef] [PubMed]

19. Ciesiolka, J.; Illangasekare, M.; Majerfeld, I.; Nickles, T.; Welch, M.; Yarus, M.; Zinnen, S. Affinity selection-amplification from randomized ribooligonucleotide pools. *Methods Enzym.* **1996**, *267*, 315–335.

20. Legiewicz, M.; Lozupone, C.; Knight, R.; Yarus, M. Size, constant sequences, and optimal selection. *RNA* **2005**, *11*, 1701–1709. [CrossRef] [PubMed]

21. Carothers, J.M.; Oestreich, S.C.; Davis, J.H.; Szostak, J.W. Informational complexity and functional activity of RNA structures. *J. Am. Chem. Soc.* **2004**, *126*, 5130–5137. [CrossRef] [PubMed]

22. Geiger, A.; Burgstaller, P.; von der Eltz, H.; Roeder, A.; Famulok, M. RNA aptamers that bind L-arginine with sub-micromolar dissociation constants and high enantioselectivity. *Nucleic Acids Res.* **1996**, *24*, 1029–1036. [CrossRef] [PubMed]

23. Irvine, D.; Tuerk, C.; Gold, L. Selexion. *J. Mol. Biol.* **1991**, *222*, 739–761. [CrossRef]

24. Levine, H.A.; Nilsen-Hamilton, M. A mathematical analysis of SELEX. *Comput. Biol. Chem.* **2007**, *31*, 11–35. [CrossRef] [PubMed]

25. Lozupone, C.; Changayil, S.; Majerfeld, I.; Yarus, M. Selection of the simplest RNA that binds isoleucine. *RNA* **2003**, *9*, 1315–1322. [CrossRef] [PubMed]

26. Connell, G.J.; Illangsekare, M.; Yarus, M. Three Small Ribooligonucleotides with Specific Arginine Sites. *Biochemistry* **1993**, *32*, 5497–5502. [CrossRef] [PubMed]

27. Yarus, M.; Caporaso, J.G.; Knight, R. Origins of the genetic code: The escaped triplet theory. *Annu. Rev. Biochem.* **2005**, *74*, 179–198. [CrossRef] [PubMed]

28. Majerfeld, I.; Yarus, M. A diminutive and specific RNA binding site for L-tryptophan. *Nucleic Acids Res.* **2005**, *33*, 5482–5493. [CrossRef] [PubMed]

29. Majerfeld, I.; Chocholousova, J.; Malaiya, V.; Widmann, J.; McDonald, D.; Reeder, J.; Iyer, M.; Illangasekare, M.; Yarus, M.; Knight, R. Nucleotides that are essential but not conserved; a sufficient L-tryptophan site in RNA. *RNA* **2010**, *16*, 1915–1924. [CrossRef] [PubMed]

30. Knight, R.; Yarus, M. Finding specific RNA motifs: Function in a zeptomole world? *RNA* **2003**, *9*, 218–230. [CrossRef] [PubMed]
31. Yarus, M.; Knight, R.D. The scope of selection. In *The Genetic Code and the Origin of Life*; Pouplana, L.R., Ed.; Landes Bioscience: Georgetown, TX, USA, 2004; pp. 75–91.
32. Uhlenbeck, O.C. Keeping RNA happy. *RNA* **1995**, *1*, 4–6. [PubMed]
33. Knight, R.; De Sterck, H.; Markel, R.; Smit, S.; Oshmyansky, A.; Yarus, M. Abundance of correctly folded RNA motifs in sequence space, calculated on computational grids. *Nucleic Acids Res.* **2005**, *33*, 5924–5935. [CrossRef] [PubMed]
34. Legiewicz, M.; Yarus, M. A more complex isoleucine aptamer with a cognate triplet. *J. Biol. Chem.* **2005**, *280*, 19815–19822. [CrossRef] [PubMed]
35. Illangasekare, M.; Turk, R.; Peterson, G.C.; Lladser, M.; Yarus, M. Chiral histidine selection by D-ribose RNA. *RNA* **2010**, *16*, 2370–2383. [CrossRef] [PubMed]
36. Chumachenko, N.V.; Novikov, Y.; Yarus, M. Rapid and simple ribozymic aminoacylation using three conserved nucleotides. *J. Am. Chem. Soc.* **2009**, *131*, 5257–5263. [CrossRef] [PubMed]
37. Coleman, T.M.; Huang, F. RNA-catalyzed thioester synthesis. *Chem. Biol.* **2002**, *9*, 1227–1236. [CrossRef]
38. Ruta, J.; Ravelet, C.; Grosset, C.; Fize, J.; Ravel, A.; Villet, A.; Peyrin, E. Enantiomeric separation using an l-RNA aptamer as chiral additive in partial-filling capillary electrophoresis. *Anal. Chem.* **2006**, *78*, 3032–3039. [CrossRef] [PubMed]
39. Sokal, R.; Rohlf, F. *Biometry: The Principles and Practice of Statistics in Biological Research*; Freeman & Co.: New York, NY, USA, 1995.
40. Illangasekare, M.; Yarus, M. Phenylalanine-binding RNAs and genetic code evolution. *J. Mol. Evol.* **2002**, *54*, 298–311. [PubMed]
41. Mannironi, C.; Scerch, C.; Fruscoloni, P.; Tocchini-Valentini, G.P. Molecular recognition of amino acids by RNA aptamers: The evolution into an L-tyrosine binder of a dopamine-binding RNA motif. *RNA* **2000**, *6*, 520–527. [PubMed]
42. Tao, J.; Frankel, A.D. Arginine-Binding RNAs Resembling TAR Identified by in Vitro Selection. *Biochemistry* **1996**, *35*, 2229–2238. [CrossRef] [PubMed]
43. Knight, R.; Birmingham, A.; Yarus, M. BayesFold: Rational secondary folds that combine thermodynamic, covariation, and chemical data for aligned RNA sequences. *RNA* **2004**, *10*, 1323–1336. [CrossRef] [PubMed]
44. Knight, R.D.; Landweber, L.F.; Yarus, M. 2003 Tests of a stereochemical genetic code. In *Translation Mechanisms*; Lapointe, J., Brakier-Gingras, L., Eds.; Kluwer Academic/Plenum: New York, NY, USA, 2003; pp. 115–128.
45. Yarus, M.; Christian, E.L. Genetic Code Origins. *Nature* **1989**, *342*, 349–350. [CrossRef] [PubMed]
46. Guo, F.; Gooding, A.R.; Cech, T.R. Structure of the Tetrahymena Ribozyme: Base Triple Sandwich and Metal Ion at the Active Site. *Mol. Cell* **2004**, *16*, 351–362. [CrossRef] [PubMed]
47. Breaker, R.R.; Atilho, R.M.; Malkowski, S.N.; Nelson, J.W.; Sherlock, M.E. The Biology of Free Guanidine As Revealed by Riboswitches. *Biochemistry* **2017**, *56*, 345–347. [CrossRef] [PubMed]
48. Ruff, K.M.; Strobel, S.A. Ligand binding by the tandem glycine riboswitch depends on aptamer dimerization but not double ligand occupancy. *RNA* **2014**, *20*, 1775–1788. [CrossRef] [PubMed]
49. Serganov, A.; Huang, L.; Patel, D.J. Structural insights into amino acid binding and gene control by a lysine riboswitch. *Nature* **2008**, *455*, 1263–1267. [CrossRef] [PubMed]
50. Ames, T.D.; Breaker, R.R. Bacterial aptamers that selectively bind glutamine. *RNA Biol.* **2011**, *8*, 82–89. [CrossRef] [PubMed]
51. Reiss, C.W.; Xiong, Y.; Strobel, S.A. Structural Basis for Ligand Binding to the Guanidine-I Riboswitch. *Struct. Lond. Engl.* **2017**, *25*, 195–202.
52. Johnson, D.B.; Wang, L. Imprints of the genetic code in the ribosome. *Proc. Natl. Acad. Sci. USA* **2010**, *107*, 8298–8303. [CrossRef] [PubMed]
53. Sievers, A.; Beringer, M.; Rodnina, M.V.; Wolfenden, R. The ribosome as an entropy trap. *Proc. Natl. Acad. Sci. USA* **2004**, *101*, 7897–7901. [CrossRef] [PubMed]
54. Illangasekare, M.; Yarus, M. Small aminoacyl transfer centers at GU within a larger RNA. *RNA Biol.* **2012**, *9*, 59–66. [CrossRef] [PubMed]
55. Kumar, R.K.; Yarus, M. RNA-catalyzed amino acid activation. *Biochemistry* **2001**, *40*, 6998–7004. [CrossRef] [PubMed]

56. Xu, J.; Appel, B.; Balke, D.; Wichert, C.; Muller, S. RNA aminoacylation mediated by sequential action of two ribozymes and a nonactivated amino acid. *ChemBioChem* **2014**, *15*, 1200–1209. [CrossRef] [PubMed]

57. Szathmáry, E. Coding coenzyme handles: A hypothesis for the origin of the genetic code. *Proc. Natl. Acad. Sci. USA* **1993**, *90*, 9916–9920. [CrossRef] [PubMed]

58. Wong, J.T.F. Coevolution of genetic code and amino acid biosynthesis. *Trends Biochem. Sci.* **1981**, *6*, 33–36. [CrossRef]

59. Freeland, S.J.; Wu, T.; Keulmann, N. The case for an error minimizing standard genetic code. *Orig. Life Evol. Biosph.* **2003**, *33*, 457–477. [CrossRef] [PubMed]

60. Wong, J.T.F. A Co-Evolution Theory of the Genetic Code. *Proc. Natl. Acad. Sci. USA* **1975**, *72*, 1909–1912. [CrossRef] [PubMed]

61. Freeland, S.J.; Hurst, L.D. The genetic code is one in a million. *J. Mol. Evol.* **1998**, *47*, 238–248. [CrossRef] [PubMed]

62. Woese, C.R.; Dugre, D.H.; Saxinger, W.C.; Dugre, S.A. The molecular basis for the genetic code. *Proc. Natl. Acad. Sci. USA* **1966**, *55*, 966–974. [CrossRef] [PubMed]

63. Di Giulio, M. Genetic code origin: Are the pathways of type Glu-tRNA(Gln)—>Gln-tRNA(Gln) molecular fossils or not? *J. Mol. Evol.* **2002**, *55*, 616–622. [CrossRef] [PubMed]

64. Caporaso, J.G.; Yarus, M.; Knight, R. Error minimization and coding triplet/binding site associations are independent features of the canonical genetic code. *J. Mol. Evol.* **2005**, *61*, 597–607. [CrossRef] [PubMed]

65. Crick, F.H.C. The Origin of the Genetic Code. *J. Mol. Evol.* **1968**, *38*, 367–379. [CrossRef]

66. Orgel, L.E. Evolution of the Genetic Apparatus. *J. Mol. Evol.* **1968**, *38*, 381–393.

67. Forterre, P. The universal tree of life: An update. *Front. Microbiol.* **2015**, *6*, 717. [CrossRef] [PubMed]

68. Benner, S.A.; Ellington, A.D.; Tauer, A. Modern metabolism as a palimpsest of the RNA world. *Proc. Natl. Acad. Sci. USA* **1989**, *86*, 7054–7058. [CrossRef] [PubMed]

Concept Paper

What Froze the Genetic Code?

Lluís Ribas de Pouplana [1,2,*], Adrian Gabriel Torres [1] and Àlbert Rafels-Ybern [1]

[1] Institute for Research in Biomedicine (IRB Barcelona), The Barcelona Institute of Science and Technology, Baldiri Reixac, 10, 08028 Barcelona, Spain; adriangabriel.torres@irbbarcelona.org (A.G.T.); albert.rafels@irbbarcelona.org (À.R.-Y.)

[2] Catalan Institution for Research and Advanced Studies (ICREA), Passeig Lluis Companys 23, 08010 Barcelona, Spain

*   Correspondence: lluis.ribas@irbbarcelona.org; Tel.: +34-934034868

Academic Editor: Koji Tamura
Received: 2 March 2017; Accepted: 3 April 2017; Published: 5 April 2017

Abstract: The frozen accident theory of the Genetic Code was a proposal by Francis Crick that attempted to explain the universal nature of the Genetic Code and the fact that it only contains information for twenty amino acids. Fifty years later, it is clear that variations to the universal Genetic Code exist in nature and that translation is not limited to twenty amino acids. However, given the astonishing diversity of life on earth, and the extended evolutionary time that has taken place since the emergence of the extant Genetic Code, the idea that the translation apparatus is for the most part immobile remains true. Here, we will offer a potential explanation to the reason why the code has remained mostly stable for over three billion years, and discuss some of the mechanisms that allow species to overcome the intrinsic functional limitations of the protein synthesis machinery.

Keywords: translation; evolution; speciation; protein folds; tRNA; ribosome

---

1. The Limits of the Genetic Code

The race to identify the structure of the Genetic Code was intense. However, the literature of the time suggests that it was, nevertheless, a collaborative exercise enriched by an intense academic debate that tried to offer explanations to the many questions that kept popping up.

In his seminal paper 'The origin of the Genetic Code', Francis Harry Compton Crick offered a good example of this dynamic as he and Leslie Orgel published their respective views on this topic in back-to-back papers [1,2]. In his paper Crick used the term 'frozen accident' to refer to the apparent inability of the code to accept new variations, and he contrasted this hypothesis with an alternative possibility: the stereochemical theory for the origin of the Genetic Code.

In the forty-nine years that have passed since the publication of this paper, we have advanced very significantly in our understanding of the molecular mechanisms that govern the Genetic Code. However, many fundamental questions regarding the origin and evolution of the code remain open, and chief among them is the reason why the system stopped incorporating new amino acids despite the obvious availability of codon sequences.

Nevertheless, progress has been made. The remarkable advances in the structural analysis of ribosomes, tRNAs, and aminoacyl-tRNA synthetases (ARS) have led to several important conclusions regarding the central roles of RNA in the early Genetic Code, which persist today in the functions of transfer RNAs and the ribosome, among others [3–5]. We now have strong support for the notion that extant proteomes functionally replaced a preceding RNA world where most, if not all, biological catalysis was performed by RNA molecules [6].

It is generally accepted that a primitive Genetic Code, using a limited number of amino acids or groups of related amino acids under a single identity, expanded through the generation of new tRNA identities that increased the number of residues being used, while allowing for a better discrimination between similar amino acid sidechains [7]. The remarkable clustering of chemically-related amino acids that can be seen in the Genetic Code possibly reflects the process of establishment of the different codon and tRNA identities, and is the basis for the coevolution theory of Wong [8,9].

It is reasonable to expect that the expansion of tRNA identities was accompanied by the evolution of tRNA-associated polypeptides (ancestors of extant ARS). Indeed, both the distribution of amino acids in the Genetic Code, as well as the structural features of tRNA, are closely mirrored by the organization of the two ARS classes [10].

It is possible that the initial interaction between primitive tRNAs and the ancestral forms of ARS was in a complex of tRNA molecules bound by a heterodimer from which the two families of ARS later would emerge [11]. It has also been proposed that these two ancestral ARS domains could be coded by complementary strands and, as such, be under tightly coupled selection [12]. This hypothesis can explain the broad internal organization of the two ARS classes, the intriguing distribution of amino acid specificities that can be seen within these same classes, and the many unexplained similarities in identity elements found between tRNAs that are aminoacylated by ARS of different classes [13,14].

2. Why Did the Genetic Code Freeze?

Given the extraordinary chemical diversity of biological amino acids, and the potential for a three-base code based on four bases to theoretically incorporate up to sixty-three amino acids, it is a priori unclear why the universal Genetic Code includes only twenty amino acids. This is even more puzzling if one considers that several additional amino acids, such as selenocysteine and pyrrolysine, are used for protein synthesis. Chemical modifications of side chains are widespread, suggesting that cells could use a larger repertoire of residues within the canonical Genetic Code. Thus, what drove the arrest in the emergence of new tRNA identities and the expansion of the Genetic Code?

Although faithful amino acid recognition is an essential feature of the Genetic Code, it is unlikely that it was a limiting factor in the growth of the system because the recognition is limited to the interactions with ARS active sites, which are extremely adaptable and supported by editing domains that can discriminate between similar side chains [15]. On the other hand, the recognition of tRNAs is a much harder challenge because the three-dimensional structures of all tRNAs are very similar, their chemical composition before modifications is more uniform, and the number of required specific interactions with protein components of the translation apparatus is much larger.

We have proposed that a functional boundary exists with regards to the ability of the translation apparatus to successfully discriminate different tRNA identities. This boundary is determined by the overall capacity of the tRNA structure to incorporate different recognition elements. The incorporation of a new amino acid (hence a new tRNA identity) greatly increases the combinatorial problem faced by the translation machinery to specifically recognize individual tRNAs. This problem applies to modification enzymes, transport systems, ARS, elongation factors, ribosomes, etc. All tRNA identity elements need to coexist in a short RNA sequence whose structure is necessarily similar among all tRNAs in the cell. Additional constraints on tRNA evolution emerging from its non-canonical functions can also be envisaged. Our proposal is that this complex recognition network reaches a limit beyond which the incorporation of new tRNA identities is impossible without generating a recognition conflict with a pre-existing tRNA [16].

We have demonstrated that the saturation of structural and identity signals in a tRNA can prevent this molecule from incorporating other identities in evolution. We investigated the reasons for the intriguing lack of tRNA$^{Gly}_{ACC}$ in eukaryotic genomes and showed that pre-existing features of the tRNA^{Gly} anticodon loop are incompatible with the presence of an adenosine at position 34, explaining why an A34-containing tRNA could not evolve and become enriched in eukaryotes [16].

At the genomic level, we observed that species with low numbers of tRNA genes have significantly more nucleotide differences between their orthologous tRNA pairs than closely related species with a larger number of tRNA genes. This is consistent with the notion that an increase in complexity of tRNA populations leads to a higher conservation of tRNA sequences. Conversely, it would be expected that tRNA sequences would evolve faster in genomes with smaller numbers of tRNA genes. This situation is evident, for example, in mitochondria whose genomes have low numbers of tRNA genes [17]. Mitochondrial genomes display abundant deviations from the canonical Genetic Code, and contain the highest known variability in the structure and identity elements of tRNAs [18–20].

3. Evolutionary Strategies to Expand the Functional Boundaries of the Translation Apparatus

The study of globular protein folds has shown that the extant universe of proteins covers a minimal area of the vast potential number of protein structures. It is likely that extant protein structures evolved from the repetition of simpler domains that were assembled gradually through mechanisms of genetic recombination [21,22].

The synthesis of proteins generated through multiple repetitions of simple sequences may encounter difficulties due to the physicochemical characteristics of such repetitive peptides, or to the inability of tRNAs to maintain fidelity and reading frame when low complexity mRNA sequences are encountered. A number of adaptations have emerged to overcome some of these limitations. For example, the structure of the mammalian mitochondrial ribosome reveals that its polypeptide exit channel has been remodeled to allow not only the synthesis of the hydrophobic proteins that constitute the mitochondrial respiratory chain, but also their insertion into the mitochondrial membrane [23]. Also, EF-P (or eIF5A in eukaryotes), is a universally distributed elongation factor required for the translation of stretches of poly-proline codons [24]. Finally, translation of the extremely codon-biased mRNA transcripts coding for sericin and fibroin (protein components of silk) in the salivary glands of some arthropods requires a unique and highly skewed pool of cellular tRNAs, specifically selected to favor the translation of these mRNAs [25,26]. Thus, certain sequence combinations are a priori inaccessible to the translation apparatus, and functional improvements are needed to translate them.

The existence of species-specific adaptations of the translation apparatus indicate that some species have access to protein structures that are inaccessible to others [27]. We envisage that adaptations of the protein synthesis apparatus that allowed a given species to assemble new types of proteins would provide these organisms with the opportunity to evolve novel and unique functions (in the aforementioned example, the production of silk resulted in a novel mechanism for the growth and development of certain arthropods). We believe that this evolutionary process could start with a simple modification of the translation apparatus, which would allow species to increase their proteome diversity and drive speciation in a punctuated manner (Figure 1A).

Two important parameters that differentiate the translation apparatuses of the three domains of life are their genomic composition of tRNA genes and the set of base modifications in their mature tRNAs [28,29]. We have shown that the divergence of eukaryotic and bacterial genomes in terms of tRNA composition is tightly linked to the evolution of different base modifications in the two kingdoms [30]. In eukaryotic genomes, a remarkable enrichment in genes coding for A34-containing tRNA isoacceptors coincided with the appearance of heterodimeric adenosine deaminases acting on tRNAs (ADAT). This enzyme deaminates A34 to inosine (I34) in tRNAs decoding for eight different amino acids [31]. The activity of this enzyme allows the tRNA pool in eukaryotic cells to match the codon composition of their genomes [30].

In the human transcriptome, codons recognized by ADAT-modified tRNAs are significantly more abundant than those that do not require these modified tRNAs, and this preference is greater in proteins that are highly enriched in the eight amino acids that can be decoded by ADAT-modified tRNAs. We have shown that, in the human proteome, the polypeptides that display the highest preference for these ADAT-modified tRNAs contain extremely biased stretches of the amino acids threonine, alanine, proline, and serine (TAPS) [32]. Figure 1B shows an example of such proteins, Syndecan 3 (SDC3), a member of a proteoglycan family unique to placentals (Figure 1C). This observation suggests that the emergence of TAPS-enriched proteins in eukaryotes was facilitated by the evolutionary emergence of ADAT, which caused an 'upgrade' of the translation machinery through the modification of the composition and the codon-pairing capacity of their tRNA pool. We propose that the capacity of bacterial- and archaeal-type translation machineries to synthesize polypeptides highly enriched in TAPS, is limited by the functional characteristics of their tRNA pools, which may be either inefficient during the elongation phase of these transcripts (causing ribosomal stalling), or prone to decoding errors in these circumstances (causing deleterious levels of mutations in the resulting polypeptides).

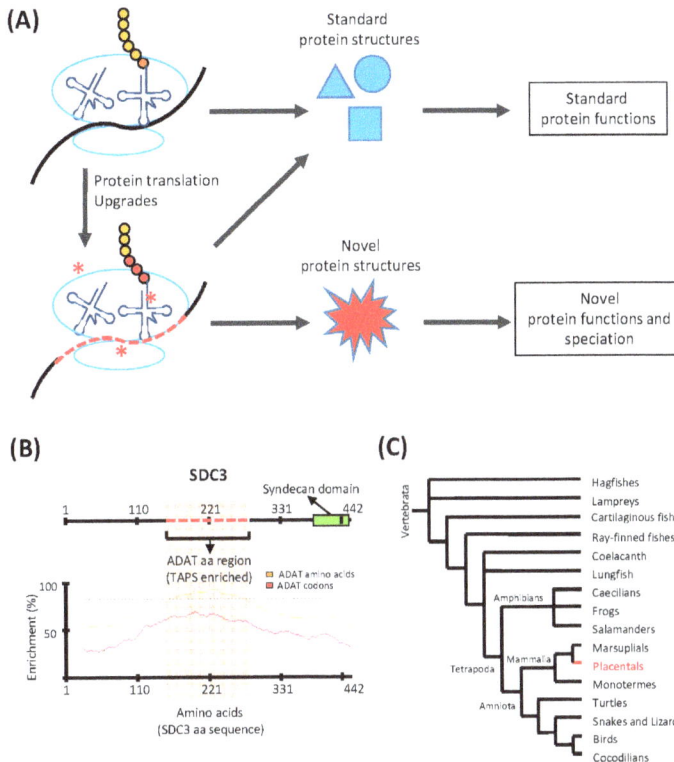

Figure 1. Translation upgrades may lead to novel protein structures and drive speciation. (**A**) The translation machinery is capable of synthesizing a finite number of standard protein structures, and translation 'upgrades' (red asterisks) such as codon usage adaptations, or modulation of the tRNA pool, allow the translation machinery to synthesize proteins with novel structures and functions. This process may drive speciation; (**B**) An example of a gene (SDC3) containing a region with a sequence highly enriched in ADAT-related amino acids (red dashed line; upper panel). The codon composition of the DNA coding for this domain is highly biased towards triplets recognized by tRNAs modified by ADAT. The lower panel shows the enrichment in ADAT-related amino acids (yellow line) and ADAT-dependent codons (red line) across the whole sequence of SDC3. The dashed line marks an enrichment level of ADAT-dependent codons of 80%; (**C**) Consensus phylogeny for Vertebrata. SDC3 belongs to the syndecan proteoglycan family found solely in placentals (highlighted region). The activity of ADAT may have contributed to the emergence of SDC3-type domains in placentals.

The number of known species-specific features of the translation apparatus continues to grow and already includes the composition and regulation of several tRNA modifications, alterations to the ribosomal structure, the differential functionality of translation factors, and the protein and RNA composition of ribosomes, among others. Some of these adaptations may have resulted in translation machinery upgrades that allowed the synthesis of proteins with novel structures and functionality.

The extent to which each of these differential features contributed to the divergence of proteomes is still unknown. However, a comparative analysis of the regions of the protein universe that are occupied by the proteomes of archaeal, bacterial, and eukaryotic organisms could shed light on this question.

In conclusion, the frozen accident that Francis Crick proposed with his characteristic genius may have been the result of the intrinsic limitations imposed by tRNA recognition, but translation has learned to overcome some of these initial limitations through additional functional adaptations that allow species to increase the range and roles of their proteins.

Acknowledgments: This work was supported by the Spanish Ministry of Economy and Competitiveness [FPDI-2013-17742] to AGT; [BES2013-064551] to AR-Y; and [BIO2015-64572] to LRdP.

Conflicts of Interest: The authors declare no conflict of interest.

References

1. Crick, F.H. The origin of the genetic code. *J. Mol. Biol.* **1968**, *38*, 367–379. [CrossRef]
2. Orgel, L.E. Evolution of the genetic apparatus. *J. Mol. Biol.* **1968**, *38*, 381–393. [CrossRef]
3. Di Giulio, M. The origin of the tRNA molecule: implications for the origin of protein synthesis. *J. Theor. Biol.* **2004**, *226*, 89–93. [CrossRef] [PubMed]
4. Noller, H.F. *On the Origin of the Ribosome: Coevolution of Subdomains of tRNA and rRNA, in The RNA World*; Gesteland, R.F., Atkins, J.A., Eds.; Cold Spring Harbor Laboratory Press: New York, NY, USA, 1993; pp. 137–156.
5. Petrov, A.S.; Gulen, B.; Norris, A.M.; Kovacs, N.A.; Bernier, C.R.; Lanier, K.A.; Fox, G.E.; Harvey, S.C.; Wartell, R.M.; Hud, N.V.; et al. History of the ribosome and the origin of translation. *Proc. Natl. Acad. Sci. USA* **2015**, *112*, 15396–15401. [CrossRef] [PubMed]
6. Pressman, A.; Blanco, C.; Chen, I.A. The RNA World as a Model System to Study the Origin of Life. *Curr. Biol.* **2015**, *25*, R953–R963. [CrossRef] [PubMed]
7. Grosjean, H.; Westhof, E. An integrated, structure- and energy-based view of the genetic code. *Nucleic Acids Res.* **2016**, *44*, 8020–8040. [CrossRef] [PubMed]
8. Wong, J.T. A co-evolution theory of the genetic code. *Proc. Natl. Acad. Sci. USA* **1975**, *72*, 1909–1912. [CrossRef] [PubMed]
9. Wong, J.T. Coevolution theory of the genetic code at age thirty. *Bioessays* **2005**, *27*, 416–425. [CrossRef] [PubMed]
10. Ribas de Pouplana, L.; Schimmel, P. Aminoacyl-tRNA synthetases: Potential markers of genetic code development. *Trends Biochem. Sci.* **2001**, *26*, 591–596. [CrossRef]
11. Ribas de Pouplana, L.; Schimmel, P. Two classes of tRNA synthetases suggested by sterically compatible dockings on tRNA acceptor stem. *Cell* **2001**, *104*, 191–193. [CrossRef]
12. Pham, Y.; Li, L.; Kim, A.; Erdogan, O.; Weinreb, V.; Butterfoss, G.L.; Kuhlman, B.; Carter, C.W., Jr. A minimal TrpRS catalytic domain supports sense/antisense ancestry of class I and II aminoacyl-tRNA synthetases. *Mol. Cell* **2007**, *25*, 851–862. [CrossRef] [PubMed]
13. Giege, R.; Sissler, M.; Florentz, C. Universal rules and idiosyncratic features in tRNA identity. *Nucleic Acids Res.* **1998**, *26*, 5017–5035. [CrossRef] [PubMed]
14. Beuning, P.J.; Musier-Forsyth, K. Transfer RNA recognition by aminoacyl-tRNA synthetases. *Biopolymers* **1999**, *52*, 1–28. [CrossRef]
15. Martinis, S.A.; Boniecki, M.T. The balance between pre- and post-transfer editing in tRNA synthetases. *FEBS Lett.* **2010**, *584*, 455–459. [CrossRef] [PubMed]
16. Saint-Leger, A.; Bello, C.; Dans, P.D.; Torres, A.G.; Novoa, E.M.; Camacho, N.; Orozco, M.; Kondrashov, F.A.; de Pouplana, L.R. Saturation of recognition elements blocks evolution of new tRNA identities. *Sci. Adv.* **2016**, *2*, e1501860. [CrossRef] [PubMed]
17. Gray, M.W.; Burger, G.; Lang, B.F. The origin and early evolution of mitochondria. *Genome Biol.* **2001**, *2*. REVIEWS1018. [CrossRef] [PubMed]
18. Chihade, J.W.; Brown, J.R.; Schimmel, P.R.; De Pouplana, L.R. Origin of mitochondria in relation to evolutionary history of eukaryotic alanyl-tRNA synthetase. *Proc. Natl. Acad. Sci. USA* **2000**, *97*, 12153–12157. [CrossRef] [PubMed]

19. Sengupta, S.; Higgs, P.G. Pathways of Genetic Code Evolution in Ancient and Modern Organisms. *J. Mol. Evol.* **2015**, *80*, 229–243. [CrossRef] [PubMed]
20. Sengupta, S.; Yang, X.; Higgs, P.G. The mechanisms of codon reassignments in mitochondrial genetic codes. *J. Mol. Evol.* **2007**, *64*, 662–688. [CrossRef] [PubMed]
21. Alva, V.; Soding, J.; Lupas, A.N. A vocabulary of ancient peptides at the origin of folded proteins. *Elife* **2015**, *4*, e09410. [CrossRef] [PubMed]
22. Alva, V.; Remmert, M.; Biegert, A.; Söding, J. A galaxy of folds. *Protein Sci.* **2010**, *19*, 124–130. [CrossRef] [PubMed]
23. Greber, B.J.; Boehringer, D.; Leitner, A.; Bieri, P.; Voigts-Hoffmann, F.; Erzberger, J.P.; Leibundgut, M.; Aebersold, R.; Ban, N. Architecture of the large subunit of the mammalian mitochondrial ribosome. *Nature* **2014**, *505*, 515–519. [CrossRef] [PubMed]
24. Lassak, J.; Wilson, D.N.; Jung, K. Stall no more at polyproline stretches with the translation elongation factors EF-P and IF-5A. *Mol. Microbiol.* **2016**, *99*, 219–235. [CrossRef] [PubMed]
25. Chevallier, A.; Garel, J.P. Differential synthesis rates of tRNA species in the silk gland of Bombyx mori are required to promote tRNA adaptation to silk messages. *Eur. J. Biochem.* **1982**, *124*, 477–482. [CrossRef] [PubMed]
26. Li, J.Y.; Ye, L.P.; Che, J.Q.; Song, J.; You, Z.Y.; Yun, K.C.; Wang, S.H.; Zhong, B.X. Comparative proteomic analysis of the silkworm middle silk gland reveals the importance of ribosome biogenesis in silk protein production. *J. Proteomics* **2015**, *126*, 109–120. [CrossRef] [PubMed]
27. Caetano-Anolles, G.; Wang, M.; Caetano-Anollés, D.; Mittenthal, J.E. The origin, evolution and structure of the protein world. *Biochem. J.* **2009**, *417*, 621–637. [CrossRef] [PubMed]
28. Chan, P.P.; Lowe, T.M. GtRNAdb: A database of transfer RNA genes detected in genomic sequence. *Nucleic Acids Res.* **2009**, *37*, D93–D97. [CrossRef] [PubMed]
29. Grosjean, H.; de Crecy-Lagard, V.; Marck, C. Deciphering synonymous codons in the three domains of life: co-evolution with specific tRNA modification enzymes. *FEBS Lett.* **2010**, *584*, 252–264. [CrossRef] [PubMed]
30. Novoa, E.M.; Pavon-Eternod, M.; Pan, T.; Ribas de Pouplana, L. A role for tRNA modifications in genome structure and codon usage. *Cell* **2012**, *149*, 202–213. [CrossRef] [PubMed]
31. Torres, A.G.; Piñeyro, D.; Rodríguez-Escribà, M.; Camacho, N.; Reina, O.; Saint-Léger, A.; Filonava, L.; Batlle, E.; de Pouplana, L.R. Inosine modifications in human tRNAs are incorporated at the precursor tRNA level. *Nucleic Acids Res.* **2015**, *43*, 5145–5157. [CrossRef] [PubMed]
32. Rafels-Ybern, A.; Attolini, C.S.; Ribas de Pouplana, L. Distribution of ADAT-Dependent Codons in the Human Transcriptome. *Int. J. Mol. Sci.* **2015**, *16*, 17303–17314. [CrossRef] [PubMed]

life

MDPI

Review

Self-Referential Encoding on Modules of Anticodon Pairs—Roots of the Biological Flow System

Romeu Cardoso Guimarães

Laboratório de Biodiversidade e Evolução Molecular, Departamento de Biologia Geral, Instituto de Ciências Biológicas, Universidade Federal de Minas Gerais, Belo Horizonte, Minas Gerais 31270-901, Brasil; romeucardosoguimaraes@gmail.com or romeucg@icb.ufmg.br; Tel.: +55-31-98897-6439; Fax: +55-31-3274-4988

Academic Editor: Koji Tamura
Received: 13 January 2017; Accepted: 26 March 2017; Published: 6 April 2017

Abstract: The proposal that the genetic code was formed on the basis of (proto)tRNA Dimer-Directed Protein Synthesis is reviewed and updated. The tRNAs paired through the anticodon loops are an indication on the process. Dimers are considered mimics of the ribosomes—structures that hold tRNAs together and facilitate the transferase reaction, and of the translation process—anticodons are at the same time codons for each other. The primitive protein synthesis system gets stabilized when the product peptides are stable and apt to bind the producers therewith establishing a self-stimulating production cycle. The chronology of amino acid encoding starts with Glycine and Serine, indicating the metabolic support of the Glycine-Serine C1-assimilation pathway, which is also consistent with evidence on origins of bioenergetics mechanisms. Since it is not possible to reach for substrates simpler than C1 and compounds in the identified pathway are apt for generating the other central metabolic routes, it is considered that protein synthesis is the beginning and center of a succession of sink-effective mechanisms that drive the formation and evolution of the metabolic flow system. Plasticity and diversification of proteins construct the cellular system following the orientation given by the flow and implementing it. Nucleic acid monomers participate in bioenergetics and the polymers are conservative memory systems for the synthesis of proteins. Protoplasmic fission is the final sink-effective mechanism, part of cell reproduction, guaranteeing that proteins don't accumulate to saturation, which would trigger inhibition.

Keywords: genetic code; (proto)tRNA Dimer-Directed Protein Synthesis; self-reference; modularity; error-compensation; metabolism chronology; protein stability; punctuation

---

Living beings are metabolic flow systems that self-construct on the basis of memories and adapt/evolve on the basis of constitutive plasticity.

Life is the ontogenetic and evolutionary process instantiated by living beings.

1. Context

Cellular structures and functions are composed by a network that may be divided into a mostly internal segment of macromolecules, the proteins and the nucleic acids, and another of micromolecules, that communicate and exchange intimately with the environment. The genetic encoding/decoding system is a key mechanism in both of these processes. It participates in the informational, the polymer sequence order-based self-referential cycle that is the cellular nucleoprotein core, the specifically endogenous and molecular identifier of living beings (Figure 1). The decoding system mediates the translation of genetic sequences—string memories—into proteins, which are the main functional working components of the cell system. The protein synthesis machinery is centered on ribosomes, while the mRNAs, tRNAs and aminoacyl-tRNA synthetases (aRS) carry the translation triplet codes, which establish the correspondences between mRNA and tRNA base triplets and the amino acids.

It may be said, with some simplification, that the main function of the other—inter-genic—sequences and of the multitude of RNAs that do not code for proteins would be regulatory, in the sense of retrieving information, processing and preparing protein-coding sequences for coordinated expression in the adequate time and space contexts [1]. There are ribozyme activities in different parts of the system, its main representative being possibly the ribosomal Peptidyl Transferase Center [2,3]. It could be said that the main catalytic function in the RNP complexes would reside in enzymes, while proteins and RNAs together help and guide each other to identify targets and substrates with precision.

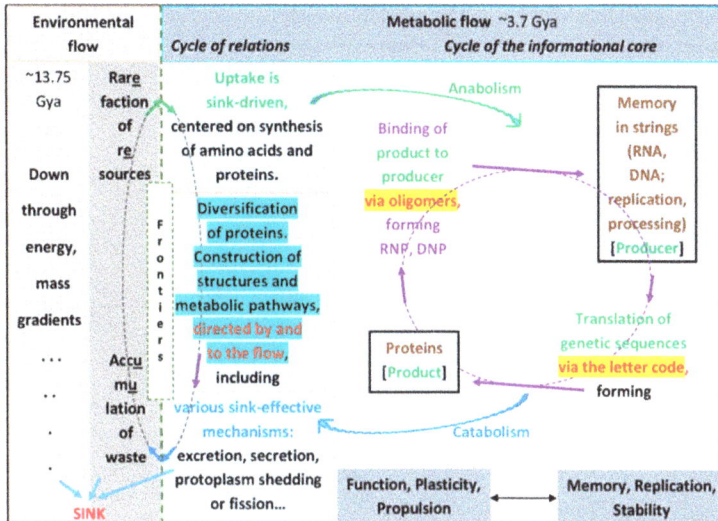

Figure 1. Fundamentals of living beings. Cellular functions are described as composed by two interconnected and interdependent self-referential cycles. The informational cycle is composed by the system where proteins and nucleic acids generate each other and are connected through the genetic code and the nucleic acid-binding proteins. The plasticity function, which is mainly of proteins and networks, is responsible for propulsion of the system through adaptations and evolution. The memory in strings and the replication function of nucleic acids is responsible for stability and identity that are conservation properties. The metabolism cycle is continuous with the environment that is transformed via uptake of resources and extrusion of waste. The cycles configure a flow system that is centered on the initial sink of protein synthesis. This is kept continually active through the addition of various other sink mechanisms, culminating in the extrusion of cytoplasmic chunks which gave origin to reproduction. Cycles are drawn with dashed lines indicating their limited regenerative capacity and duration, and dependence on contexts.

The other portion of this cycle relies upon activities of the nucleic acid-binding oligomeric segments of proteins. These are involved with stabilization of the genetic molecules and the regulation of their activities through formation of nucleoprotein complexes. A useful metaphor for the self-referential aspect of the system would be of a producer-product ensemble that can only work efficiently and along the time, when the products—proteins—are able to recognize and feedback positively upon the producers—RNA, DNA. In case such informational closure would not be effective, the producers would eventually stop acting in those directions which only resulted in wasted investment, and may attempt functions in other directions that could accomplish the self-stimulating process. It is assumed that such processes occurred in the proto-biotic realm of events, involving oligomers, and led to the encodings. They are akin to autocatalysis [4], but differ from those by the composition of distinct agents that integrate among themselves to form a system.

In the genetic encoding process, the initial recognition of the two components—tRNA and amino acid, that is mediated by the enzyme aRS—could simply involve the binding of the protein, product of the proto-, pre-(t)RNA dimer, to this, which is the producer, and the result is the formation of an ensemble that is stabilized (Figure 2B). Stabilization *per se* has a (self)stimulatory effect upon the system. The stabilized couple would evolve into a aRS-tRNA cognate set given that (1) this initial binding would at least not harm the protein synthesis activity and in case (2) the specificity would show benefits to the chemical system in which they are immersed, starting with e.g., facilitation of the flows through reduction of turbulence and of mass and energy gradients, among other effects. The two components of this cycle produce the compositional (various recognition sites in the enzyme and in the tRNA) to discrete (one triplet—the anticodon—and one amino acid) correspondences and also the self-stimulatory properties that are at the core of the living systems [5,6].

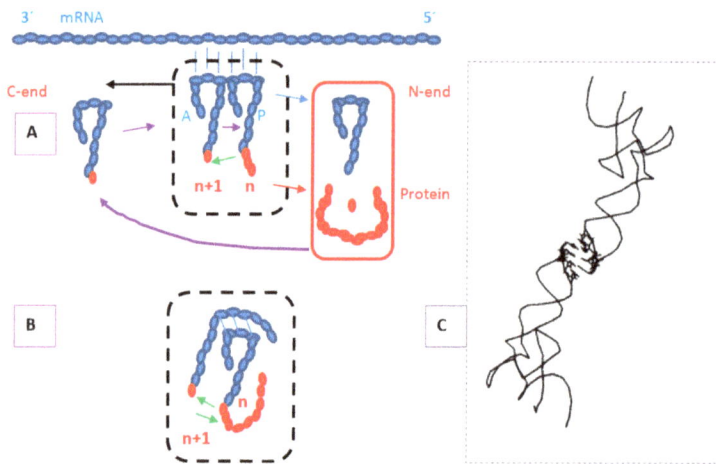

Figure 2. Protein synthesis directed by tRNA couples at ribosomal translation and directed by dimers of (proto)tRNAs. (**A**) Ribosomal translation of mRNA is made by couples of tRNAs whose tails, carrying an amino acid or a peptide, are placed in contact that facilitates formation of a peptide bond. Successive couples scan the mRNA from the 5′ to the 3′ termini that correspond to the protein N-end and C-ends, respectively. There are two cycles of the substrates that are fed by the aRSs. A tRNA receives an amino acid and the aminoacyl-tRNA enters the ribosomal acceptor A-site. This receives the transfer, from the P-site, of the initiator Methionine or of the nascent peptide and a peptide bond is made by the transferase activity (green arrow) at the ribosomal Peptidyl Transferase Center. The ribosome moves forward while the A-site tRNA with the n + 1 peptidyl tail is translocated to the P-site, which releases its empty tRNA that may be later re-charged. The amino acids (red elipses) derive from uptake or are recycled from protein degradation and are polymerized into proteins. These may participate in the translation system, therewith establishing a self-stimulating process (the aRSs are examples of such class of proteins) or may be released to build other structures and functions; (**B**) The self-referential model proposes that the encoding process derives from Dimer-Directed Protein Synthesis, which mimics the tRNA couples of the translation process. The anticodons are at the same time codons for each other and the dimer structure would also be a proto-ribosome, holding two tRNAs together and facilitating the transferase reaction. The tRNAs composing the dimers and the direction of the reaction could, in principle, vary along iterations of the process. Encoding would result from the binding and stabilization of the dimer structures by selected kinds of peptides produced, adequate also for keeping their functions; (**C**) A dimer of uncharged tRNAs observed experimentally [7], which survived drastic purification processes. The acceptor tails are far apart. The anticodons are held together in spite of the central base mismatch, due to the self-complementariness of the Asp anticodons $\frac{GUC}{CUG}$.

Another self-referential cycle couples with these endogenous, uniquely cellular identity structures and establishes (hetero-referential) relations with the environment. The two cycles together comprise the definition of the cell (the living being) as a metabolic flow system. The relational cycle is composed by elements of the nucleoprotein system—enzymes, ribozymes, and its products or subcomponents, such as cofactors and carriers—and is devoted to its maintenance. The scheme of the cell is fully self-referential. The 'auto or self' aspect relies upon memories, guaranteeing the maintenance of identity, conservation and stability. They are of two kinds: the replicative, stored in genetic strings [8] plus the systemic, dynamically conserved in positive feedback cycles that are constructed with the products [9]. Functions are of proteins and networks, highly plastic and diversified, whose restlessness and constructiveness propel and drive adaptations and evolution [10–12], in concert with the metabolic flow directionality. Both also contribute to the metabolic sink effectiveness.

The setting for the origins of cells inside geochemical systems (discussion in [13,14]) considers plain continuity between the universal flow where the sink is almost virtual, that is, the entropic direction given by thermodynamic degradation, which becomes real in its application to matter, where the degradative processes are energetic and material, that is, molecular. In living beings, the sink is first identified in the consumption of amino acids at the protein synthesis system, which also involves the genetic memories. A schema of the steps would be as follows. There is (a) a push to keep the metabolic flow going, driven by the gradients of concentrations of amino acids and energy, among other organics, which is relieved and at the same time reinforced by creating the (b) protein synthesis sink (pulling), and this is kept continuously active by the (c) constitutive plasticity of the components, especially the proteins, which add diversification of kinds and (d) efficiency through the development of specificity, which may be identified with the genetic code.

The sketch in Figure 1 considers the molecular bases of the cell systems, which are the constitutive grounds or bottom layers of the living—metabolic flow systems that self-construct through memories and adapt/evolve through plasticity. It is expected that this definition would be useful to accommodate other aspects of the life process—the entire set of activities (from ontogenesis to evolution) instantiated by living beings—as evolutionary additions to the basic biomolecular. The constitutive functions in the terrestrial life process are: propulsion through the plasticity properties, which is most extensive in proteins, second in RNA and less in DNA, and conservation through the replication properties of the string memory structures, especially DNA. The other propulsion subsystem is the directional metabolic flow, which is also morphogenetic and coupled concertedly with the informational subsystem. This self-referential or cyclical panorama [15] is at the basis of the model for the genetic code structure that is based on encoding directed by dimers of tRNAs (Figure 2). The bi-functional ensemble, combining propulsion/innovation/change (that are mostly in proteins) with conservation/identity/stability (mostly in replication of nucleic acids) is a basic attribute of life in general; the application of this 'balance in mutuality' to human affairs might consider it a source of tension and possible conflicts.

The intermingled structure of the system also means that all components are involved with the formation of a network where it is difficult to establish linear chronologies for sectors of the system. The main paradigm becomes of fully integrated coevolution [16]. When an amino acid is in the process of being encoded its availability should be guaranteed, which means that its production pathway should also be at least in the process of being fixed, as well as those routes of its utilization by the cell, and all aspects should fit each other for general shared benefits.

Some selected aspects of the model are recalled to compose this review, without much focus on technical details but attempting to encourage the reader to examine the original papers and to submit the model to further tests. It has survived already some empirical checks. We will concentrate in a few key references from which others can be reached [17].

2. Modular Structure in the Genetic Code and Layout of the Text

Study of the structure and formation of the genetic code has to concentrate on its elements—tRNAs with their anticodon triplets, amino acids and aRSs—that are independently moving in the cellular fluids and have to compose and aggregate for generation of function (e.g., [18,19]). We will not treat much of the work based on codons in mRNAs and genes. These are strings of triplets adjusted for translation speed, where much of the characters that were important for the encoding process do not have specific functions anymore and are not manifest in the string context. The regularities in the matrix of anticodes that indicate a structure based on dimers or pairs are here sometimes introduced in a simplified mode, through utilization of the '*principal dinucleotides* (pDiN) + *wobble*' structure of the triplets, which is established by the aRSs and identify the 16 boxes; the wobble position may not be specified, becoming N. Some data are available only for dinucleotides, e.g., the hydropathies, due to low solubility of the triplets that limit the chromatographic experiments [20]. The simplification of triplets to doublets of bases is informative enough for some comparisons. The 16 boxes are organized through the formation of dimers in four modules, each with two pairs of boxes (Table 1).

Table 1. Structure of the anticode organized through modules of dimers and the chronology of encoding. The two sectors and four modules are identified by the principal dinucleotides. The principal dinucleotides also dictate the pairing possibilities. (**A**) Bases in the 5′ position (wobble) are chosen among the variety (indicated by N), according to the generic complementariness R:Y, which is dictated by the base in the 3′ position. The central base pair is strictly of the Watson:Crick type, A:U, G:C. (**B**) Amino acids of the Glutamate family are in bold, in the homogeneous sector plus the Arg that expands into the mixed sector; Phenylalanine is the first amino acid encoded that derives from sugar precursors, also bold. Lysine biosynthesis has pathways derived from Asp or Glu. Punctuation is indicated to form after the elongation amino acid encodings; its three codes are in blue to indicate their mechanistic relatedness. Further details on the chronology will be shown later in the text.

(A)				
Wobble	Principal dinucleotide			→
5′ N	Central			*Anticodon*
3′	Central		5′ N	*Anticodon*
Principal dinucleotide			Wobble	←
Generic R:Y	Standard G:C, A:U		Generic R:Y	*Base pair*

(B)				
Sector	**Homogeneous Principal Dinucleotides**			
Modules	**Central G:C**		**Central A:U**	
Pairs Elongation	1a	1b	2a	2b
	NGG	NGA	NAG	NAA
	CCN	UCN	CUN	UUN
1st occupier (5′N)	Gly	Ser	Leu	Leu
	Gly	Ser	Asp	Asn
Present (5′N or 5′G/5′Y)	Pro	Ser	Leu	Phe/Leu
	Gly	Ser/Arg	Asp/Glu	Asn/Lys
Sector	**Mixed Principal Dinucleotides**			
Pairs Elongation	3a	3b	4a	4b
	NCG	NCA	NUG	
	CGN	UGN	CAN	NUAUAN
1st occupier (5′N)	Arg	Cys	His	Tyr
	Ala	Thr	Val	Ile
Present (5′N or 5′G/5′Y)	Arg	Cys/Trp	His/Gln	Tyr
	Ala	Thr	Val	Ile/Met
Punctuation				/Met, iMet
		/Trp, X		Tyr/X

The results that compose the Self-Referential Model (SRM) for the structure and the formation of the genetic code are summarized as follows. Dimers involving the triplets form network systems that are different for the self-complementary (SC) and the nonself-complementary (NSC) kinds (Table 2 shows the kinds of triplets; the networks are shown further ahead, Section 3.2). These are: SC with an

R and a Y in the lateral positions of the triplet (RNY, YNR), NSC with Rs or Ys in both lateral positions (RNR, YNY).

Table 2. Splitting the boxes into 5′R and 5′Y halves and the base size 'topographic landscape' of the triplets. Anticodon triplets are composed of the 5′ base—the wobble position—plus the principal dinucleotide (pDiN). The 16 pDiN define the 16 boxes. The 5′ position fills boxes with options that may be undefined (N) or made explicit as G, C or Y, since the standard set contemplates the absence of 5′A [21]. Purines (R) are two-ring bulky bases and pyrimidine (Y) one-ring small bases. When they are together in a duplet or triplet, rugged topographies are presented for interactions; when the duplet or triplet is composed homogenously by one kind—all R or all Y, the landscape is smooth or planar. The self-complementary (SC) triplets may allow for self-dimerization and possibly transient internal looping since the bases in the 5′ and 3′ positions are complementary to each other [7,22]. The nonself-complementary (NSC) triplets are maintained in the extended or open configuration, and only accept hetero-dimerization—the minihelices accommodate obligatorily different kinds of triplets, obeying complementariness. The mixed pDiN is always part of rugged topography triplets; the homogeneous pDiN may lose the planar landscape configuration when it is part of a SC triplet; only the NSC triplets of the homogeneous sector maintain the full planar character. Only the NSC triplets, irrespective of sectors, maintain a symmetry center.

Sector of Principal Dinucleotide (pDiN)		Homogeneous pDiN		Mixed pDiN	
Base size configuration of the anticodon triplets		NRR	Nyy	NRy	NyR
Self-complementary triplet	Triplet	yRR	Gyy	GRy	yyR
SC	Purine (R)	••	•	••	•
lateral bases one R the other y	Pyrimidine (y)	•	••	•	••
Nonself-complementary triplet	Triplet	GRR	yyy	yRy	GyR
NSC	Purine (R)	•••		•	• •
lateral bases both R or both y	Pyrimidine (y)		•••	• •	•

The networks formed by dimers of nonself-complementary anticodon triplets, after the known *en bloc* 5′A elimination [21], become asymmetric and provide an easy way to start the encoding in the modules (Section 3.2). Initiation and termination anticodes are proposed to be connected through a mechanism of competition for pairing with the initiation codons, which resulted in the deletion of the triplets corresponding to the termination codons (Section 5).

The amino acid and protein aspects of the coding system are presented with respect to the amino acid composition of functional sequence segments or motifs (Section 4), and with respect to protein metabolic stability—half-life (Section 4.3). A synthesis of the information on both components of the system—triplets and amino acids—generated a chronology of encoding whose understanding required close examination of pathways of amino acid biosynthesis. The resulting model says that the code was formed entirely on biological-metabolic grounds, starting with one-carbon unit substrates to form amino acids, followed by gluconeogenesis (Section 6). Relevant aspects of prebiotic chemistry are discussed, facing the complexity of abiotic organics and the apparent simplicity of the C1 substrates. Models for the genetic code that are based on the sets of prebiotic abundant amino acids are considered proto-codes, possibly active in contexts where different oligomers could be adjusting their structures among themselves inside aggregates. The modules in the structure of the whole set of triplet dimers correspond to subsets of amino acids devoted to specific functions with the consequent regionalization or clustering of the functions, which would be at the basis of the property of error-reduction in the system (Section 7).

A schematic on the encoding process on the basis of proto-tRNA dimers follows. (a) In the proto-biotic realm of events the interacting compounds are oligomers. There are various options for their structures, still to be defined. (b) These can function as pre/proto tRNAs: carriers of monomers, among which there would be the amino acids. (c) They would also be apt for dimerization, through complementary sites, and the stability of the dimer structure would propitiate the transferase reaction. (d) In accordance with the types of monomers carried, they would produce of a variety of oligomers/polymers, including peptides/proteins, which is Dimer-Directed Protein Synthesis (DDPS). (e) Among the products some would be able to bind to the producers, or better, would stay bound to the producers, forming pre/protonucleoproteins. (f) When these, at the least, do not harm but maintain the producer activity and stabilize the complex, which lasts longer, the result is equivalent to stimulating the activity. (g) This is the root of a positive feedback or self-stimulating system, which is a kind of auto-catalytic network. (h) Some of these products would be structural, analogous to ribosomal proteins, others enzymatic, such as the aRSs or members of biosynthesis pathways, and so forth. (i) These compounds would manifest cohesiveness through binding protein-protein and protein-nucleic acids etc., in the route to building proto-cell globules. (j) The encoding process would require repetition of the production cycles where mutual adjustment of producers (dimers) and products (peptides) would result in specificity of binding and of catalysis, which is the beginning of fixation of the tRNA-aRS-amino acid systems. (k) It is envisaged that at some point in the process the dimers would be separated by the intromission of exogenous RNA. This would become later the messenger RNA and the process of development of individualized charging systems is enforced.

3. Functionality in the Modular Organization

3.1. Hydropathy Correlation

The graph on the hydropathy correlation is an empirical demonstration that the set of anticodons is organized in modules [23,24], therewith providing a structure for the code system (Figure 3). The correlation is known for a long time (see [20]) but it was not conducive to modeling due to having been based on data from amino acid molecules in solution. Our reexamination utilized data on amino acid residues in proteins together with the principal dinucleotides (pDiN) of anticodons [20,23], which was fruitful in generating the self-referential model.

The rationale for the separation of modules is based on complementariness of the codes. They are seen here in a simplified way, through the pDiN (see Table 1), since data on triplets are not available due to their limited solubility. We treat further down the complexity added with consideration of the wobble base, together with other questions posed by this initial presentation of the modules, especially on the chronology of amino acid encoding and of the associated functions.

The modular structure shown in Figure 3 does not separate the two parts of the mixed sector of attributions, where the pDiN have one R and one Y base and pDiN hydropathies are intermediate. The homogeneous sector, pDiN with both bases R or both Y and hydropathies extreme, is separated into the central G:C and central A:U modules, with distinct behaviors. The central G:C should have been encoded earlier than the central A:U due to the greater thermal stability, which coincides with its simpler constitution, such as the lower number of attributions that corresponds to higher average degeneracy (Section 7.1). This rationale is extended with the correlation data: encodings start producing the non-correlated attributions (the central G:C module 1 of the homogeneous sector), then develop a moderate inclination regression line (the central A:U, module 2), and end with the steeper regression line of the mixed sector (modules 3 and 4).

Installation of the correlation in module 2 means that proteins composed by the set of ten amino acids of the homogeneous sector would be complex enough to be able to construct enzyme pockets and to develop specificity in interactions. One of such specific products is the correlation. The early aRS pockets would accommodate amino acids and proto-tRNA segments (proto-anticodons) with each other in a manner that would be more stable when they were coherent with respect to the hydropathy behavior of the partners. When both were hydrophobic they would cooperate in expelling water out of their interactive surfaces; when both were hydrophilic they would cooperate in organizing eventual water molecules in the interactive sites. When such hydropathic coherence between partners would not be established, cooperativity would fail leading to instability in the associations inside the pockets, which would hamper the fixation of a correspondence.

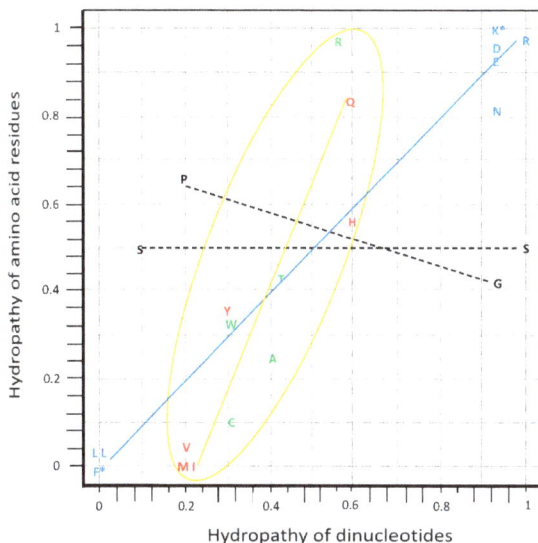

Figure 3. Correlation between hydropathies of amino acid residues in proteins and of dinucleotides. The dinucleotides are considered the principal dinucleotides (pDiN) of anticodons. The regularity in the distribution of anticodon types is clear [23]. The sector of mixed pDiN (circled in yellow and with a steep regression line), composed of one R and one Y, has intermediate hydropathies but the modules, one with the central bases G:C (amino acids in green), the other A:U (amino acids in red), are not resolved. The homogeneous pDiN sector (strands RR or YY) has extreme pDiN hydropathies and is separated into the two modules: hydropathies in the central G:C are not correlated (amino acids in black) but those in the central A:U are correlated (amino acids in blue). Arginine is the only amino acid belonging to both sectors. Note the extreme hydropathies of Lys and Phe; the * labels refer to the atypical character of their synthetases, commented further down.

The constitution of the proto-tRNA involved in the associations in the non-correlated module 1 remains an open question. The lack of correlation could be due to poorness in the amino acid constitution of the module 1 set or to a non-RNA or a pre-RNA constitution of the dimers. The lack of correlation refers to the dinucleotides but there might have been a correlation between the amino acids and whatever were the compounds in the dimers, which belong to the pre- or proto-biotic chemical systems (some non-RNA options are in [25,26]).

Exclusion of Arg from module 1 brought clearness to the interpretation based on dimers of anticodons. The hydrophilic Arg attributions traverse the sectors and are correlated coherently with the triplet kinds but do not belong to the hydroapathetic group. Difficulties in rationalization were reduced when only the Ser, Gly and Pro attributions could be interpreted as a coherent group. The 'enigma' of the two Ser codes, which could not be explained in any of the previous studies (see [27,28]), now receives an understanding based on dimer complementariness. The indication that the Gly and Pro codes would also have belonged to an unforeseen pair (Table 1) found support only when the interpretation provided by the metabolic pathways of amino acid biosynthesis entered the argument. Both Pro and Arg biosyntheses derive from Glu so that their encoding followed that of Glu, occurring in or after module 2. Therefore, the Pro-NGG correspondence would have been a full concession from a previous Gly-NGG, as well as Arg-YCU would have been a concession from a previous Ser-NCU that retained finally only the Ser-GCU anticode. It is indicated that the first Arg correspondence was made by the ArgRS with the YCU anticodes, in the homogeneous sector, thereafter expanding the specificity of the recognitions also to accommodate the NCG, of the mixed sector, that is, developing the 3′ U/G ambiguity. The third hexacodonic attribution shows a similar but simpler 3′R ambiguity, LeuRS for NAG and YAA.

3.2. Networks of Dimers in the Modules and the Encoding Process

We chose to examine the configuration of the dimer sets by accepting only the standard kind of base pair in the central position [29] and the unrestricted R:Y kinds in the lateral positions. Exclusion of base A from the 5′ position (Figure 4) is considered typical of the standard matrix of anticodes [21].

The networks distinguished, among the sets of triplets that compose the boxes (these defined by the pDiN), the self-complementary from the nonself-complementary. These kinds of triplets present different thermodynamic parameters [22] and a main differentiating property is highlighted, considered relevant for choosing among them for the protein synthesis activity of the dimers: pairs composed by molecules bearing the SC kind would be neither always nor entirely free for the hetero-referential dimerization due to the possibility of getting involved with self-dimerization [7] (Figure 2).

A reason for the chronological precedence of the homogenous over the mixed sector of triplets would be the simplicity of the former over the complexity of the latter (Table 2). An intuitive image would take the 'topographic landscape' of the interaction surfaces that, in the minihelices, are in the minor grooves. The mixed sector would be presenting for interactions a non-repetitive rugged surface—bases in a string have different sizes, with bumps and depressions—in comparison with the NSC triplets of the homogeneous sector where the topology is planar, monotonically repetitive, therefore indicating a corresponding repetitiveness in the interacting partner. The rationale assumes that repetitiveness in the structures would mean simplicity both in the mechanisms for producing them and in their interactions, which would not require precise location of sites inside the repetitive segments, therewith allowing for slippages [30]. The non-repetitive structures would require some degree of specificity in the ordering or organization, and in both partners. Together with the above, there is also the symmetric structure of the NSC triplets, irrespective of the sectors, which should enlighten on the character of the primordial aRS pockets; the mature (present day) triplets are not structurally symmetric but 'wobble + pDiN' and the aRSs follow this organization.

Another positive attribute of the NSC dimers for the protein synthesis activity and for the encoding is consequent to the exclusion of 5′ A, which reduced the size and produced an asymmetric topology in the networks, and is repeated identically in the four modules. There are in each module two triplets with 5′ G connected to four with 5′ Y, totaling eight dimers per module. Without the exclusion of a base from the set in the 5′ position, the network would contain 16 dimers, which is the size of the SC 5′Y networks (dashed lines in blue, Figure 4A, top half). The other kind of SC dimers, with 5′G, connect between themselves through a different topology, with a total of four dimers (dashed lines in blue, Figure 4A, bottom half). Both the SC 5′G and the SC 5′Y are repeated twice and are internally fully symmetric, another component of their disadvantage in comparison with the NSC. The process of deciding on which dimer to choose for the first encoding would take longer in symmetric networks, and the degree of difficulty increases factorially with the size of the network. The encoding of the second pairs, in each of the modules, is also facilitated by the asymmetry in the network of pairs, which is a manifestation of the flow from the 1st to the 2nd members of the modules.

The suggested process of encoding (Figure 5) does not involve a typical decision process due to being almost 'automatic': the formation of the highest ΔG pair in the module (in the central rows of the matrix, GNG:CNC [5]) is followed by lowered concentrations of five other pairs that members of the first pair would be involved with. Two dimers are left at high concentrations, for the second encoding round, the low ΔG pairs (GNA:YNU, in the top and bottom rows of the matrix).

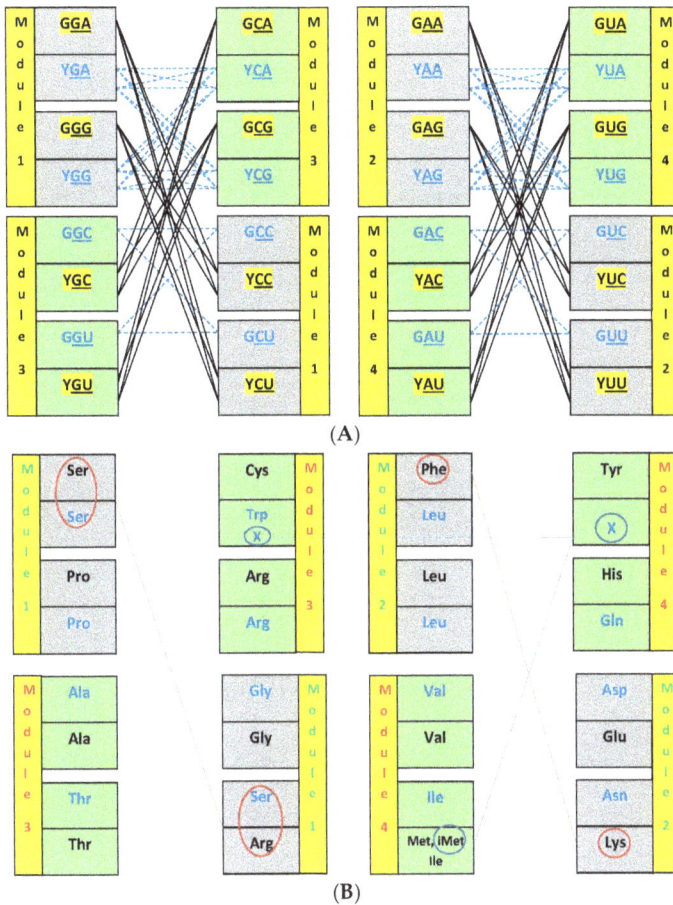

Figure 4. (**A**) Dimer networks. The nonself-complementary triplets (highlighted in yellow) pair among themselves forming four identical networks, combining the central G:C triplets or the central A:U triplets in each of the homogenous or mixed principal dinucleotide (pDiN) sectors. Connections in this graph follow the diagonal black lines. These networks are asymmetrical due to the exclusion of the base A from the 5′ position, which is typical of the standard anticode [21], combining two 5′G with four 5′ Y triplets (8 pairs). The self-complementary triplets (blue font) pair among themselves (connections follow the rows, blue dashed lines) forming two kinds of networks: the 5′ Y triplets are untouched by the 5′ A exclusion and the network of pairs among them is symmetric of size 16 (4 × 4); the network of the 5′G triplets is also symmetric but of size 4 (2 × 2). Chronological numbering of the modules (see text) follows both triplet kinds (homogeneous before mixed and, in each, central G:C before A:U) and amino acid characters (metabolic pathways and protein properties; see text); (**B**) Meanings of the triplets in the dimer network format. The three instances where the meanings would indicate the dimer-directed diagonal organization (in circles) are either partial (the Ser-NGA : Ser-NCU pair is corrupted by the Arg-YCU 'invasion'; the initiation iMet-CAU would pair with the termination X-YUA, which would require some peculiar wobble-pairing between pyrimidines, but not with the X-UCA) or not obvious to the non-informed reader in the case of the Phe-GAA: Lys-YUU pair, where the link in the pair is through the respective aRSs that are both atypical (see text).

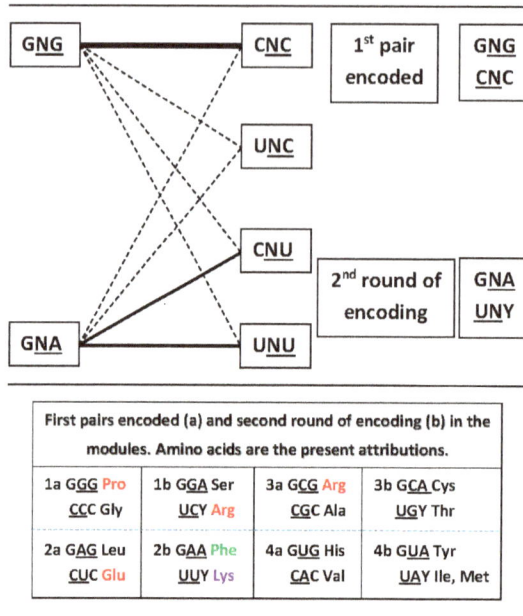

| | | 1st pair encoded | GNG CNC |

| | | 2nd round of encoding | GNA UNY |

First pairs encoded (a) and second round of encoding (b) in the modules. Amino acids are the present attributions.

1a GGG Pro	1b GGA Ser	3a GCG Arg	3b GCA Cys
CCC Gly	UCY Arg	CGC Ala	UGY Thr
2a GAG Leu	2b GAA Phe	4a GUG His	4b GUA Tyr
CUC Glu	UUY Lys	CAC Val	UAY Ile, Met

Figure 5. Mechanism of encoding the NSC modules. Two steps are sufficient for the description: the high ΔG pair would be stable enough to propitiate the transferase reaction and to accommodate the binding of the peptide product to itself; the five other pairs these triplets would also be involved with, of lower ΔG, become scarce, so that the two pairs left free to form are now in high concentration, which should facilitate the second round of encoding. Note that some of the present attributions in the homogeneous sector do not coincide with the constraints imposed by the metabolic pathway adopted in our scheme (see below) and require the proposition of concessions from an earlier occupier of the triplet (proposed by the metabolic scheme) to the present occupier. Three of these are related to the late entrance of the Glu family of amino acids (in red font), which does not belong to the proposed first metabolic pathway—the Glycine-Serine Cycle: in module 1 pair a, GGG-Gly conceded to Pro and in pair b YCU-Ser conceded to Arg; in module 2 pair a, CUC-Asp conceded to Glu. The other is Phe, whose biosynthesis requires a sugar (green) component. There are two pathways for the biosynthesis of Lys, involving either Asp or Glu derivations. After this stage of encodings, there are no more metabolic constraints.

We could not find any interesting possibility for participation of the SC subnetworks in the encoding process. It is possible that they would be involved in integration of the networks of dimers (the NSC, diagonal, with the SC, horizontal) or in regulatory processes but this has not been specifically tested, except graphically. An integrative role is plainly acceptable for the protein associations in the Multi-aRS Complexes [31], and the aRSs in the complex would be further associated through the dimers they bring in together with them ([5,17]; an update is available as Supplementary Information 1), including the SC subnetworks.

The encoding process can be followed through studies on the aRS/tRNA interactions. We compiled data reviewed by Beuning and Forsyth [32] and displayed them according to a succession from high to low degeneracy (Table 3). The data are consistent with an evolution of the interactions from simple to complex, with respect to involvement of the anticodon bases. The high degeneracy aRSs, for the hexa- and tetracodonic amino acids, do not bind the base in the 5' position (nucleotide 34). The aRS/tRNA recognition involves nucleotides outside the anticodon or, when they interact with the anticodon, it is done through the principal dinucleotide (nucleotides 35–36). The low degeneracy aRSs bind the three anticodon bases. There are only two exceptions to the rule, which are amino acids in the 5′G of the neighbor boxes of Tyr and His.

Table 3. Evolution of aRS/anticodon interactions. The data, compiled from [32] are consistent with the proposition that the initial encoding of a (proto)tRNA is dictated by the protein (aRS) binding to it, which may or may not involve the anticodon. When it involves the anticodon, it is directed to the principal dinucleotide, which means a full box degeneracy, the 5′ base being N. Evolution of proteins with the enlarged amino acid 'alphabet', required creation of some multi-meaning boxes, which is accompanied by interaction of the aRS with all three positions in the anticodon. The six stages are 1 = module 1; 2 = module 2; 3 = maturation of the homogeneous sector (modules 1 + 2); 4 = module 3; 5 = module 4; 6 = amino acids in the initiation NAU and the double-termination sign NUA pair of boxes.

aRS	Stage	Anticodon 34	Anticodon 35–36	Nucleotide 73	Acceptor Arm Pair 1:72	2:71	3:70	4:69

aRS	Stage	34	35–36	Nucleotide 73	1:72	2:71	3:70	4:69
Hexa-, Tetracodonic (ancestral states): no interaction with the anticodon or binding through the principal dinucleotide (35–36) only								
Ser	1	-	-	-	+	+	+	+
Leu	2	-	-	+	-	-	-	-
Arg	3	-	+	+Weak	-	-	-	-
Ala	4	-	-	+	+	+	+	-
Gly	1	-	+	+	+	+	+w	-
Thr	4	-	+	-	+	+	-	-
Pro	3	-	+	+	+	-	-	-
Val	5	-	+	+	-	-	-	+w
Mono-, Dicodonic (derived states, multi-meaning boxes): binding through the three anticodon positions								
Gln	5	+	+	+	+	+	+	-
Cys	4	+	+	-	+	+	-	-
Met	6	+	+	-	+	+	-	-
Glu	3	+	+	+	+	-	-	-
Asp	2	+	+	+w	+w	-	-	-
Ile	6	+	+	-	-	-	-	+w
Trp	4	+	+	-	-	-	-	-
Asn	2	+	+	-	-	-	-	-
Lys	3	+	+	-	-	-	-	-
Phe	3	+	Weak	-	-	-	-	-
Ambiguous: dicodonic, binding in the ancestral modes								
Tyr	6	-	+	+	+	-	-	-
His	5	+Weak (w)		+	+	-	-	-

3.3. Network Symmetry-Breaking

An important step in the encoding process was the creation of asymmetry inside the originally symmetric networks of dimers. If the encoding were to be done in the latter, some decision-type process would be required but it would take long to resolve via iterative cycles of many trials and errors with few hits, as if they were 'stagnant', with many weak and fluctuating bindings in the network. The flow through the system would be impaired via turbulence, continuous back and forth reactions, that is undecidability. There would be many factors involved in establishing a gradient type push-pull dynamics favoring the symmetry-breaking process, not to be detailed here.

In the asymmetric network the solution comes up almost automatically, besides the expediency resulting from the reduced size (from 16 to eight dimers) of the network: the first encoding is in the pair with the highest thermal stability and the second encodings are directed to the pairs left behind at high concentration (Figure 5). Asymmetry was created by elimination of a base from the set in the 5′ position. Any base would do the job but one had to be chosen. We could not find a major component that directed the choice to A but a list of factors is offered, still to be adequately weighed and combined.

(1) Keep the (a) high stability of the G:C pair plus the (b) mono-specificity of C and bi-specificity of G. The choice becomes between A and U, where A is tri-specific and U tetra [21].

(2) Avoid the very weak A:C pair at the laterals of the triplets, in favor of the G:U, which is frequent in RNA. Among the eight dimers in a module, the structure is exactly repetitive. Counting the numbers of lateral base pairs, excluding 5′A makes G:C 6, *G:U 6*, *A:C 2*, A:U 2 while excluding 5′U would make G:C 6, *G:U 2*, *A:C 6*, A:U 2.

(3) Keep U due to its being the base most used for modifications, provided that these would already be present at the times of developing the 5′ elimination [33,34].

(4) Keep U to profit from its tetra-fold ambiguity at decoding, as used in full in today's vertebrate mitochondria [35].

(5) Presence of A at the 5′ position in the P-site anticodon would, in a not well explained manner, destabilize the codon-anticodon pair in the A-site [36].

In view of the weakness in any of the single choices offered above, we prefer to adopt a rationale inspired on protein folding studies [37]: (a) The pressure for creating the asymmetry came via the flow. Any loss of a base would facilitate the flow from the first to the second round of encodings in a module, which was beneficial in reducing turbulence in the system caused by excess of substrates. (b) The choice for A utilized the strategy of 'minimum harm', which would be greater at elimination of any other base. It is rationalized that the avoidance of 5′A should be en bloc in view of the repetitive mutational formation of such kinds of tRNAs, which would be located anywhere in the space of tRNAs. This rationale is similar to that explaining the formation of eventual termination suppressors, which are continually scrutinized for deletion. Genomic data show that the absence of 5′A anticodes is at the gene level [38–40].

A global logic key for the encoding process produces the following summary. The homogeneous sector is encoded before the mixed sector. In each sector, the central G:C module is encoded before the central A:U module. According to the asymmetry produced by the 5′A exclusion, the high ΔG pair—whose members reside in the two central rows of the matrix, nonself-complementary GNG:CNC—is encoded first, leaving the lower ΔG pairs for the last encodings, whose members reside in the upper and lower rows of the matrix, nonself-complementary GNA:YNU. The self-complementary triplets are encoded afterwards, via expansion of the degeneracy of the first encodings in the box, which is directed to the pDiN. In case a box receives new attribution(s), triplets are conceded to them, which are usually the 5′Y anticodons to aRSs class I; exceptions are the two atypical aRSs of class II that will be treated further down.

4. Protein Secondary Structure, Nucleic Acid Binding and Composition of Termini

Some properties of peptides or proteins were examined with respect to amino acid composition, with the purpose of attributing functional meaning to the modules and helping in the establishment of their chronology.

4.1. Secondary Structure

The homogeneous sector codes for the non-periodic protein secondary structures (Table 4; original data in [41]). All amino acids preferred in protein non-periodic segments (including coils and turns) belong to the homogeneous sector (GNPSD); those preferred in α-helices are distributed in the two sectors, namely, homogeneous (ELKR) and mixed (AMQRH); those preferred in β- strands are mostly in the mixed sector (VIYCWT), only one in the homogeneous sector (F). The same trend is built by the data on amino acids builders of Intrinsically Disordered segments of proteins (data from [42]). The latter is especially important for the model, indicating the role of non-structured segments of proteins in not constraining the kinds of structures that may be accepted for interactions and in offering open choices for development of structures after the interactions, besides the possibility of maintaining the disordered character even after the interactions [43–45].

Table 4. Amino acids preferred in the types of protein conformation structures.

Modules of tRNA Pairs	Sectors of Anticode Dimers (Wobble + Principal Dinucleotide)							
	Homogenous Sector, RNP Realm				Mixed Sector, DNP Realm			
	1 G S	2 L D N	Sector mature E P F K R	Total wRR YYw	3 R A T C W	4 V H Q I M Y	Punctu ation iM, X	Total wRY RYw
Non-periodical, [34] Coils, Turns	G S	D N	P	5				0
Helices		L	E K R	3–4	R A	H Q M		4–5
Strands			F	1	T C W	V I Y		6
Disorder [35]	S		E P K R	4–5	R	Q M		2–3
Borderline, Neutral	G	D N		3	A T	H		3
Order		L	F	2	C W	V I Y		5

4.2. RNA and DNA Binding

The homogeneous sector codes for the RNP realm (Table 5). Amino acids preferred in conserved positions of RNA-binding motifs are 75% in the homogeneous sector (GPLKFS), with only VM in the mixed sector. Amino acids belonging to the DNP realm may be preferred exclusively in DNA-binding motifs (80% in the mixed sector: AHCT; with only Glu in the homogeneous sector), or preferred in both DNA- and RNA-binding motifs (IYRQW, Arg belongs to both sectors of the code). Asp and Asn were not preferred in any nucleic acid-binding motifs. The compilation [46] does not consider highly basic motifs owing to their non-specificity for bases, interacting mostly with the sugar-phosphate backbones.

Table 5. Amino acids preferred in RNA and DNA binding motifs of proteins.

Modules of tRNA pairs	Sectors of Anticode Dimers (Wobble + Principal Dinucleotide)							
	Homogenous Sector, RNP Realm				Mixed Sector, DNP Realm			
	1 G S	2 L D N	Sector mature E P F K R	Total wRR YYw	3 R A T C W	4 V H Q I M Y	Punctu ation iM, X	Total wRY RYw
RNA binding [39]	G S	L	P K F	6		V M		2
Both			R	0–1	R W	I Y Q		4–5
DNA binding			E	1	A C T	H		4

The two criteria above are linked. Radivojac et al.'s review [42] is nicely consistent with our proposition for the self-referential function of early proteins, explicitly citing the intrinsically disordered regions as characteristic of ribosomal proteins and of the splicing complex proteins. These sets encompass proteins bearing the RRM (RNA Recognition Motif, Gly-rich); the intrinsically disordered regions are also rich in Ser and Arg, which sum to the Module 1 subset (GS) of our previous list of amino acids typical of RNA-binding motifs [46].

4.3. Protein Termini and the N-End Rule

Data from the N-end rule of protein stabilization against degradation [47–50] and from statistical frequency of amino acids in the N- and C-terminal segments of proteins [51] (respectively left and right in Figure 6) are shown to superpose coherently upon each other. There is overall consistency between the two modes of examining protein strings, indicating that properties of amino acids generating the N-end rule were utilized by cells to locate stabilizing amino acids at the heads and destabilizing at the tails, which could be considered a primitive 'punctuation' system [52].

The circularly 'closed' mode of presenting the code system derives from an informational description. Both the N-end rule and the protein termini regularities, plus the link between the punctuation signs (see below) conjoin to produce the informational closure drawn as a circle. Amino acids that are stabilizers of proteins against degradation are all at the left side. The drawing also requires that the codes in module 1, pairs 1 and 2, follow each other *in tandem*, initiating a central loop structure. Otherwise, amino acids of modules 3 and 4, when corresponding to central R triplets, are added backwards in the string, following from module 1 pair 1 (the initial head) to the direction of the future head (Met); the neck keeps growing until the set of amino acids is exhausted. Amino acids that are destabilizers of proteins when added to the N-ends, therefore being preferred to compose the C-ends (tails), are all at the right side of the circle. The early loop is completed by adding, still *in tandem*, module 2 to pair 2 of module 1. By the end of this loop, the amino acid Arg that fills up the homogeneous sector and initiates the mixed sector is added as the primitive tail. This is then elongated forward with addition of the amino acids in modules 3 and 4, taken from the central Y triplets, in the direction of the future tail and the termination codes. The closure is accomplished by the mechanistic and informational link between initiation and termination (topic below).

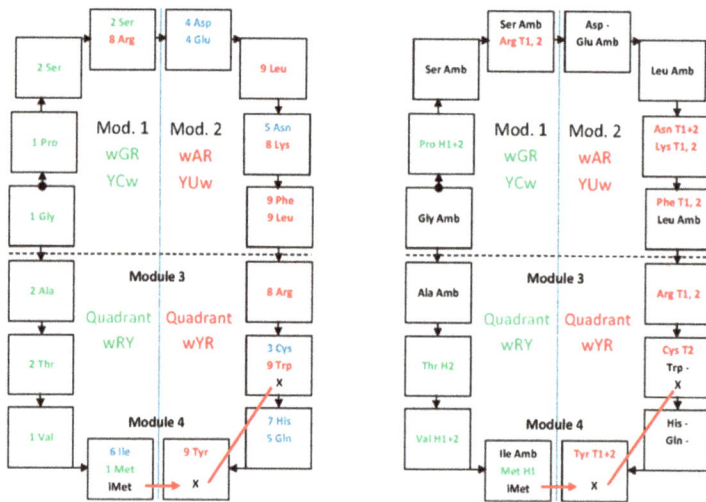

Figure 6. Amino acids displayed according to the N-end rule (left) and to their preferential location in protein termini (right). The N-end rule classifies amino acids as strong stabilizers of proteins against catabolism (grades 1, 2; respectively GPMV, SAT) when the half-life of proteins bearing those amino acids at the N-terminus is long; strong destabilizers (grades 8, 9; RK, LFWY) when the half-life is drastically shortened; and intermediate (grades 3–7; CDENQIH). Data on preferential location of amino acids at protein terminal segments suffer from database-dependency and from frequent ambiguities, when amino acids are preferred or avoided in both locations (Amb). H (head), significant statistical preference at the N-terminus (H1), at the second position (H2), or when the two first positions are summed as a dipeptide (H1 + 2); T (tail), significant statistical preference at the C-terminus (T1), at the second position (T2), when the preference is significant at both last positions individually (T1, 2), or when the two last positions are summed as a dipeptide (T1 + 2). In spite of the care taken by the authors with the sample sizes examined and the statistical processing, some results may suffer from deficiencies, such as the difficulties with Pro, significantly preferred in heads only in fungi, repeating the problems at trying to experimentally add it to the heads of proteins [47], and with the overall low frequency of Trp.

5. Location of the Punctuation Codes

The puzzle set forth by the indication that the boxes where the initiation Met and the termination double signs are located, respectively NAU:NUA, form a pair is only accentuated at closer examination.

The real triplets do not pair due to incompatible wobble bases CAU:YUA. The other termination box, bearing the single sign UCA would not pair with the iMet but shows partial non-aligned identity, of the CA doublets. We were unable to find previous studies relating the initiation and termination codes or mechanisms. Such complication prompted a full examination of all possible connections between triplets of the related boxes: identity or complementariness, alignment or the different possible slippages, codons and anticodons. The demarcation criterion is that a mechanism should accommodate all entities involved in one same explanation: the initiation Met, alone or in combination with the second triplet, and the three X together with their neighbors in the respective boxes. The meaningful resulting combination is presented in Figure 7. The consensus constitution of the mostly virtual X anticodes, which are not present except for the suppressor instances, is YYA, minus the Trp-CCA.

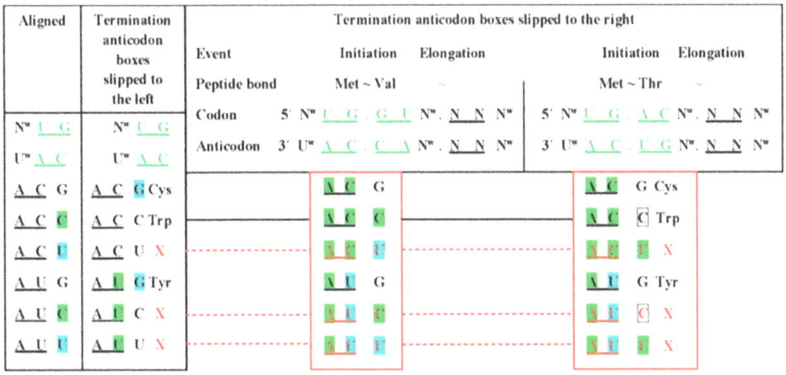

Aligned	Termination anticodon boxes slipped to the left	Termination anticodon boxes slipped to the right					
		Event	Initiation	Elongation		Initiation	Elongation
		Peptide bond	Met ~ Val	~		Met ~ Thr	~
Nᵂ U G	Nᵂ U G	Codon	5′ Nᵂ U G . G U Nᵂ. N N Nᵂ			5′ Nᵂ U G . A C Nᵂ. N N Nᵂ	
Uᵂ A C	Uᵂ A C	Anticodon	3′ Uᵂ A C . C A Nᵂ. N N Nᵂ			3′ Uᵂ A C . U G Nᵂ. N N Nᵂ	
A C G	A C G Cys		A C G			A C G Cys	
A C C	A C C Trp		A C C			A C C Trp	
A C U	A C U X		A C U			A C U X	
A U G	A U G Tyr		A U G			A U G Tyr	
A U C	A U C X		A U C			A U C X	
A U U	A U U X		A U U			A U U X	

Figure 7. Localization of termination codes and elimination of the correspondent tRNAs directed by the initiation mechanism. **Initiation.** The principal dinucleotide of the initiation anticodon is slipped so that the wobble position is the 3′ (alternative codons would be Val-GUG and Leu-YUG). The consequence is a strengthening of the initiation mechanism where there is no wobbling interposed between the two first coding triplets. The two frequent kinds of initiation couple of triplets is presented, Met-Val and Met-Thr. Note the peculiar arrangements of the contiguous principal dinucleotides, respectively, an inverted repeat wUG.GUw codons or a direct complement wUG.ACw. **Termination.** Our search for connections between the initiation and termination codes tested all possibilities of identity or complementariness and with alignment or the two kinds of slippages. Only the configuration displayed satisfied the condition of involving coherently the three termination codes: the principal dinucleotides of the boxes containing the termination codes align with the principal dinucleotide of the initiation Met and the wobble base aligns with first base of the second triplet. A conflict is instantiated where the 5′Y triplets of the cited boxes compete with the initiation Met-Val or Met-Thr anticodon doublets. This would have been the reason for eliminating the conflicting anticodons while keeping the initiator Met, which created the termination codons, void of decoding tRNAs. It is as yet not known whether the tRNA$_{Trp}$^{CCA} was previously encoded, probably tRNA$_{Trp}$^{YCA}, therefore retained, or was encoded afterwards, being recoded from X^{YCA} which conceded X^{CCA} to Trp^{CCA}; in both cases, mechanisms should be added to protect initiation from the conflict. It is highlighted the crucial determinant 3′A of the X anticodes, which conserves the Watson:Crick type pair through all tests and should have been a main guide for the X tRNA exclusions. Key: N = bases of the principal dinucleotides of the initiation triplets; N = standard complementariness of the termination anticodon base to the codon base, which is identity with the iMet anticodon; N = G:U base pair, N = termination anticodons that were deleted; N = first anticodon base whose complementariness to the first base of the second amino acid (Val or Thr) indicates the mechanism of competition with the initiation anticodon but gives no indication as to how it would have contributed to the retention or concession of the Trp anticode.

The obvious message, which nonetheless deserves highlighting is that, in all steps of the protein synthesis system, the basis is the construction of a peptide bond, which involves a couple of tRNAs. These are laterally associated in the case of translation and dimerized through an anticodon minihelix in our DDPS model for the code origin (Figure 2). Installation of initiation involved a reconfiguration of the elongation mechanisms to eliminate a wobble position between the two first codes, which may be described as a functional 'inversion' of the first triplet. Such inversion would be analogous to the capping process that is observed in mRNAs of eukaryotes. Our results indicate that this mechanism, at the same time, created conflicts with some anticodons that had to be eliminated therewith originating the Stop (X) codons (Figure 7).

The connection between initiation and termination is most probably only in trans. The conflictive termination tRNAs would meet the interference sites through diffusion, not involving direct physical contacts between rostral and caudal portions of the mRNAs. Observations on circular RNA configurations pop up eventually in the bibliography but they are neither typical of nor compelling for [53] application in mRNA conformations. After the mechanism of termination tRNA exclusion was installed, some other tRNAs may mutate to acquire that constitution, then called termination or non-sense suppressors [54,55]. Such mutational events should be continually purged by purifying selection. Otherwise, the circular drawing of the code seems to be plainly justified on the basis of the informational closure character, which is also esthetically appealing to indicate that the encoding process probably reached completion in extent, while not prohibiting further evolutionary modifications inside the system.

It could be envisaged that the punctuation system was installed upon a full set of elongation encodings. In this context, initiation would have to recode a previous triplet, which was achieved through a principal dinucleotide 'slippage' from the elongation ^{W}CAU to the initiation CAU^W. Another immediate consequence of the mechanism would be the creation of conflicts with competing tRNAs, whose identity, to become the terminators, would only depend on the identity of the chosen initiator. In this case, when the choice of one depends also on the possible conflictive consequences, it is suggested that the process involved some multifactorial iterative variety of trials, errors and hits. The natural case was constrained by the amino acid properties that gave rise to the N-end rule, therewith reducing the amplitude of the window of trials: initiation was directed to an N-end stabilizer, located at the N-end segment, whose conflictive tRNAs would correspond to an N-end destabilizer, located at the C-end segment. It should be interesting to investigate two related points that could test the credibility of the mechanism proposed: (a) which would be the characters of the Trp system that allowed its retention without interfering with the initiation system, and (b) is there some kind of toxicity directed to the initiation subsystem when there are nonsense suppressor tRNAs or recoded X codons [56] in the system?

RNP World Instead of RNA World

The existence of protein translation factors that mimic the shapes of tRNAs, which include the Release Factors (RFs), would easily suggest that the latter came into the play as substitutes for the tRNAs complementary to the X codons. There are many other instances of reduction of the tRNA set, most appealing the generally forbidden 5'A anticodons. This rationale would be part of the RNA World proposition [57–59], indicating that RNAs started the construction of the biosystem and at some point developed the proteins as functional helpers for structures and catalysis, and developed the DNA as more stable genomic helpers, thereafter receding to the job of intermediates between those. These propositions seem to be now being made more flexible in favor of the coevolutionary panorama [16], that is, some kind of an RNP World preceding the present cellular DNP World, which is in entire consistency with the proposal of the Self-Referential Model (SRM), as sketched in Figure 2.

The cases of protein-nucleic acid mimicry are now known to be more extensive and multifaceted [60–62]. The emergent consensus for the cases of tRNA mimicry in translation factors is that it refers mostly to the shapes of the tRNAs, necessary to fit the pockets and tunnels of the ribosome, while some of the functional details may differ from the expected if they would be substituting the tRNA functions [63,64]. It is realized that tRNAs are not capable of terminating protein synthesis, which requires protein activities. Termination may be achieved through a variety of ways, including the traditional absence of X anticodons, but it may occur even in the absence of the X codons, where it may rely upon directions given by other features of the mRNA 3' end [65,66]. Even one of the paradigmatic proposed ribozyme activities, the ribosomal Peptidyl Transferase Center, has been drastically challenged [67,68] and is under reevaluation.

6. Metabolic Pathways

The chronology offered by the self-referential model (SRM) clashes with the traditional studies on the origins of the code [69–78]. These are centered on the set of amino acids that are produced in relative abundance under prebiotic conditions, possibly representing what would be expected from early Earth geochemical systems. Even the coevolution hypothesis, saying that much of the code organization is derived from amino acid biosynthesis pathways, starts with the geochemical set [70,71,78]. This is centered on the amino acids in the 3'C row of the matrix Val Ala Gly Asp, which could also profit from high stability of some triplet pairs such as codons Ala-GCC:GGC-Gly, less so in Val-GUC:GAC-Asp. This coincidence between an all-G:C triplet pair carrying the most abundant prebiotic amino acids gave origin to the proposals [72,76,77] based on the Ala and Gly correspondences. Note that this choice through rows has to rely upon self-complementary triplets. In the coevolution hypothesis, the hybrid molecules aminoacyl-tRNAs would be able to generate concertedly amino acid families of biosynthetic derivations and tRNA variants. The biosynthesis pathways recalled to be involved in structuring the code were the traditional heterotrophic and catabolic, centered on glycolysis, the pentose shunt and the citrate cycle.

Our starting amino acids Gly and Ser did not fit any of the propositions above. The couple of the most abundant prebiotic is Gly and Ala. The derivation of Gly and Ser from the glycolysis route starts with a phosphorylated compound and follows the degradative path from C6 to C3 and then to C2, which is not an easy path for initial evolution to work with; the last transformations are 3-phosphoglycerate → 3-phosphohydroxypyruvate → 3-phosphoserine → Ser → Gly. A main problem with respect to the encoding process relying upon prebiotically synthesized amino acids would be the frequent and drastic fluctuations in precursor availability, approaching the 'feast and famine' regimes, which would not be adequate for fixation of encodings; these would require reliable and reasonably continuous sources of the substrates. In the panorama of the SRM, propositions based on the sets of prebiotic abundant amino acids could have corresponded to proto-codes where proto-tRNAs and peptides might have possibly passed through processes of mutual adjustments.

We found a simple pathway at pan-searching inside microbial biochemical diversity, the *Glycine-Serine Cycle* (Figure 8), more frequently called the *Serine Cycle* [79–81], which fits the SRM modular scheme. It belongs to the C1 realm of metabolism and is typical of Type II Methylotrophs (α-proteobacteria, the same group where mitochondria originated from), while having sectors of compounds that overlap other pathways and other groups of organisms. It is the simplest among central metabolic routes, starting with the C2 glyoxylate and reaching the maximum C4. It is said that nowadays its main function would be of producing AcetylCoA from C1 units. These are one CO_2 and one reduced kind of C1, brought in by the TetraHydroFolate (H_4F, THF) or the correlate and older H_4MethanoPTterine (H_4MPT) carrier. Such kind of pathway is already at the bottom row of simplicity with respect to the substrates, at the fuzzy borders between methylotrophy and autotrophy. Methylotrophs and methanotrophs are not considered autotrophs because they have wider metabolic abilities in the assimilation realm; they are able to incorporate some of the partially oxidized C1 compounds into cellular carbon before they are completely oxidized to CO_2, this being directed to energy production.

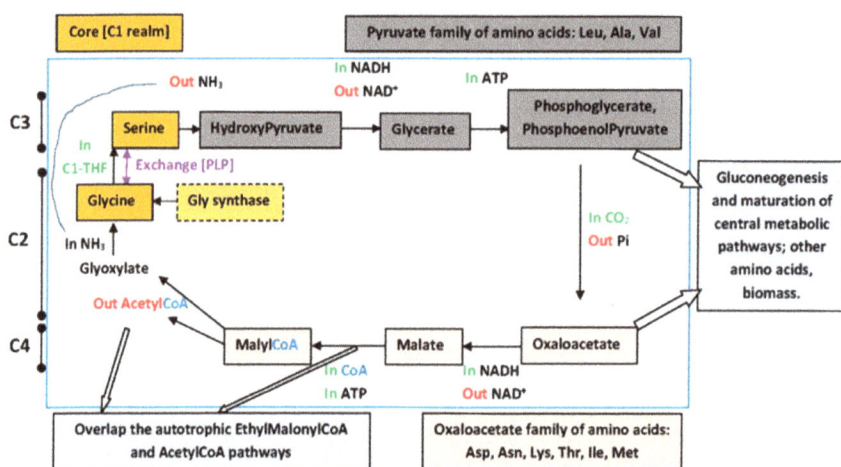

Figure 8. The Glycine-Serine Cycle. This is at the bottom layer of complexity among the assimilation central metabolic pathways. Just one step above the basal synthesis of Acetate, it reaches C4 compounds through additions of a reduced carbon (C1-H$_4$F) and a CO$_2$. Glyoxylate is aminated to Glycine and this receives a –OH-methyl to form Serine. These amino acids can feed directly the first module of the code. Various C3 derive from Ser, which can feed gluconeogenesis and synthesis of the Pyruvate family of amino acids. Addition of the CO$_2$ forms the C4 compounds where the oxaloacetate family of amino acids derive, gluconeogenesis is again nourished, other key connections are established from Malate and MalylCoA. This breaks into two C2 compounds that regenerate the cycle starter and generate AcetylCoA.

The Gly-Ser Cycle can be divided into (**a**) a core portion, which is the C2 + Ser cut that could be considered a nearly autonomous subcycle, and (**b**) the more complex set of the C3 derivatives from Ser plus the C4 compounds, some of which may feed directly on to the gluconeogenesis and the Citrate Cycle. Elements in the core are connected via the Gly $\longleftrightarrow$ Ser interconversion through the H$_4$Folate carrier of the hydroxymethyl radical –CH$_2$OH and the enzyme SHMT (Serine HydroxyMethylTransferase, SHMT, which is dependent on Pyridoxal-phosphate, PLP). This enzyme belongs to a peculiar family whose reactions and functions are much varied, sometimes called even 'promiscuous', which seems to fit adequately the requirements for early processes [82]. A deamination/amination reaction further unites, respectively, the Ser $\rightarrow$ HydroxyPyruvate conversion to the Glyoxylate $\rightarrow$ Glycine conversion. This core portion is nearly universally distributed and at the roots of autotrophy [83], therefore more relevant to support the earliest stage (module 1: Gly and Ser) of the genetic code. The relevance of the C3 and C4 portion for the next stage (module 2: Leu, Asp and Asn) of the code fits nicely our scheme but biochemists might question the validity of the proposal mainly in view of the supposed involvement of those reactions with oxygenated environments while it is also assumed that the code would have been formed under anoxic conditions. It is recalled that there are many possible variations on the insertion of the simple pathways of Serine synthesis, including some anaerobic [84], while the phosphorylated catabolic pathway became predominant at heterotrophy [85].

The homogeneous sector of codes contains presently other five amino acids, four of them in the Glu family plus the Phe, whose biosynthesis requires a sugar compound coming from the pentose shunt. We therefore add a stage of 'metabolic maturation of most of the central pathways', necessary for the completion of the homogeneous sector encodings. These pathways develop from the serine cycle C3 and C4 compounds into gluconeogenesis and then glycolysis, the pentose shunt and the citrate cycle. The key role of Glu in this stage of encodings, just at the border between the two sectors of the code, is reminiscent of the discussion asking for the causes of obligate autotrophy and obligate methanotrophy, which is supposed to reside in the sensitive step involving the loss of the α-keto glutarate dehydrogenase activity [86]. Such deficiency divides the citrate cycle into 'halves', the cycle assuming a 'horseshoe' shape: the C4 portion in one side and the C6-C5 in the other side. Such situation is also called an 'incomplete' citrate cycle, indicating that the cycle would have been formed or 'completed' through the fusion of the two sides promoted by the C5→C4 processing enzyme. It is tempting to propose that this maturation step was limiting also to the process of traversing between the code sectors via $Arg^{YCU} \rightarrow Arg^{NCG}$ since Arg is derived from Glu.

The phylometabolic approach [83] also reached the $H_4Folate/H_4MPT$ roots of autotrophic pathways. The bioenergetics group of William Martin pointed to the Ljungdahl-Wood autotrophic pathway, which also has some steps involving $H_4Folate/H_4MPT$ carriers but is very complex with respect to the variety of cofactors required for the biosynthesis of AcetylCoA [87,88]. A portion of this route is shared with the methanogenic pathways. Despite the simplicity of the Glycine-Serine Cycle and in face of the basal autotrophic routes for production of AcetylCoA being complex, there is the appealing possibility of biochemistry having started from geochemical C2 units, acetate and derived, which could be a reliably abundant source in some adequate environments, utilizing the H_2 reductant produced at serpentinization [87,89,90] (I thank Yoshi Oono for this suggestion).

An apparent inconsistency in our chronology of amino acid encoding is the precocious presence of the C6 Leu, together with the C4 of module 2. Searches for proposals to substitute Leu by a simpler amino acid reach the repetitive suggestions for the central A column having been initiated with the C5 Val. Even if this suggestion is adopted, the inconsistency remains more or less of the same magnitude. The fact of Leu biosynthesis incorporating two molecules of pyruvate, the same occurring with Val biosynthesis, might make the discrepancy more acceptable in the sense of not becoming too complex with respect to sources of substrates. It might also be relevant to biochemistry the prebiotic tendency of pyruvate (the same with other α-keto acids, reactive compounds) to engage in highly complex derivations, including its dimerization into parapyruvate and formation of up to C8 compounds, that are obtained from pyruvate of carbonaceous meteorites in water [60]. Otherwise, it is possible that Ala might have been the precursor to Leu in module 2, as indicated by the existence of a variant of this kind in mitochondria of a few filamentous Saccharomycetacea and Debasyomyces [91].

Amino Acid Hydrodynamic Size and the aRS Classes

A synthetic description of some characters of the development of encoding (Figure 9) by the aRSs can be obtained from the correlation between the increasing amino acid size [92] and biosynthetic pathway complexity (number of steps and diversity of compounds). A trend is shown, starting (module 1) with small amino acids and increasing sizes steadily to the end of the homogeneous sector, thereafter maintaining large average sizes. Variability is high in intermediate stages and lower in the sets of modules 1 and 4. Enzymes class II are typical of small and class I of large amino acids. ArgRS traverses the sectors. The Figure 9 graph can be paralleled with the hydropathy graph (Figure 3) to form groups of amino acids for the modules: the starting amino acids are (module 1) small, hydroapathetic and class II; these amino acids are among the constant and frequent oligomers of putative earliest proteins [93]. The last amino acids (module 4) are large, hydrophobic and mostly class I. Intermediate stages explore the entire hydropathy range.

The atypical aRSs are class II for the only large amino acids of the class, and of extreme hydropathies: PheRS, acylating in the class I mode (2′), and LysRS, class I or II in different organisms, the class II being the only in the class to occupy a 5′Y set of anticodes. These occupy the last pair of triplets of the homogeneous sector GAA:YUU. The complete set of atypical aRSs includes amino acids beyond the standard 20. There are two additions, Selenocysteine (Sec, the 21st amino acid) and Pyrrolysine (Pyl, 22nd), which makes the anticodon pair $\frac{5GAA}{3UUY}\frac{PheRS}{LysRS}$ only the prototype for the atypical systems. These remain all class II affecting the two Stop boxes. The class II typical acylation site is 3′ (Gly Pro Ser Ala Thr His), while Asp Asn are variable; the class I typical acylation site is 2′ (Leu Arg

Glu Val Ile Met), while Cys Trp Gln Tyr are variable [94]. Distribution of the aRS classes according to the utilization of the 5′ bases in the multi-meaning boxes obtains: 5′ R is variable (class II Phe Ser^{CU} His Asp Asn/class I Ile Cys Tyr) while 5′ Y is homogeneously class I or punctuation (Leu^{AA} Ile Met Trp X Arg Gln Glu), plus the atypical Lys. There are some organisms that charge Lys via a class I enzyme, which is typical [95].

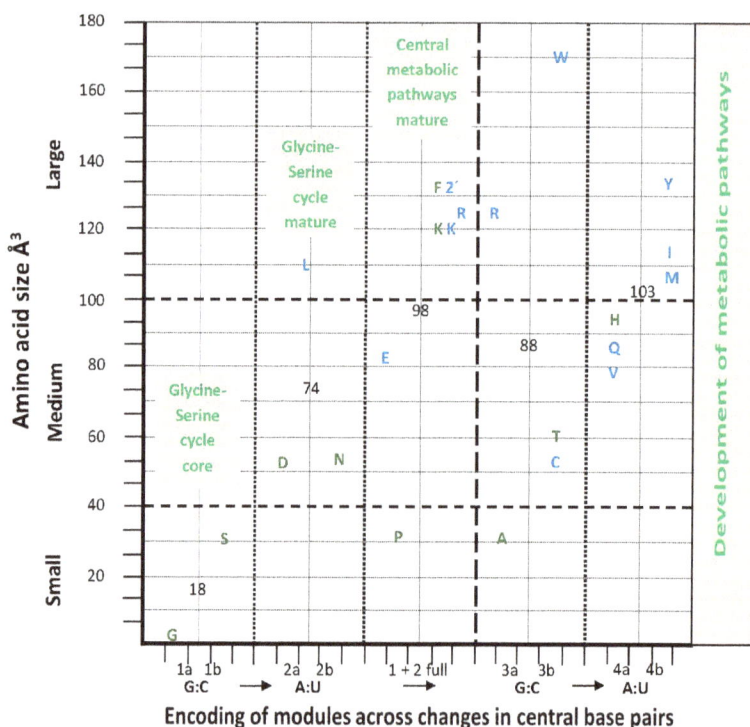

Figure 9. Amino acid hydrodynamic size, aRS classes and metabolic maturation. Enzymes class I in blue, class II green. Average amino acid side chain size [92] for the module in black. LysRS is uniquely class I or II (atypical) in different organisms. PheRS is also atypically a class II that acylates in the class I mode (2′). The metabolic pathways relevant to the encoding chronology are the Glycine-Serine Cycle, whose core coincides with module 1, while the C3 derivatives of Ser and the C4 compounds coincide with precursors to the amino acids in module 2. Other amino acids had to wait for development of the other central metabolism pathways, after gluconeogenesis, glycolysis, the pentose shunt and the citrate cycle. There are no metabolic constraints upon the mixed sector encodings.

The other atypical systems share aspects with the prototypes. Charging of both amino acids Sec and Pyl repeat the LysRS class II atypical attribution to 5′Y anticodes, but in tRNAs (anticodons, UCA and CUA, respectively) that were introduced to decode Stop codons, whose tRNAs are not present (X) in the standard anticode. These cases are called recoding of X codons and differ from the termination suppression in being internal and functional in protein sequences. Termination suppression is deleterious to protein functions due to the extended anomalous C-termini.

There is an alternative route for the charging of Cys, via the O-phosphoseryl charging of tRNA^{Cys} by a PheRS homolog (SepRS), which maintains the 2′ acylation site. The Sep-tRNA^{Cys} is thereafter transformed into Cys-tRNA^{Cys}. A tRNA^{Sec} with anticodon U<u>CA</u> (principal dinucleotide underlined), which is not present in the standard anticode set (it corresponds to the X codon <u>UGA</u>), is serylated by the standard SerRS. The seryl moiety is then phosphorylated (to Sep) and this transformed to selenocysteine (Sec), to get the final product Sec-tRNA^{Sec}. The pathway to Pyl starts with its synthesis from two molecules of Lys. The PylRS is homologous to PheRS but the acylation site is the 3′, typical of the class II enzymes, indicating that PylRS retains the original character of the class II that later developed the atypical function, which is shared by PheRS and SepRS [96,97].

7. Symmetries and Error-Reduction in Modularity

Nucleic acids should be a prime locus for the generation of symmetric structures at the acquisition of secondary or higher order organizations in view of their basic complementariness property. There have been many attempts at reaching the depths of the genetic code structure through the symmetry-search approaches (e.g., [98]) but most of them utilized codons and strings as the subject matter and are still waiting for obtaining wider support. The appeal of symmetric arrangements would reside on the possibility of facilitation of functional processes due to the repetitiveness that symmetries offer, be they derived from duplications or convergence, from direct, inverted or complementary arrangements, among other possibilities.

Our approach is centered on the elements proper to the encoding/decoding machinery—tRNAs and their anticodons, aRSs and amino acids, considering the strings of codons later developments, either enchained triplets derived from pre-encoded tRNAs [99–103] or exogenous strings that acquired the ability of being translated through evolutionary adjustments. Such strings and the decoding machinery would have coevolved with focus on e.g., speed while maintaining accuracy in translation, in a process that might have not preserved the details of the process of encoding. This would be a reason for the lack of success of the studies based on the strings of codons and we hope to have uncovered some aspects of the encoding process through studies of anticodons.

7.1. Central Bases Compose Standard Base Pairs, Columns Are Divided into Hemi-Columns

The structure of tRNA pairs or dimers is the basal symmetric: the central base pair is of the standard Watson: Crick type while the lateral base pairs allow for the generic R:Y, keeping fixed the base in the 3′ position and choosing among the variation in the wobble position (Table 1). The primacy of the central base [29] is reflected in some regularities that are recognized in the code matrix since the time of its deciphering about half a century ago, especially the correlation of the most hydrophobic amino acids coinciding with the most hydrophobic central purine A, which is the first column.

Our approach keeps this organization, improves the correlation (Figure 3) and adds the correction that the vertical organization (in columns) derives from the juxtaposition of distinct hemicolumns. These are formed in different sets of encoding events (numbered and colored in Figure 10c), one for the homogeneous sector boxes (e.g., NAG, NAA; the central R of module 2) the other for the mixed sector boxes (e.g., NAC, NAU; the central R of module 4). The horizontal organization (in rows) is also a juxtaposition of distinct sets of encoding events. The first pairs of the four modules (1a, 2a, 3a, 4a) that start with the nonself-complementary dimers (GNG:YNC) compose the two central rows, one of them coinciding with the most traditional component of models for the code origins, which are centered on the GADV row. The self-complementary YNG:YNG and GNY:GNY of the two rows just cited are encodings secondary to the nonself-complementary. The orthogonal organization of the matrix, through columns and rows, is now the result of overlapping crisscrossed hemi-columns whose elements are pairs or dimers of triplets with a disposition that follows the diagonals inside the matrix. The full symmetry in the distribution of mono- and multi-meaning boxes (see below) is a consequence of this design.

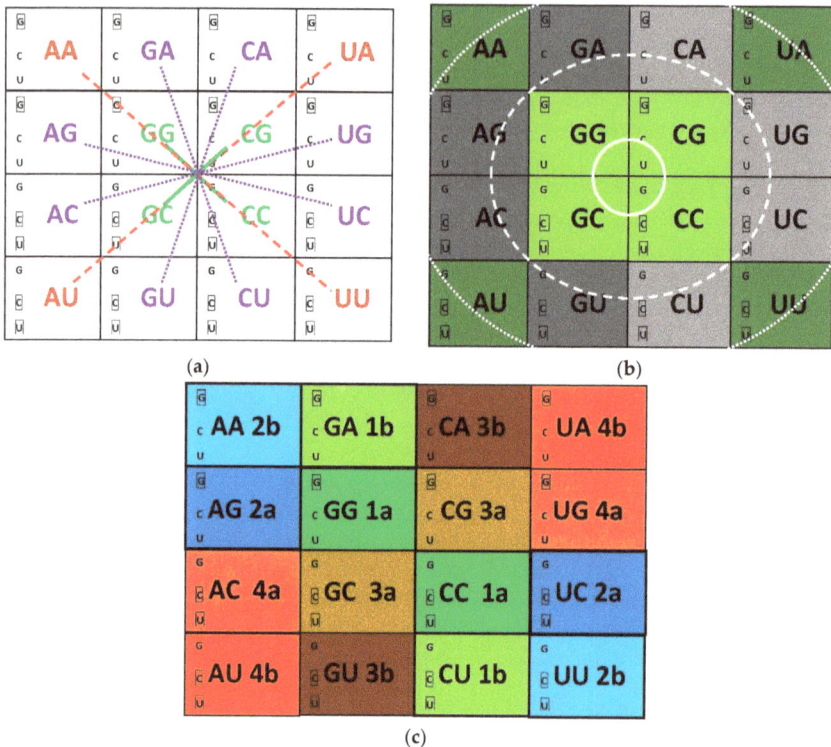

Figure 10. Symmetries in the organization of the matrix of anticodon dimers. They are easier to follow through the pairs of principal dinucleotides but are composed only by the nonself-complementary triplets (with the 5′ bases boxed). (**a**) Eight separate pairs are shown united by lines of different colors and shapes; (**b**) The pairs are united according to the composition of the degeneracy in the boxes: light green, inner circle, two pairs of high ΔG, principal dinucleotides with Gs or Cs only, boxes mono-meaning; dark green, outer circle, two pairs of low ΔG, principal dinucleotides with As and Us only, boxes multi-meaning; intermediate circle, principal dinucleotides with one G or C and one A or U, four pairs of intermediate ΔG, boxes mono-meaning when central base is R (dark grey), multi-meaning when central base is Y (light grey). The process ran through the organization in pairs and sectors but the final result is fully symmetric so that no hint of a deeper organization is given to the observer; (**c**) The pairs are united into modules and sectors. The format of presentation of the modules is different from the matrix in Figure 4a.

Symmetries are not expected to be constructed on the basis of properties of amino acids. When symmetric arrangements are observed on the basis of amino acid characters, they would probably be following the symmetry that was constructed by the triplets. In fact, we could detect three such instances (Figure 4b; one per module, except module 3, and all in the upper and lower rows) but none of them would be compelling enough to have directed research projects focusing on those arrangements. These molecules would, to the contrary, follow functional dictums of the protein segments they would be constituents of, that were examined in Sections 3.1 and 4–6.

The central base primacy in dimerization of anticodons finds a different albeit convergent counterpart in the codon-anticodon pairing. The ribosomal decoding center has an asymmetric 'principal doublet-plus-wobble' structure; the doublet of bases is contiguous but the third base comes from a distant segment. The contacts made by the central base pair with the ribosomal center are more numerous than those of the other bases [104]. Evidence from the aRS-tRNA interactions [34] (Table 3) indicates that the situation is highly complex. While bases are all contiguous in a triplet, there are curvatures in the anticodon loop which would be behind the development of the asymmetric interaction structure. We could attempt to suggest that the asymmetry would have derived from the process of tightly packing the adjacent tRNA anticodon loops inside the ribosome at translation of a continuous string. The forced accommodation of the loops required one of them to curve and yield space to the other. The base that was dislocated became the wobble position. Later adjustments created, among a variety of base selection and modifications, the peculiar U-turn that helps to stabilize the loop when the principal dinucleotide is only moderately stable but has a purine central base [19,105].

7.2. Symmetry in the Distribution of Mono- and Multi-Meaning Boxes

The picture composed by dimers in the matrix is fully symmetric and smooth (Figure 10a). The distribution of single- and multiple-meaning boxes follows the complete symmetry without any deviation (Figure 10b). These characters did not catch the attention of previous observers beyond the level of decoding where the explanation involves formation of codon: anticodon pairs [19,72,91]. The participation of anticodon dimers did not need to be invoked since questionings were satisfied by the decoding function only.

The matrix presentation is clear enough to display the global result as the triple-circle concentric arrangement (Figure 10b), relying upon the interpretation that base composition is accompanied by the ΔG values (data in Supplementary Information 2). There is no influence of the modular structure (Figure 10c) in the final picture of the distribution of degeneracies. The inner circle is the core of the matrix, composed by single-meaning and entirely G or C principal dinucleotides. These are able to reach high thermal stability from three G:C base pairs with the correct choice of the wobble base to form nonself-complementary triplets. It is indicated that such high ΔG offered high resistance to being split into multiple attributions. The outer circle is composed by the tips of the matrix, all A or U principal dinucleotides and multiple-meaning boxes. These lower thermal stability boxes offered the least resistance to being split into multi-meanings. The principal dinucleotides in the intermediate circle have one G or C and one A or U base, therefore reaching intermediate ΔG values. A specific stabilizing factor is now recognized to be at work, namely a U-turn structure that can be formed between a central anticodonic purine base (not with a central pyrimidine base) and the U-base at position 33 [19,105]. Boxes remained with single-meanings when their triplets had a central R and developed multiple-meanings when they were central Y. The full symmetry indicates that the ΔG effects were obeyed strictly and that the affordances were profited from to saturation. Further informational necessities were fulfilled with added post-translational effects.

The reasoning above only makes sense when applied to the anticodon dimers and the aRSs that are encoding the tRNAs. Base composition is accompanied by the ΔG values and the correlation with the distribution of degeneracy per box is perfectly linear (Supplementary Information 2). The possible participation of aRSs in the dimer structures has not been studied, as well as any possible influence on thermodynamic parameters [22] between self- and nonself-complementary dimers.

7.3. Symmetry in Modules of Dimers and the Error-Reduction Property

The property of the encoding/decoding system of reducing the consequences of errors [106] is indicated to derive from a regionalization of the attributions or correspondences in the modules and sectors [107]. The conserved structure of the sets of triplets across modules would facilitate the encoding process from the second module onward, after the 'learning' at construction of module 1. The main difficulty presented to the aRS duplications would reside in adapting to different central bases and in the change from the homogeneous to the mixed sector. At the same time, each new step would benefit from the expanded repertoire of amino acids encoded in the preceding step, that is, the process incorporates cumulative and self-feeding characteristics.

Our limited list of protein properties examined supports the assertion. We detected (i) the clustering of attributions dedicated to the construction of different protein motifs (ii) inside the modules of dimers. The characters examined are coherent with the modules, which may be examined in different formats of presentation, from the graphs in Figures 3 and 9 to the Tables 4 and 5 to the circular structure of Figure 6, which may be extended into a string. The informational circle is read from the central and ancient core of elongation codes—the homogeneous sector, to both extensions, backward and forward to reach, respectively, the present configuration of the initiation and the termination sites. The circle is closed through the informational connections between initiation and termination.

The main character that oriented decisively the configuration of the circular model (Figure 6) was the N-end rule. The core of the process of building strings required tandem ligation of the first codes—module 1 (the N-ends of the cores) to the immediate followers—module 2 (the C-ends). Modules 3 and 4 are then added by staggered ligation, extending the N-ends backwards and the C-ends forwards. The attributions are shown in the order of encoding (Table 1). Functional characters of the attributions are in Tables 4 and 5, Figures 6 and 9. The strict module 1 attributions (GPS) form a conserved cluster through all characters: no hydropathy correlation, preferred in non-periodic conformations, in RNA-binding motifs and contributing to protein stabilization. At the encoding of all attributions of the homogeneous sector, amino acids preferred in nonperiodic conformations are completed and a half of the preferred in protein α-helices are added, together with some others preferred in RNA-binding motifs. Amino acids forming preferentially β-strands and DNA-binding motifs are typical of the mixed sector (modules 3 and 4). The components of the NRY quadrant of the mixed sector complete the sets of amino acids preferentially forming RNA-binding motifs and contributing to N-end stabilization. Amino acids that destabilize proteins when residing in the N-ends are all located in module 2 and in the C-end extension.

7.4. Metabolic Perspectives

The self-referential (SR) process, of protein products looping back to stabilize and stimulate the nucleic acids—producers, thereafter memories in RNPs—has already received contributions from diverse sources, beyond the proposition of the SRM [108,109]. Ribosomal proteins belong to the category of Intrinsically Disordered [42], which is a character of the early encodings in the SRM. Recent studies have even detected the possibility of ribosomal proteins being coded for by ribosomal RNA [102], extending and updating our older propositions [99,100] on tRNA-rRNA homologies that are consistent also with [110].

Main functional components in the self-referential processes are obviously the aRSs, whose richness and diversity of activities range much more widely than the mere role in amino acid activation and tRNA charging (e.g., [31]). It may be highly significant to the SRM propositions to find that also some of the known instances of non-ribosomal protein or peptide synthesis (NRPS) [111–114] share aspects showing similarities to what the SRM advocates. [I thank Gustavo Caetano Anollés for bringing this to my attention]. Some proteins involved in NRPS show homologies to synthetases class II, especially the SerRS, but are small and monodomain, not bearing the tRNA binding activity while preserving the amino acid activation. Other instances of proteins active in NRPS show homologies to synthetases class I, closer to the Tyr- and TrpRS, and do not activate amino acids but utilize a couple of aminoacyl-tRNAs to promote sequential reactions for peptide synthesis and cyclo-dehydration. These proteins are dated earlier than the translation machinery [110,115]. The diversity in NRPS is large but the two examples collected here suggest an intriguing analogy with the bipartite structure of tRNAs—one segment for the aminoacylation activity, the other for the anticodon. Tests are to be devised to check whether these possible ´halves´ of the synthetase activities or domains are only fortuitous results of deletions on the present-day genes or might be relics of an evolutionarily process of independent origins of the segments that were joined into the present genes. The evolutionary process of linking pieces might also leave fragile sites for genetic truncation, which will complicate the investigation.

8. The Flow Is the Logic

The scheme for the development of the biological flow system draws a chain of internal sinks (Figure 1) that is initiated at the protein synthesis subsystem. All other pathways converge here, the bioenergetics plus the amino acid, nucleobase, nucleotide and nucleic acid polymer biosynthesis pathways. It reaches an end at the cytoplasmic fission process which is part of cell reproduction. The sink function keeps protein synthesis active, which drains and provides a 'suction force' (enzyme pockets, carrier and receptor sites empty and 'avid'—presenting high affinity—for substrates) that shall not let the system in danger of stagnation or blockade, and of the associated toxicities. Cytoplasmic components should not accumulate to the point of saturation [116] that would inhibit protein synthesis. When some of them are difficult to get rid of, through degradation or extrusion [117–120], cytoplasmic fission is a solution. Some of the waste that cannot be degraded will aggregate into clumps that are allocated to one of the daughter cells, leaving the other healthy and clean [121].

If we consider that the picture described in the sections above answers some questions on the origins of the encoding/decoding system—the origins of biological information [122,123], the most difficult aspect to be rationalized would be on the constitution of the geochemical pre- and proto-biotic context that allowed the encoding system to be installed. After the start, the forward process—building genetic strings—can at the least be sketched with some grounding. With respect to the backward sightings, directly relevant to the triggering conditions for the initial encodings, we can only list some indications derived mostly from the hydropathy correlation and the chronology of encodings.

Considering that the encodings required guaranteed amino acid availability from metabolic pathways, that the chronology pinpointed the pathway of assimilation of C1 units into amino acids, and assuming continuity of the prebiotic with the biologic realms, that is, the latter substituting the former and conserving some records of it, we are allowed to suggest that the steps immediately prior to the installation of the biologic process also included the production of amino acids from C1 units. Such substrates would have been abundant and relatively constant, supporting both prebiotic and biotic processes, which coincide with respect to the pathways of generating C2 compounds. It is a common theme among the abiotic organics the higher concentration of acidic compounds, which include the moderately reactive α-keto acids whose amination dampens their reactivity through the formation of amino acids. These are obtained in meteorites at concentrations in parts-per-million while nucleobases are in the range of parts-per-billion [69,124]. These became an important kind of monomeric substrate for the dimer-directed synthesis of chains, including peptide chains (DDPS). Our focus on the biomass aspect, via Gly and Ser, coincides with the bioenergetic aspect that focuses on AcetylCoA [79,83,87–89]. Our identification of continuity with respect to Glycine seems justified but we cannot indicate the same for Serine since prebiotic concentrations of this are generally low.

It is tempting to propose that the mechanism of dimer-directed synthesis of chains would have had a prebiotic counterpart. A reasonable prebiotic source of oligomers would be the mineral surfaces such as the crystals that compose clays [125]. Stability of the template would create repetitiveness of production which is a result equivalent to the reproduction of the oligomers. Oligomerization of slightly different monomers, such as purines and pyrimidines, or from complementary surfaces of the mother crystal layers, could create the possibility of dimerization sites via complementariness. Our hydropathy correlation data on the first encodings leaves open the question on the character of the dimerized oligomers. They might have been RNA and the lack of correlation due to the poorness of the set of encoded amino acids, but the suspicion stays that they might not have been RNA. The RNA world community has been able to offer bench modes of synthesis of nucleotides that are supposed to be possible to have happened prebiotically [126], but there is no idea of its quantification or localization, while the real observations on chondritic meteorites says of amino acids being recovered in the range of one thousand-fold more than nucleobases [69,124].

In face of the persisting doubts on the feasibility of a pure RNA world—in spite of the beauty in the RNA technology, our studies are plainly consistent with an early RNP world, late DNP. The nucleic acid and the protein component structures would have coevolved. The self-referential process depicted in Figure 2 suggests starting with the formation of RNP globules that have all components held together via protein-protein and protein-RNA binding, not requiring too much from external compartmentalization. The code is mute with respect to membrane structure or composition but it would be adequate to accept the simple proposition that the surface of globules might have been initially proteinic [127] that accommodated lipids around and in between the proteins [128], thereby generating the composite membrane structures.

In the globules, the in versus out distinction would come, respectively, from the internal crowded macromolecular gel structures, adhered to each other, in face of the external fluid. The macromolecules develop intense motility when imbibed in water, more intensely in proteins, less in DNA. The strings are continually challenged by the polar and reactive character of water molecules, hitting either the polar exposed sites of the monomers or the bonds between the monomers of the backbones of the polymers, peptide and phosphodiester, always at the danger of hydrolysis. The trembling at the inside of the spongy structures creates continuous exchanges of materials with the environments. Those that are more necessary in the interior would be trapped there through binding. This process became more easily controllable with the help of the sequestration of some compounds by the developing lipid components.

A main component of the internal drive that keeps the flow always active is the creativity provided for by plasticity [10–12], which is a property most salient in proteins, less in DNA than in RNA. The starting challenge that keeps activity restless would be water and its provocation of hydrolysis which is followed by re-ligation or re-synthesis. These challenges are the sources and triggers of plasticity when they may (1) generate functional novelties, at modification of conformations and activities of the polymers and (2) generate structural and functional novelties, when re-ligation and re-synthesis produce new and different kinds of proteins and genes. Plasticity creates diversity, which are the basic richness of biosystems. The diversity is organized into structures under the influence, among many factors, of the flow itself. The flow is morphogenetic, the structures become compliant with and facilitators of the flow. I recall having heard in the early 1970s from the late David P. Bloch and read from Stuart A. Kauffman [129], respectively, that the life process is an accelerator of the universal evolutionary flow, and that all forms of organization are modes of processing energies, to which we can add many others e.g., [130,131].

9. Coda and Direct Tests

A proposed test for the model, as shown in Figure 2, should not be difficult to the benchwork. Since the prebiotic oligomers proposed to have started the DDPS are not known, a proxy would be the readily available small RNAs than can self-aminoacylate [132,133] or the kinds of mini-tRNAs or mini-helices that have been utilized as substitutes for the acceptor arm of tRNAs [32]. In both cases, the oligomer sets should contain segments or loops that would provide the dimerization ability and tails that would carry the amino acid. The terminal segments should be a main subject of creativity in investigating structures and contexts adequate to facilitate the transferase reaction and mimic the ribosomal, instead of the already observed terminal addition of amino acids. Evolution of the DDPS process could be investigated through utilization of different compositions of the oligomers that will dimerize and of the amino acids offered for acylation and peptide synthesis, plus the observation of the rounds of syntheses and of the activity of precursor-product binding.

Transition to Biological Specificity

In the pre- or proto-biotic realm, the oligomers that would function as substrate carriers, which include amino acids, and would be able to dimerize, the dimers functioning as peptidyl transferase, are considered of unknown constitution. This indication derives from the unique lack of hydropathy correlation between the first set of encoded amino acids (Gly and Ser, module 1) and their correspondent RNA anticodons (Figure 3). They are called pre- or proto-tRNAs only because the present day molecules doing those jobs are the tRNAs in the ribosome. The term does not imply that they were structurally RNA-like but just says that they functionally preceded the tRNAs. The tRNAs would have entered the system in module 2, together with enzyme specificity.

Any incursion into the pre-biotic geochemical realm faces great difficulties in view of the distant time and the lack of fossil evidence. At least two kinds of hypotheses may be envisaged. The RNA World conjecture [16] presents a main problem of requiring the prebiotic existence of the highly complex nucleotides. Phosphate is not easy to obtain [134]. Sugars may be obtained from formaldehyde oligomerization, but the outcome of the reactions is a complex mixture of many kinds of compounds [135]. Nucleobases are obtained in concentrations about one thousand-fold lower than amino acids [69,124]. Composition of the nucleotide structures with the precise RNA-type links would almost certainly require enzyme catalysis, that is, they would have happened in already developed (proto)cells. There is for sure a need for starting with simpler compounds. In this case, any hypothesis that would lead to experimentation would at the same time create the additional problem of how to move from the simpler oligomer to RNA. This transition should ideally be gradual, to attempt to circumvent the problems of another transliteration or 'translation' system, as if reversing what happens in cells.

In view of the difficulties in taking sides with respect to the RNA World proposition and together with the possibility of the lack of hydropathy correlation with the dinucleotides being derived only from the poorness of the encoded amino acid set—just two, probably amidst other amino acids or other compounds polymerized together with them due to catalytic non-specificity—we prefer to say of an unknown constitution of the carriers. We are not committed with the RNA World proposition and cannot offer a justified alternative. Some of the commonly accepted pre-biotic supports and guides for oligomerization reactions, e.g., from mineral surfaces such as clays [125], would be non-specific enough to accept diverse kinds of monomers. Any choice among the possibilities would have to rely upon robust experimentation.

How confident could we be on the validity of the rationale for the origins of the metabolic maze? The path from simple to complex is not to be considered 'the' paradigm of evolution, but it is appealing in view of having been reached here as a result, not having been taken as a premise or assumption. There is apparently no way to get simpler, in both C1 sources and in the simplicity of the C1-C4 Glycine-Serine Cycle. Methylotrophs bearing this pathway and mitochondrial ancestors trace back to the same group, the α-proteobacteria, but this is not considered among extant relatives to primitive or root organisms. Amino acids are precursors to sugars, through gluconeogenesis, and are included in pathways for biosynthesis of nucleobases. Origins of bioenergetics pathways is presently being directed to the autotrophic AcetylCoA pathway, which converges with the Glycine-Serine Cycle core. Key co-factors in these are the (H_4)Pterins, main among these being the MethanoPterin and Folate, while the AcetylCoA pathway is much richer in cofactor requirements. In this possibly highly conflictive area, we remain.

Supplementary Materials: They are available online at www.mdpi.com/2075-1729/7/2/16/s1.

Acknowledgments: To the fruitful discussions with the UNICAMP—CLE group, especially Alfredo Pereira Jr, with the *Complex Cognitio* group of PUC—Belo Horizonte, especially Hugo Mari and Celson Diniz Pereira, and with Savio Torres de Farias, UFPB. To Yoshi Oono, for the suggestion of initiating biologic assimilation with prebiotic C2 instead of C1 compounds. To Gustavo Caetano Anollés, for the indication of incorporating data on non-ribosomal peptide synthesis to our discussions.

Conflicts of Interest: The author declares no conflict of interest.

References

1. Guimarães, R.C. Essentials in the life process indicated by the self-referential genetic code. *Orig. Life Evol. Biosph.* **2014**, *44*, 269–277. [CrossRef] [PubMed]
2. Farias, S.T.; Rêgo, T.; José, M.V. Origin and evolution of the Peptidyl Transferase Center from 160 proto-tRNAs. *FEBS Open Bio.* **2014**, *4*, 175–178. [CrossRef] [PubMed]
3. Tamura, K. Origins and early evolution of the tRNA molecule. *Life* **2015**, *5*, 1687–1699. [CrossRef] [PubMed]

4.   Hordijk, W.; Steel, M. Chasing the tail: The emergence of autocatalytic networks. *Biosystems* **2017**, *152*, 1–10. [CrossRef] [PubMed]

5.   Guimarães, R.C. Mutuality in discrete and compositional information: Perspectives for synthetic genetic codes. *Cogn. Comput.* **2012**, *4*, 115–139. [CrossRef]

6.   Vitas, M.; Dobovisek, A. On a quest for reverse translation. *Found. Chem.* **2016**, 1–17. [CrossRef]

7.   Moras, D.; Dock, A.C.; Dumas, P.; Westhof, E.; Romby, P.; Ebel, J.-P.; Giegé, R. Anticodon-anticodon interaction induces conformational changes in tRNA: Yeast tRNA^{ASP}, a model for tRNA-mRNA recognition. *Proc. Natl. Acad. Sci. USA* **1986**, *83*, 932–936. [CrossRef] [PubMed]

8.   Pascal, R.; Pross, A. The logic of life. *Orig. Life Evol. Biosph.* **2016**, *46*, 507–513. [CrossRef] [PubMed]

9.   Ptashne, M. Epigenetics: Core misconcept. *Proc. Nat. Acad. Sci. USA* **2013**, *110*, 7101–7103. [CrossRef] [PubMed]

10. Des Marais, D.L.; Hernandez, K.M.; Juenger, T.E. Genotype-by-environment interaction and plasticity: Exploring genomic responses of plants to the abiotic environment. *Annu. Rev. Ecol. Evol. Syst.* **2013**, *44*, 5–29. [CrossRef]

11. Kenkel, C.D.; Matz, M.V. Gene expression plasticity as a mechanism of coral adaptation to a variable environment. *Nat. Ecol. Evol.* **2016**, *1*, 0014. [CrossRef]

12. Murren, C.J.; Auld, J.R.; Callahan, H.; Ghalambor, C.K.; Handelsman, C.A.; Heskel, M.A.; Kingsolver, J.G.; Maclean, H.J.; Masel, J.; Maughan, H.; et al. Constraints on the evolution of phenotypic plasticity: Limits and costs of phenotype and plasticity. *Heredity* **2015**, *115*, 293–301. [CrossRef] [PubMed]

13. Jackson, J.B. Natural pH gradients in hydrothermal alkali vents were unlikely to have played a role in the origin of life. *J. Mol. Evol.* **2016**, *83*, 1–11. [CrossRef] [PubMed]

14. Wächtershäuser, G. In praise of error. *J. Mol. Evol.* **2016**, *82*, 75–80. [CrossRef] [PubMed]

15. Füllsack, M. Circularity and the micro-macro difference. *Constructivist Foundations* **2016**, *12*, 1.

16. Lehman, N. The RNA world: 4,000,000,050 years old. *Life* **2015**, *5*, 1583–1586. [CrossRef] [PubMed]

17. Guimarães, R.C. Formation of the Genetic Code—Review and Update as of November 2012. 2013. Available online: http://www.icb.ufmg.br/labs/lbem/pdf/GMRTgeneticodeNov12.pdf. All original publications are available online: https://www.researchgate.net/profile/Romeu_Guimaraes (both sites accessed on December 2016).

18. Caetano-Anollés, G.; Wang, M.; Caetano-Anollés, D. Structural phylogenomics retrodicts the origin of the genetic code and uncovers the evolutionary impact of protein flexibility. *PLoS ONE* **2013**, *8*, e72225. [CrossRef] [PubMed]

19. Gulik, P.T.S.; Hoff, W.D. Anticodon modifications in the tRNA set of LUCA and the fundamental regularity in the standard genetic code. *PLoS ONE* **2016**, *11*, e0158342.

20. Lacey, J.C., Jr.; Mullins, D.W., Jr. Experimental studies related to the origin of the genetic code and the process of protein synthesis—A review. *Orig. Life Evol. Biosph.* **1983**, *13*, 3–42. [CrossRef]

21. Osawa, S. *Evolution of the Genetic Code*; Oxford University Press: Oxford UK, 1995.

22. Xia, T.; SantaLucia, J., Jr.; Burkard, M.E.; Kierzek, R.; Schroeder, S.J.; Jiao, X.; Cox, C.; Turner, D.H. Thermodynamic parameters for an expanded nearest-neighbor model for formation of RNA duplexes with Watson-Crick base pairs. *Biochemistry* **1998**, *37*, 14719–14735. [CrossRef] [PubMed]

23. Farias, S.T.; Moreira, C.H.C.; Guimarães, R.C. Structure of the genetic code suggested by the hydropathy correlation between anticodons and amino acid residues. *Orig. Life Evol. Biosph.* **2007**, *37*, 83–103. [CrossRef] [PubMed]

24. Guimarães, R.C. Anti-complementary order in the genetic coding system. *Int. Conf. Orig. Life* **1996**, *26*, 435–436.

25. Egholm, M.; Buchardt, O.; Nielsen, P.E.; Berg, R.H. Peptide nucleic acids (PNA). Oligonucleotide analogs with an achiral backbone. *J. Am. Chem. Soc.* **1992**, *114*, 1895–1897. [CrossRef]

26. Francis, B.R. The Hypothesis that the Genetic Code Originated in Coupled Synthesis of Proteins and the Evolutionary Predecessors of Nucleic Acids in Primitive Cells. *Life* **2015**, *5*, 467–505. [CrossRef] [PubMed]

27. Bernhardt, H.S. Clues to tRNA evolution from the distribution of class II tRNAs and serine codons in the genetic code. *Life* **2016**, *6*, 10. [CrossRef] [PubMed]

28. Rogozin, I.B.; Belinsky, F.; Pavlenko, V.; Shabalina, S.A.; Kristensen, D.M.; Koonin, E.V. Evolutionary switches between two serine codon sets are driven by selection. *Proc. Natl. Acad. Sci. USA* **2016**, *113*, 13109–13113. [CrossRef] [PubMed]

29. Johansson, M.; Zhang, J.; Ehrenberg, M. Genetic code translation displays a linear trade-off between efficiency and accuracy of tRNA selection. *Proc. Natl. Acad. Sci. USA* **2012**, *109*, 131–136. [CrossRef] [PubMed]

30. Guimarães, R.C.; Erdmann, V.A. Evolution of adenine clustering in 5S ribosomal RNA. *Endocyt. Cell Res.* **1992**, *9*, 13–45.

31. Lee, E.Y.; Lee, H.C.; Kim, H.K.; Jang, S.Y.; Park, S.J.; Kim, Y.H.; Kim, J.H.; Hwang, J.; Kim, J.H.; Kim, T.H.; et al. Infection-specific phosphorylation of glutamyl-prolyl tRNA synthetase induces antiviral immunity. *Nat. Immunol.* **2016**, *17*, 1252–1262. [CrossRef] [PubMed]

32. Beuning, P.J.; Musier-Forsyth, K. Transfer RNA recognition by aminoacyl-tRNA synthetases. *Biopolymers* **1999**, *52*, 1–28. [CrossRef]

33. Jackman, J.E.; Alfonzo, J.D. Transfer RNA modifications: Nature's combinatorial chemistry playground. *WIREs RNA* **2013**, *4*, 35–48. [CrossRef] [PubMed]

34. A Database of RNA Modifications. Available online: modomics.genesilico.pl/modifications (accessed on 10 November 2016).

35. Seligmann, H. Natural mitochondrial proteolysis confirms transcription systematically exchanging/eleting nucleotides, peptides coded by expanded codons. *J. Theor. Biol.* **2017**, *414*, 76. [CrossRef] [PubMed]

36. Lim, V.I. Analysis of action of wobble nucleoside modification on codon-anticodon pairing within the ribosome. *J. Mol. Biol.* **1994**, *240*, 8–19. [CrossRef] [PubMed]

37. Tzul, F.O.; Vasilchuk, D.; Makhatadze, G.I. Evidence for the principle of minimal frustration in the evolution of protein folding landscapes. *Proc. Natl. Acad. Sci. USA* **2017**, E1627–E1632. [CrossRef] [PubMed]

38. Marck, C.; Grosjean, H. tRNomics: Analysis of tRNA genes from 50 genomes of Eukarya, Archaea, and Bacteria reveals anticodon-sparing strategies and domain-specific features. *RNA* **2002**, *8*, 1189–1232. [CrossRef] [PubMed]

39. Targanski, I.; Cherkasova, V. Analysis of genomic tRNA sets from Bacteria, Archaea and Eukarya points to anticodon-codon hydrogen bonds as a major determinant of tRNA compositional variation. *RNA* **2008**, *14*, 1095–1109. [CrossRef] [PubMed]

40. Agris, P.F. Decoding the genome: A modified view. *Nucleic Acids Res.* **2004**, *32*, 223–238. [CrossRef] [PubMed]

41. Creighton, T.E. *Proteins: Structures and Molecular Properties*; WH Freeman: New York, NY, USA, 1993.

42. Radivojac, P.; Iakoucheva, L.M.; Oldfield, C.J.; Obradovic, Z.; Uversky, V.N.; Dunker, A.K. Intrinsic disorder and functional proteomics. *Biophys. J.* **2007**, *92*, 1439–1456. [CrossRef] [PubMed]

43. Dagliyan, O.; Tarnawski, M.; Chu, D.H.; Shirvanyants, D.; Schlichting, I.; Dokholyan, N.V.; Hahn, K.M. Engineering extrinsic disorder to control protein activity in living cells. *Science* **2016**, *354*, 1441–1444. [CrossRef] [PubMed]

44. Gruszka, D.T.; Mendonça, C.A.T.F.; Paci, E.; Whelan, F.; Hawkhead, J.; Potts, J.R.; Clarke, J. Disorder drives cooperative folding in a multidomain protein. *Proc. Natl. Acad. Sci. USA* **2016**, *113*, 11841–11846. [CrossRef] [PubMed]

45. Tuite, M.F. Remembering the past: A new form of protein-based inheritance. *Cell* **2016**, *167*, 302–303. [CrossRef] [PubMed]

46. Guimarães, R.C.; Moreira, C.H.C. Genetic code—A self-referential and functional model. In *Progress in Biological Chirality*; Pályi, G., Zucchi, C., Caglioti, L., Eds.; Elsevier: Oxford, UK, 2004; pp. 83–118.

47. Varschavsky, A. The N-end rule: Functions, mysteries, uses. *Proc. Natl. Acad. Sci. USA* **1996**, *93*, 12142–12149. [CrossRef]

48. Meinnel, T.; Sereno, A.; Giglione, C. Impact of the N-terminal amino acid on targeted protein degradation. *Biol. Chem.* **2006**, *387*, 839–851. [CrossRef] [PubMed]

49. Wang, K.H.; Hernandez, G.R.; Grant, R.A.; Sauer, R.T.; Baker, T.A. The molecular basis of N-end rule recognition. *Mol. Cell* **2008**, *32*, 406–414. [CrossRef] [PubMed]

50. Kim, M.K.; Oh, S.J.; Lee, B.G.; Song, H.K. Structural basis for dual specificity of yeast N-terminal amidase in the N-end rule pathway. *Proc. Natl. Acad. Sci. USA* **2016**, *113*, 12438–12443. [CrossRef] [PubMed]

51. Berezovsky, I.N.; Kilosanidze, G.T.; Tumanyan, V.G.; Kisselev, L.L. Amino acid composition of protein termini are biased in different manners. *Prot. Eng.* **1999**, *12*, 23–30. [CrossRef]

52. Guimarães, R.C. Two punctuation systems in the genetic code. In *First Steps in the Origin of Life in the Universe*; Chela-Flores, J., Owen, T., Raulin, F., Eds.; Kluwer: Dordrecht, The Netherlands, 2001; pp. 121–124.

53. Chen, L.L. The biogenesis and emerging roles of circular RNAs. *Nature Rev. Mol. Cell Biol.* **2016**, *17*, 205–211. [CrossRef] [PubMed]

54. Beier, H.; Grimm, M. Misreading of termination codons in eukaryotes by natural nonsense suppressor tRNAs. *Nucl. Acids Res.* **2001**, *29*, 4767–4782. [CrossRef] [PubMed]

55. Roy, B.; Friesen, W.J.; Tomizawa, Y.; Leszyk, J.D.; Zhuo, J.; Johnson, B.; Dakka, J.; Trotta, C.R.; Xue, X.; Mutyam, V.; et al. Ataluren stimulates ribosomal selection of near cognate tRNAs to promote nonsense suppression. *Proc. Natl. Acad. Sci. USA* **2016**, *113*, 12508–12513. [CrossRef] [PubMed]

56. Chen, Y.; Ma, J.; Lu, W.; Tian, M.; Thauvin, M.; Yuan, C.; Volovitch, M.; Wang, Q.; Holst, J.; Liu, M.; et al. Heritable expansion of the genetic code in mouse and zebrafish. *Cell Res.* **2016**, *27*, 294. [CrossRef] [PubMed]

57. Eltschinger, S.; Bütikofer, P.; Altmann, M. Translation elongation and termination—Are they conserved processes? In *Evolution of the Protein Synthesis Machinery and Its Regulation*; Hernández, G., Jagus, R., Eds.; Springer: Cham, Switzerland, 2016; pp. 277–311.

58. Ma, C.; Kurita, D.; Li, N.; Chen, Y.; Himeno, H.; Gao, N. Mechanistic insights into the alternative termination by ArfA and RF2. *Nature* **2016**, *541*, 550. [CrossRef] [PubMed]

59. Petrov, A.S.; Gulen, B.; Norris, A.M.; Kovacs, N.A.; Bernier, C.R.; Lanier, K.A.; Fox, G.E.; Harvey, S.C.; Wartel, R.M.; Hud, N.V.; et al. History of the ribosome and the origins of translation. *Proc. Natl. Acad. Sci. USA* **2015**, *112*, 15396–15401. [CrossRef] [PubMed]

60. Colussi, T.M.; Constantino, D.A.; Hammond, J.A.; Ruehle, G.M.; Nix, J.C.; Kieft, J.S. The structural basis of tRNA mimicry and conformational plasticity by a viral RNA. *Nature* **2014**, *511*, 366–369. [CrossRef] [PubMed]

61. Loc'h, J.; Bloud, M.; Réty, S.; Lebaron, S.; Deschamps, P.; Barreille, J.; Jombart, J.; Paganin, J.R.; Delbos, L.; Chardon, F.; et al. RNA mimicry by the Fap7 adenylate kinase in ribosome biogenesis. *PLoS Biol.* **2014**, *12*, e1001860. [CrossRef] [PubMed]

62. Marzi, S.; Romby, P. RNA mimicry, a decoy for regulatory proteins. *Mol. Microbiol.* **2012**, *83*, 1–6. [CrossRef] [PubMed]

63. Nakamura, Y.; Ito, K. tRNA mimicry in translation termination and beyond. *WIREs RNA* **2011**, *2*, 647–668. [CrossRef] [PubMed]

64. Suetsuzu, K.H.; Sekine, S.I.; Sakai, H.; Takemoto, C.H.; Terada, T.; Unzai, S.; Tame, J.R.H.; Kuramitsu, S.; Shirouzu, M.; Yokoyama, S. Crystal structure of elongation factor P from *Thermus thermophilus* HB8. *Proc. Natl. Acad. Sci. USA* **2004**, *101*, 9595–9600.

65. James, N.R.; Brown, A.; Gordiyenko, Y.; Ramakrishnan, V. Translational termination without a stop codon. *Science* **2016**, *354*, 1437–1440. [CrossRef] [PubMed]

66. Zinshteyn, B.; Green, R. When stop makes sense. *Science* **2016**, *354*, 1106. [CrossRef] [PubMed]

67. Anderson, R.M.; Kwon, M.; Strobel, S.A. Toward ribosomal RNA catalytic activity in the absence of protein. *J. Mol. Evol.* **2007**, *64*, 472–483. [CrossRef] [PubMed]

68. Voorhees, R.M.; Weixlbaumer, A.; Loakes, D.; Kelley, A.C.; Ramakrishnan, V. Insights into substrate stabilization from snapshots of the peptidyl transferase center of the intact 70S ribosome. *Nat. Struct. Mol. Biol.* **2009**, *16*, 528–533. [CrossRef] [PubMed]

69. Cooper, G.; Reed, C.; Nguyen, D.; Carter, M.; Wang, Y. Detection and formation scenario of citric acid, pyruvic acid, and other possible metabolism precursors in carbonaceous meteorites. *Proc. Natl. Acad. Sci. USA* **2011**, *108*, 14015–14020. [CrossRef] [PubMed]

70. Di Giulio, M. An extension of the coevolution theory of the origin of the genetic code. *Biol. Direct* **2008**, *3*, 37. [CrossRef] [PubMed]

71. Di Giulio, M. An autotrophic origin of the coded amino acids is concordant with the coevolution theory of the genetic code. *J. Mol. Evol.* **2016**. [CrossRef] [PubMed]

72. Hartman, H.; Smith, T.F. The evolution of the ribosome and the genetic code. *Life* **2014**, *4*, 227–249. [CrossRef] [PubMed]

73. Higgs, P.G. A four-column theory for the origin of the genetic code: Tracing the evolutionary pathways that gave rise to an optimized code. *Biol. Direct* **2009**, *4*, 16. [CrossRef] [PubMed]

74. Ikehara, K. Evolutionay steps in the emergence of life deduced from the bottom-up approach and 'GADV hypothesis (top-down approach)'. *Life* **2016**, *6*, 6. [CrossRef] [PubMed]

75. Sengupta, S.; Higgs, P.G. Pathways of genetic code evolution in ancient and modern organisms. *J. Mol. Evol.* **2015**, *80*, 229–243. [CrossRef] [PubMed]

76. Trifonov, E.N. Consensus temporal order of amino acids and evolution of the triplet code. *Gene* **2000**, *261*, 139–151. [CrossRef]

77. Trifonov, E.N. The triplet code from first principles. *J. Biomol. Struct. Dyn.* **2004**, *22*, 1–11. [CrossRef] [PubMed]

78. Wong, J.T.F.; Ng, S.K.; Mat, W.K.; Hu, T.; Xue, H. Coevolution theory of the genetic code at age forty: Pathway to translation and synthetic life. *Life* **2016**, *6*, 12. [CrossRef] [PubMed]

79. Guimarães, R.C. Metabolic basis for the self-referential genetic code. *Orig. Life Evol. Biosph.* **2011**, *41*, 357–371. [CrossRef] [PubMed]

80. Chistoserdova, L.; Kalyuzhnaya, M.G.; Lidstrom, M.E. The expanding world of methylotrophic metabolism. *Annu. Rev. Microbiol.* **2009**, *63*, 477–499. [CrossRef] [PubMed]

81. Chistoserdova, L. Modularity of methylotrophy, revisited. *Environ. Microbiol.* **2011**, *13*, 2603–2622. [CrossRef] [PubMed]

82. Florio, R.; Salvo, M.L.; Vivoli, M.; Contestabile, R. Serine hydroxymethyltransferase: A model enzyme for mechanistic, structural, and evolutionary studies. *Biochim. Biophys. Acta* **2011**, *1814*, 1489–1496. [CrossRef] [PubMed]

83. Braakman, R.; Smith, E. The emergence and early evolution of biological carbon-fixation. *PLoS Comput. Biol.* **2012**, *8*, e1002455. [CrossRef] [PubMed]

84. Xiong, W.; Lin, P.P.; Magnusson, L.; Warner, L.; Liao, J.C.; Maness, P.C.; Chou, K.J. CO2-fixing one-carbon metabolism in a cellulose-degrading bacterium *Clostridium thermocellum*. *Proc. Natl. Acad. Sci. USA* **2016**, *113*, 13180–13185. [CrossRef] [PubMed]

85. Kottakis, F.; Nicolay, B.N.; Roumane, A.; Karnik, R.; Gu, H.; Nagle, J.M.; Boukhali, M.; Hayward, M.C.; Li, Y.Y.; Chen, T.; et al. LKB1 loss links serine metabolism to DNA methylation and tumorigenesis. *Nature* **2016**, *539*, 390–395. [CrossRef] [PubMed]

86. Wood, A.P.; Aurikko, J.P.; Kelly, D.P. A challenge for the 21st century molecular biology and biochemistry: What are the causes of obligate autotrophy and methanotrophy? *FEMS Microbiol. Rev. RNA* **2004**, *28*, 335–352. [CrossRef]

87. Russell, M.J.; Martin, W. The rocky roots of the acetyl-CoA pathway. *Trends Bioch. Sci.* **2004**, *29*, 358–363. [CrossRef] [PubMed]

88. Weiss, M.C.; Neukirchen, S.; Roettger, M.; Mrujavac, N.; Sathi, S.N.; Martin, W.F.; Sousa, F.L. Reply to 'Is LUCA a thermophilic progenote?'. *Nat. Microbiol.* **2016**, *1*, 16230. [CrossRef] [PubMed]

89. Huber, C.; Wächtershäuser, G. Activated acetic acid by carbon fixation on (Fe, Ni)S under primordial conditions. *Science* **1997**, *276*, 245–247. [CrossRef] [PubMed]

90. Holm, N.G.; Oze, C.; Mousis, O.; Waite, J.H.; Lepoutre, A.G. Serpentinization and the formation of H2 and CH4 on celestial bodies (planets, moons, comets). *Astrobiology* **2015**, *15*, 587–600. [CrossRef] [PubMed]

91. Grosjean, H.; Westhof, E. An integrated, structure- and energy-based view of the genetic code. *Nucleic Acids Res.* **2016**, *44*, 8020–8040. [CrossRef] [PubMed]

92. Grantham, R. Amino acid difference formula to help explain protein evolution. *Science* **1974**, *185*, 862–864. [CrossRef] [PubMed]

93. Sobolevsky, Y.; Guimarães, R.C.; Trifonov, E.N. Towards functional repertoire of the earliest proteins. *J. Biomol. Struct. Dyn.* **2013**, *31*, 1293–1300. [CrossRef] [PubMed]

94. Sprinzl, M. Chemistry of aminoacylation and peptide bond formation on the 3′ terminus of tRNA. *J. Biosci.* **2006**, *311*, 489–496. [CrossRef]

95. Ambrogelly, A.; Söll, D.; Nureki, O.; Yokoyama, S.; Ibba, M. Class I lysyl-tRNA synthetases. NCBI Bookshelf. Landes Bioscience: Madame Curie Bioscence Database: Austin TX, USA, 2013. Available online: http://www.ncbi.nlm.nih.gov/books/NBK6444 (accessed on December 2016).

96. Xu, X.M.; Carlson, B.A.; Mix, H.; Zhang, Y.; Saira, K.; Glass, R.S.; Barry, M.J.; Gladyshev, V.N.; Hatfield, D.L. Biosynthesis of selenocysteine on its tRNA in eukaryotes. *PLoS Biol.* **2007**, *5*, e4. [CrossRef] [PubMed]

97. Englert, M.; Moses, S.; Hohn, M.; Ling, J.; O'Donohue, P.; Söll, D. Aminoacylation of tRNA 2′- or 3′-hydroxyl by phosphoseryl- and pyrrolysyl-tRNA synthetases. *FEBS Lett.* **2013**, *587*, 3360–3364. [CrossRef] [PubMed]

98. Rodriguez, L.M.; Erdogan, O.; Rodriguez, M.J.; Rivera, K.G.; Williams, T.; Li, L.; Weinreb, V.; Collier, M.; Chandrasekaran, S.N.; Ambrozzio, X.; et al. Functional class I and class II amino acid activating enzymes can be coded by opposite strands of the same gene. *J. Biol. Chem.* **2015**, *290*, 19710. [CrossRef] [PubMed]

99. Bloch, D.P.; McArthur, B.; Widdowson, R.; Spector, D.; Guimarães, R.C.; Smith, J. tRNA-rRNA sequence homologies: A model for the generation of a common ancestral molecule and prospects for its reconstruction. *Orig. Life Evol. Biosph.* **1984**, *14*, 571–578. [CrossRef]

100. Bloch, D.P.; McArthur, B.; Guimarães, R.C.; Smith, J.; Staves, M.P. tRNA-rRNA sequence matches from inter- and intraspecies comparisons suggest common origins for the two RNAs. *Braz. J. Med. Biol. Res.* **1989**, *22*, 931–944. [PubMed]

101. Kanai, A. Disrupted tRNA genes and tRNA fragments: A perspective on tRNA gene evolution. *Life* **2015**, *5*, 321–331. [CrossRef] [PubMed]

102. Root-Bernstein, R.; Root-Bernstein, M. The ribosome as a missing link in prebiotic evolution II: Ribosomes encode ribosomal proteins that bind to common regions of their own mRNAs and rRNAs. *J. Theor. Biol.* **2016**, *397*, 115–127. [CrossRef] [PubMed]

103. Macé, K.; Gillet, R. Origins of tmRNA: The missing link in the birth of protein synthesis? *Nucl. Acids Res.* **2011**, *44*, 8041–8051. [CrossRef] [PubMed]

104. Ogle, J.M.; Brodersen, D.E.; Clemens Jr, W.M.; Tarry, M.J.; Carter, A.P.; Ramakrishnan, C.V. Recognition of cognate transfer RNA by the 30S ribosomal subunit. *Science* **2001**, *293*, 897–902. [CrossRef] [PubMed]

105. Lehmann, J.; Libchaber, A. Degeneracy of the genetic code and stability of the base pair at the second position of the anticodon. *RNA* **2008**, *14*, 1264–1269. [CrossRef] [PubMed]

106. Massey, S.E. The neutral emergence of error minimized genetic codes superior to the standard genetic code. *J. Theor. Biol.* **2016**, *408*, 237–242. [CrossRef] [PubMed]

107. Guimarães, R.C. The Self-Referential Genetic Code is Biologic and Includes the Error Minimization Property. *Orig. Life Evol. Biosph.* **2015**, *45*, 69–75. [CrossRef] [PubMed]

108. Guimarães, R.C.; Moreira, C.H.C.; Farias, S.T. A self-referential model for the formation of the genetic code. *Theory Biosci.* **2008**, *127*, 249–270. [CrossRef] [PubMed]

109. Guimarães, R.C.; Moreira, C.H.C.; Farias, S.T. Self-referential formation of the genetic system. In *The Codes of Life—The Rules of Macroevolution*; Barbieri, M., Ed.; Springer: Dordrecht, The Netherlands, 2008; pp. 68–110.

110. Caetano-Anollés, D.; Caetano-Anollés, G. Piecemeal buildup of the genetic code, ribosomes, and genomes from primordial tRNA building blocks. *Life* **2016**, *6*, 43. [CrossRef] [PubMed]

111. Fung, A.W.S.; Payoe, R.; Fahlman, R.D. Perspectives and insights into the competition for aminoacyl-tRNAs between the translational machinery and for tRNA-dependent non-ribosomal peptide bond formation. *Life* **2016**, *6*, 2. [CrossRef] [PubMed]

112. Goudry, M.; Saugnet, L.; Belin, P.; Thai, R.; Amoureux, R.; Tellier, C.; Tuphile, K.; Jacqet, M.; Braud, S.; Courçon, M.; et al. Cyclodipeptide synthases are a family of tRNA-dependent peptide bond-forming enzymes. *Nat. Chem. Biol.* **2009**, *5*, 414–420.

113. Mocibob, M.; Ivic, N.; Bilokapic, S.; Maier, T.; Luic, M.; Ban, N.; Durasevic, I.W. Homologs of aminoacyl-tRNA synthetases acylate carrier proteins and provide a link between ribosomal and nonribosomal peptide synthesis. *Proc. Natl. Acad. Sci. USA* **2010**, *107*, 14585–14590. [CrossRef] [PubMed]

114. Moutiez, M.; Schmidt, E.; Seguin, J.; Thai, R.; Favry, E.; Belin, P.; Mechulan, Y.; Goudry, M. Unravelling the mechanism of non-ribosomal peptide synthesis by cyclopeptide synthases. *Nat. Commun.* **2014**, *5*, 5141–5146. [CrossRef] [PubMed]

115. Caetano-Anollés, D.; Caetano-Anollés, G. The phylogenomic roots of translation. In *Evolution of the Protein Synthesis Machinery and its Regulation*; Hernandez, G., Jagus, R., Eds.; Springer: Cham, Switzerland, 2016; pp. 9–30.

116. Dill, K.A.; Ghosh, K.; Schmit, J.D. Physical limits of cells and proteomes. *Proc. Natl. Acad. Sci. USA* **2011**, *108*, 17876–17882. [CrossRef] [PubMed]

117. Deatherage, B.L.; Cookson, B.T. Membrane vesicle release in bacteria, eukaryotes, and archaea: A conserved yet underappreciated aspect of microbial life. *Infect. Immun.* **2012**, *80*, 1948–1957. [CrossRef] [PubMed]

118. Errington, J. L-form bacteria, cell walls and the origins of life. *Open Biol.* **2013**, *3*, 120–143. [CrossRef] [PubMed]

119. Makarova, K.S.; Yutin, N.; Bell, S.D.; Koonin, E.V. Evolution of diverse cell division and vesicle formation systems in archaea. *Nat. Rev. Microbiol.* **2010**, *8*, 731–741. [CrossRef] [PubMed]

120. Mercier, R.; Kawai, Y.; Errington, J. Excess membrane synthesis drives a primitive mode of cell proliferation. *Cell* **2013**, *152*, 997–1007. [CrossRef] [PubMed]

121. Nyström, T. Spatial protein quality control and the evolution of lineage-specific ageing. *Philos. Trans. R. Soc. Lond. B* **2011**, *366*, 71–75. [CrossRef] [PubMed]

122. Guimarães, R.C. Emergence of information patterns: In the quantum and biochemical realms. *Quantum Biosyst.* **2015**, *6*, 148–159.

123. Wills, P.R. The generation of meaningful information in molecular systems. *Philos. Trans. R. Soc. A* **2016**, *374*, 20150066. [CrossRef] [PubMed]

124. Callahan, M.P.; Smith, K.E.; Cleaves, H.J., 2nd; Ruzicka, J.; Stern, J.C.; Glavin, D.P.; House, C.H.; Dworkin, J.P. Carbonaceous meteorites contain a wide range of extraterrestrial nucleobases. *Proc. Natl. Acad. Sci. USA* **2011**, *108*, 13995–13998. [CrossRef] [PubMed]

125. Ertem, G.; O'Brien, A.M.S.; Ertem, M.C.; Rogoff, D.A.; Dworkin, J.P.; Johnston, M.V.; Hazen, R.M. Abiotic formation of RNA-like oligomers by montmorillonite catalysis: Part II. *Int. J. Astrobiol.* **2008**, *7*, 1–7. [CrossRef]

126. Powner, M.W.; Gerland, B.; Sutherland, J.D. Synthesis of activated pyrimidine ribonucleotides in prebiotically plausible conditions. *Nature* **2009**, *459*, 239–242. [CrossRef] [PubMed]

127. Santoso, S.; Hwang, W.; Hartman, H.; Zhang, S.G. Self-assembly of surfactant-like peptides with variable glycine tails to form nanotubes and nanovesicles. *Nano Lett.* **2002**, *2*, 687–691. [CrossRef]

128. Rizzotti, M. *Early Evolution*; Birkhäuser: Basel, Switzerland, 2000.

129. Kauffman, S.A. *The Origins of Order: Self-Organization and Selection in Evolution*; Oxford University Press: Oxford, UK, 1993.

130. Russell, M. Thinking about life: Adding (thermo)dynamic aspects to definitions of life. In *Evolution and Transitions in Complexity—The Science of Hierarchical Organization in Nature*; Springer: Cham, Switzerland, 2016; pp. 199–202.

131. Morowitz, H.; Smith, E. Energy flow and the organization of life. *Complexity* **2007**, *13*, 51–59. [CrossRef]

132. Illangasekare, M.; Yarus, M. A tiny RNA that catalyzes both aminoacyl-RNA and peptidyl-RNA synthesis. *RNA* **1999**, *5*, 1482–1489. [CrossRef] [PubMed]

133. Yarus, M. The meaning of a minuscule ribozyme. *Philos. Trans. R. Soc. Lond. B* **2011**, *366*, 2902–2909. [CrossRef] [PubMed]

134. Goldford, J.E.; Hartman, H.; Smith, T.F.; Segrè, D. Remnants of an ancient metabolism without phosphate. *Cell* **2017**, *168*, 1126–1134. [CrossRef] [PubMed]

135. Ricardo, A.; Carrigan, M.A.; Olcott, A.N.; Benner, S.A. Borate Minerals Stabilize Ribose. *Science* **2004**, *303*, 196. [CrossRef] [PubMed]

Concept Paper

Frozen Accident Pushing 50: Stereochemistry, Expansion, and Chance in the Evolution of the Genetic Code

Eugene V. Koonin

National Center for Biotechnology Information, National Library of Medicine, National Institutes of Health, Bethesda, MD 20894, USA; koonin@ncbi.nlm.nih.gov; Tel.: +1-301-435-5913

Academic Editor: Koji Tamura
Received: 10 March 2017; Accepted: 20 May 2017; Published: 23 May 2017

Abstract: Nearly 50 years ago, Francis Crick propounded the frozen accident scenario for the evolution of the genetic code along with the hypothesis that the early translation system consisted primarily of RNA. Under the frozen accident perspective, the code is universal among modern life forms because any change in codon assignment would be highly deleterious. The frozen accident can be considered the default theory of code evolution because it does not imply any specific interactions between amino acids and the cognate codons or anticodons, or any particular properties of the code. The subsequent 49 years of code studies have elucidated notable features of the standard code, such as high robustness to errors, but failed to develop a compelling explanation for codon assignments. In particular, stereochemical affinity between amino acids and the cognate codons or anticodons does not seem to account for the origin and evolution of the code. Here, I expand Crick's hypothesis on RNA-only translation system by presenting evidence that this early translation already attained high fidelity that allowed protein evolution. I outline an experimentally testable scenario for the evolution of the code that combines a distinct version of the stereochemical hypothesis, in which amino acids are recognized via unique sites in the tertiary structure of proto-tRNAs, rather than by anticodons, expansion of the code via proto-tRNA duplication, and the frozen accident.

Keywords: genetic code evolution; frozen accident; stereochemical theory; coevolution theory; error minimization; RNA world; proto-tRNAs

1. Introduction

The time of this writing, early 2017, falls between two notable dates, the 100th anniversary of Francis Crick's birth and the 50th anniversary of Crick's 1968 classic paper on the evolution of the genetic code [1,2]. Compared to Crick's momentous contribution to the understanding of DNA structure and replication [3,4], and then the principles of protein translation [5,6], the code evolution paper might seem to be almost inconsequential. Yet, this masterpiece of conceptual thinking presents two ideas that have shaped the subsequent developments in the study of the code evolution and more generally, the study of the early evolution of life. These ideas are the frozen accident perspective on the code evolution and the inference of a RNA-only translation system.

The genetic code that defines the rules of translation from the 4-letter nucleic acid alphabet to the 20-letter alphabet of proteins is arguably the single central informational invariant of all life forms [6–9]. Indeed, notwithstanding multiple code variants that continue to emerge through the study of protein coding in diverse life forms, the basic structure of the code and the majority of the codon assignments are truly universal [10,11]. When the codon table was settled in 1965, distinct patterns in the code begging for explanation became immediately apparent [9,12]. The 64 triplet codons are neatly organized in sets of four or two, with the third base of a codon typically being synonymous. The assignment of codons to amino acids across the code table is clearly non-random: related amino acids typically occupy contiguous areas in the table. The second position of a codon is the most important specificity determinant, and three of the four columns of the table encode related, chemically similar amino acids. For example, all codons with a U in the second position correspond to hydrophobic amino acids. It is obvious from the table itself that the code is robust to error, i.e., mutational and translation errors in synonymous positions (typically, the third position in a codon) have no effect on the protein, whereas substitutions in the first position typically lead to incorporation of an amino acid similar to the correct one, thus decreasing the damage [9]. Quantitative analyses of the code using cost functions derived from physico-chemical properties of amino acids or their evolutionary exchangeability have confirmed the exceptional robustness of the standard genetic code (SGC): the probability to reach the same level of error minimization as in the SGC by random permutation of codons is below 10^{-6} [10,13–15]. However, the SGC is far from being optimal because, given the enormous overall number of possible codes ($>10^{84}$), billions of variants are even more robust to error [10].

During the 49 years since the publication of Crick's code evolution paper, extensive experimental and theoretical research effort had aimed to develop a definitive scenario for the evolution of the code that would account for its notable properties. Even if rarely coached that way, these studies can be considered as attempts on falsification of the frozen accident scenario. Although the features of the code have been explored in detail and some aspects of its evolution seem to have been elucidated, these efforts appear to fall short of a compelling refutation of the frozen accident.

In this concept article, I briefly review the current understanding of the factors that determined the universality of the SGC and the three main scenarios of the code evolution. I then sketch a new scenario for code evolution, informed by comparative genomic analyses, that combines the idea of a primordial stereochemcial code with the frozen accident perspective. As a disclaimer, I should note that no attempt is made here on anything close to a comprehensive review of the research on the code origin and evolution let alone the origin of life. The goal is to place the frozen accident concept into the context of latest efforts in the field and briefly discuss some new ideas.

2. Why the Universal Code?

The genetic code is nearly universal in all extant life forms although limited deviations from the SGC have been detected in many groups of organisms, particularly in organelles and parasitic or endosymbiotic bacteria with highly reduced genomes [11,16,17]. The changes to the SGC follow three distinct patterns: (i) reassignment of codons within the canonical set of 21 (including the stop signal); (ii) loss ("unassignment") of codons, and (iii) incorporation of new amino acids. Stop codons are strongly over-represented among the code modifications. Of the 23 non-standard code variants listed in a recent survey [11], there are 8 cases of stop codons being reassigned or acquired, 8 cases of codon loss, and 10 reassignments of a codon from one amino acid to another. Given that there are only three stop codons, the same changes to the code occurred in parallel in different groups of organisms.

At least two well-characterized non-canonical amino acids have been co-opted into the code, namely, selenocysteine that is represented in varying sets of proteins in diverse organisms from all three domains of life [18,19], and pyrrolysine, currently detected only in some archaea. The mechanisms that lead to the incorporation of these two amino acids are completely different. Pyrrolysine is accommodated via reassignment of a stop codon analogous to code changes within the canonical amino acid set; in contrast, inclusion of selenocysteine involves recoding whereby a stop codon directs selenocysteine incorporation only in the presence of a distinct, regulatory sequence element [20–22].

The evolutionary mechanisms that lead to codon reassignment and emergence of deviant codes involve changes in tRNA specificity and/or evolution of new specificities in the case of stop codon recruitment. These mechanisms fit the general 'gain and loss' framework, where 'gain' refers to acquisition of a new tRNA specificity, typically resulting from duplication of a tRNA gene, whereas 'loss' is elimination of a tRNA specificity, typically via deletion [11,23,24]. There is no evidence that any modern code modifications are adaptive. Most likely, these change evolved neutrally, through genetic drift and mutational pressure that drives small genomes toward high AT-content. The single key feature to be emphasized regarding the extant code variants is that they are all minor, i.e., involve one or at most two reassignments per variant, and typically concern rare amino acids, such as reassigning tryptophane (the least abundant amino acid of all) to a stop codon, the most common change that occurred in parallel in several deviant codes. The variant code does not venture far from the SGC at all.

Notably, new methods of synthetic biology have recently allowed substantial artificial alteration of the code in bacteria [25–28]. Although the fitness of the bacteria with altered codes has not been thoroughly studied, their viability itself seems to support the view that the fitness of different codes might not differ dramatically, so that the (near) universality of the SGC is likely to stem from high fitness barriers separating it from other codes, or in other words, low fitness of the intermediates [11].

To account for the universality of the code, Crick came up with the frozen accident argument: "This theory states that the code is universal because at the present time any change would be lethal, or at least very strongly selected against. This is because in all organisms (with the possible exception of certain viruses) the code determines (by reading the mRNA) the amino acid sequences of so many highly evolved protein molecules that any change to these would be highly disadvantageous unless accompanied by many simultaneous mutations to correct the "mistakes" produced by altering the code. This accounts for the fact that the code does not change. To account for it being the same in all organisms one must assume that all life evolved from a single organism (more strictly, from a single closely interbreeding population). In its extreme form, the theory implies that the allocation of codons to amino acids at this point was entirely a matter of "chance"." [1].

Using the language of fitness landscapes, the frozen accident perspective implies that there can be many different codes occupying fitness peaks, but they are separated by deep valleys of low fitness [15] (Figure 1). As indicated above, all discovered codon reassignments in extant organisms are indeed quite limited in scope. Moreover, in agreement with Crick's reasoning, these modifications have been identified primarily in organelles and bacteria with small genomes where the damage from reassigning rare codons could have been tolerable (contrary to what Crick thought, the code apparently does not change in viruses because these employ the host translation machinery for decoding their mRNAs). The frozen accident argument does not necessarily require that the original choice of codon assignment is literally and strictly random. Various factors could have contributed to the initial codon assignments (see discussion below) but once the choice is made, it gets frozen, i.e., only rare and minor changes may be allowed.

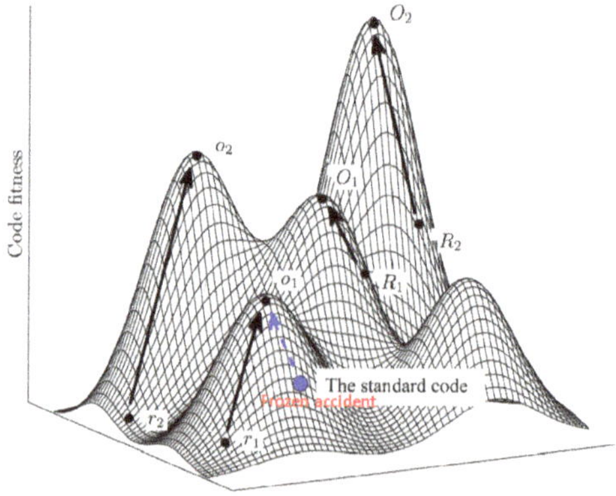

Figure 1. The code fitness landscape. The figure is a cartoon illustration of peaks of different heights separated by low fitness valleys on the code fitness landscape. The summit of each peak corresponds to a local optimum (O). Evolution towards local peaks is shown by arrows and starts either from a random code (R) or from the SGC. Modified from [14], under Creative License.

As Crick points out, the universality of the code is perhaps the strongest evidence of the existence of LUCA (Last Universal Cellular Ancestor), along with the universal conservation of the translation machinery [29]. However, LUCA certainly was not the first life form and most likely, not even the first cellular organism, only an evolutionary bottleneck. The early stages of cellular and especially pre-cellular evolution must have been dramatically different from the post-LUCA evolution of cellular organisms. This early evolution is thought to have involved competition between ensembles of "virus-like" genetic elements and selection at the level of such collectives [30–32]. Translation hardly could have evolved literally within a single such ensemble, and if it emerged on multiple occasions in different ensembles, then, initially, numerous, different codes must have existed. Then, why only one frozen accident, assuming that the actual codon assignments are indeed (quasi) accidental? To put it even more generally, why a single LUCA only?

A strikingly simple but, I believe, compelling answer was given by Vetsigian, Woese, and Goldenfeld in a breakthrough 2006 paper [33]: only one code survived because extensive horizontal gene transfer (HGT) was an aspect of early evolution without which the transition to the cellular level of complexity would not have been possible. Even a small change in the code has a prohibitive effect on HGT. Vetsigian and colleagues developed a simple mathematical model to track evolution with and without HGT. Starting with a random ensemble of codes, their simulated evolution experiments led to increased coded diversity in the absence of HGT but to survival of only a few code variants when HGT was allowed. In the original study, Vetsigian and coworkers explored a deterministic model in an infinite population approximation [33]. Notably, recent modeling efforts that took into account stochastic effects of the finite population size produced a single, universal code, with a structure similar to that of the SGC, within a range of HGT rates [34,35].

Given that HGT, even if limited by both physical and selective barriers, remained a key factor of microbial evolution [36–40], and apparently, a condition for long term survival of microbial populations [41,42], it stands to reason that it played a key role not only in the primordial universalization of the code but also in its maintenance through nearly four billion years of cellular evolution. The demonstration of the essential role of the code universality in the exchange of genetic information that promotes evolution of life arguably was one of the most important insights in the code evolution field since Crick's hypothesis. Conceivably, frozen accident and the requirement of HGT were major, complementary factors that have kept the code universal.

3. The Three Principal Scenarios of the Code Origin and Evolution: Achievements, Limitations, and Compatibility

The structure of the SGC, i.e., the mapping of 64 codons to 20 amino acids and the stop signal, is clearly non-random by multiple criteria [9,10]. This non-randomness of the code seems to require an explanation(s). The three major concepts that strive to explain the regularities in the code are the stereochemical, coevolution, and error minimization 'theories' [10,43,44]. A detailed review of the current state of these scenarios for code evolution is presented elsewhere [45]; here, I only give a brief synopsis and assessment of their contributions to our current understanding of the code evolution.

The stereochemical theory postulates that the structure of the code is determined by physicoo-chemical affinity between amino acids and codons or anticodons. The initial attempts on direct experimental demonstration of interaction between amino acids and the cognate triplets have been generally unconvincing [46]. In these early days of the study of code evolution, molecular modeling has been used to propose a variety of 'stereochemical codes' based on interactions of amino acids with cognate codons [47], reversed codons [48], anticodons [49], codon-anticodon duplexes [50], or a complex of four nucleotides including the anticodon and an additional base [51]. The difficulty of discriminating between different models and the lack of solid experimental support prevented any of these schemes from becoming the explanatory framework for the evolution of the code.

The stereochemical theory has been given a new life by the progress of the aptamer technology. At least some amino acids have been shown to select sequences significantly enriched for either codons or anticodons [52–55]. These results were taken as evidence of existence of an early stereochemical era in the code evolution during which the majority of the modern amino acids have been co-opted into the code [56]. However, as discussed elsewhere [10,45], the stereochemical evidence does not appear to be compelling. The principal problems are two-fold. First, statistically significant affinity for cognate triplets has been demonstrated only for five amino acids, and the triplet could be either codon or anticodon [55]. Second, and more damning, all the results of aptamer experiments that appear compatible with the physico-chemical affinity between amino acids and cognate triplets pertain to complex amino acids that are unlikely to have been available at the time of the code emergence (see next Section).

The second theory, known as coevolution, holds that the genetic code is shaped by the precursor-product relationships between amino acids [57–60]. Under the coevolution scenario, the code evolved from an ancestral version that included only the simple amino acids that can be produced abiogenically (see next section) by expanding to incorporate the more complex amino acids, in parallel with the evolution of the respective biosynthetic pathways (hence coevolution—between the code and the amino acid biosynthesis pathways). The importance of biosynthetic pathways for the code evolution is effectively obvious because amino acids could not be incorporated into the code unless they were available. The coevolution theory holds that the code evolved by subdivision of large blocks of codons that in the ancestral code encoded the same amino acid but were split to encode two (or more) amino acids upon the evolution of the respective metabolic pathways. The specific pattern of codon reassignment would be determined by the precursor-product relationships between amino acids. Thus, the coevolution theory cannot be reduced to the self-evident statement on the importance of biosynthetic pathways for the inclusion of the late amino acids into the code. Rather, the theory makes specific and readily falsifiable predictions on the subdivision of the primordial blocks of codons assignments following the evolution of metabolic pathways. As shown within the framework of the third theory of code evolution, these inferences do not seem to hold very well.

The grouping of similar amino acids within the same column of the code table immediately indicates that the code has a degree of robustness to mutational and translational errors. In other words, the codon assignments are organized in such a way as to minimize the deleterious effect of such errors. Hence the error minimization theory of code evolution, under which the structure of the code is determined by selection for robustness to errors. Extensive quantitative analyses that employed cost functions differently derived from physico-chemical properties of amino acids have shown that the code is indeed highly resilient, with the probability to pick an equally robust random code being on the order of 10^{-7}–10^{-8} [14,15,61–68]. Obviously, however, among the ~10^{84} possible random codes, there is a huge number with a higher degree of error minimization than the SGC. Furthermore, the SGC is not a local peak on the code fitness landscape because certain local rearrangements can increase the level of error minimization; a quantitatively, the SGC is positioned roughly halfway from an average random code to the summit of the corresponding local peak [15] (Figure 1).

The typical conclusion of the error minimization theorists is that the code evolved under selective pressure for robustness. However, this conclusion is not necessarily justified. Most if not all code optimization analyses focus on rearrangements of the standard code table, which allows formal (and in itself, valid, given the cost function) estimation of robustness but not the evolutionary processes that lead to it. A more biologically sound approach involves reconstruction of the routes of code expansion that might produce error minimization as a selectively neutral byproduct of evolution driven by other factors [64,67,68].

To conclude this brief discussion of the three major scenarios for the evolution of the code, it is useful to note that, although stemming from widely different premises, there is no reason why these scenarios should be mutually exclusive. Quite the contrary, stereochemistry, biochemical coevolution, and selection for error minimization could have contributed synergistically at different stages of the evolution of the code [43]—along with frozen accident.

4. Primordial Expansion of the Code

Little as we can claim to actually know about the evolution of the code, it appears most likely that the earliest proteins contained fewer amino acids than the modern set of 20. The canonical amino acids differ substantially in terms chemical complexity and stability. Ten amino acids have been consistently identified in prebiotic chemistry experiments as well as in in meteorites, in the following order of abundance: Gly, Ala, Asp, Glu, Val, Ser, Ile, Leu, Pro, Thr [69–72]. Notably, the recent, perhaps most promising at this time prebiotic chemistry experiments, based on hydrogen cyanide photochemistry, that yield precursors of ribonucleotides and amino acids, also primarily produce amino acids from the above list, namely, Gly, Ala, Ser, and Thr [73,74]. The ranks of amino acids in the 'early' list positively and significantly correlate with the free energy of their synthesis: the amino acids that are on top of the list are the 'cheapest' energetically [75]. An independent approach that involved analysis of the fluxes of amino acids in recent evolution of proteins in diverse organisms shows that the concentrations of the putative early amino acids in the above list are mostly decreasing, whereas those of the late amino acids are increasing [76]. This convergence of widely different methods of inference suggests that the above 10 amino acids can be confidently considered old, i.e., were represented already in the first proteins [77].

Given the confidently derived consensus set of ancient amino acids and the unavailability of the late amino acids before complex biosynthetic pathways evolved (see discussion of the coevolution theory above), early expansion of the code appears to be inevitable. The idea of such expansion again goes back to Crick's seminal paper: "The next general point about the primitive code is that it seems likely that only a few amino acids were involved" [1]. More importantly, Crick proposed, if not quite explicitly, that the error minimization property of the code could have evolved as consequence of the code expansion: "the net effect of a whole series of such changes would be that similar amino acids would tend to have similar codons, which is just what we observe in the present code" [1].

The theme of evolution of error minimization as a neutral consequence, or by-product, of the code expansion has been thoroughly developed in several recent studies [64,67,68]. Massey explored three distinct scenarios of the code expansion: the "2-1-3" model [78], which is similar to expansion schemes independently proposed by Higgs [79] and Francis [80], the ambiguity reduction model [81] and precursor-product scenario specified by the coevolution theory [57,59]. The 2-1-3 model and similar schemes posit that the code started from an ancestral stage, in which only the second base of the codon was informative, and expanded by assigning specificity to the first, and then, in some codon series, the third bases (another idea that goes back to Crick's 1968 paper). Under the ambiguity reduction model, in the ancestral code, codon series ambiguously encoded groups of amino acids, such that the subsequent evolution involved gradual increase in the specificity of the codon-amino acid mapping. The evolutionary simulations with both the 2-1-3 model and the ambiguity reduction model readily yield codes with error minimization levels exceeding that of the SGC. In contrast, the coevolution model, although producing some level of error minimization, was substantially inferior to the other scenarios, as also demonstrated in an earlier analysis by Higgs [79]. These findings cast doubt on the direct relevance of coevolution model [57,59] and imply that, although substantial error minimization is a key property of the SGC, this feature likely evolved as a by-product of code expansion rather than by direct selection for code robustness.

5. Protein Evolution Paradox, Extinct Primordial Stereochemical Code, Expansion of the Code, and Frozen Accident: A Coevolution Scenario for the Code and the Translation System

The origin and evolution of the translation system is a forbiddingly difficult problem, and therefore, in many studies on the code evolution, it is formally treated as a separate issue and approached almost like a mathematical puzzle [9]. Ultimately, however, it appears virtually certain that evolution of the code can be understood only in the context of the evolution of translation, as presciently noted by Crick: "It is almost impossible to discuss the origin of the code without discussing the origin of the actual biochemical mechanisms of protein synthesis" [1].

The translation system is universally conserved among the extant cellular life forms but many protein components of the translation apparatus, as well as the tRNAs, are paralogs, and furthermore, belong to large paralogous families. Phylogenetic analysis of such families can provide clues to the pre-LUCA phase of evolution. The results of such analyses reveal a 'protein evolution paradox'. The aminoacyl-tRNA synthetases (aaRS), the enzymes that are responsible for the accurate matching between amino acids and the cognate codons in the modern translation system, belong to two classes of paralogs, with 10 specificities in each [82,83]. As convincingly demonstrated for the Class I aaRS that contain a Rossmann-fold catalytic domain, the diversification of the aaRS occurred at a late stage in the evolution of the Rossmann-fold protein superfamily [84,85]. By the time the 10 Class I aaRS specificities evolved via radiation from a common ancestor, the evolution of the Rossmann fold superfamily had already produced a substantial diversity of other enzymatic and nucleotide-binding domains. A similar scenario holds for the evolution of the Class II aaRS, which belong to the biotin synthase superfamily, although the evolution of these proteins has not been explored as thoroughly as that of Class I [86,87]. For this early protein evolution to occur, high-fidelity (although possibly not to the level of the modern system) translation was certainly essential, and given that the different aaRS specificities have not yet evolved, the conclusion inevitably ensues that the specificity of amino acid-codon correspondence was determined by RNA molecules [88].

The conclusion that the mRNA decoding in the early translation system was performed by RNA molecules, conceivably, evolutionary precursors of modern tRNAs (proto-tRNAs) [89], implies a stereochemical model of code origin and evolution, but one that differs from the traditional models of this type in an important way (Figure 2). Under this model, the proto-RNA-amino acid interactions that defined the specificity of translation would not involve the anticodon (let alone codon) that therefore could be chosen arbitrarily and fixed through frozen accident. Instead, following the reasoning outlined previously [90], the amino acids would be recognized by unique pockets in the tertiary structure of the proto-tRNAs. The clustering of codons for related amino acids naturally follows from code expansion by duplication of the proto-tRNAs; the molecules resulting from such duplications obviously would be structurally similar and accordingly would bind similar amino acids, resulting in error minimization, in accord with Crick's proposal (Figure 2). A variant of this scenario would involve initial imprecise recognition of groups of amino acids (e.g., several bulky hydrophobic ones) followed by duplication and subfunctionalization of the proto-tRNAs, recapitulating Woese's statistical protein hypothesis [91]. This part of the scenario represents code expansion discussed above that could have been driven by the benefits of diversification of the repertoire of protein amino acids and would yield robustness as a byproduct.

The transition from the primordial, RNA only translation system to the modern RNA-protein machinery would follow the previously outlined path [90,92], in which the first translated peptides would serve as cofactors for the ribozymes, in particular, the proto-tRNAs (Figure 3). At the next stage, the ancestral catalytic domains of the aaRS would evolve and take over from the proto-tRNA as the catalysts for the aminoacylation reaction. Initially, each of these domains would indiscriminately catalyze aminoacylation of 10 proto-tRNAs. The subsequent evolution would involve duplication of the catalytic domains and accretion of accessory domains, resulting in the emergence of protein-mediated determination of aminoacylation specificity (Figure 3).

Once the amino acid specificity determinants shifted from the proto-tRNAs to the aaRS, the amino acid-binding pockets in the (proto) tRNAs deteriorated such that modern tRNAs showed no consistent affinity to the cognate amino acids. This scenario implies that attempts to decipher the primordial stereochemical code by comparative analysis of modern translation system components are largely futile. The best hope for reconstructing the ancient code could lie in experiments on in vitro evolution of specific aminoacylating ribozymes that can be evolved quite easily and themselves seem to recapitulate a key aspect of the primordial translation system [93–95]. It seems interesting to note that, although direct recognition of amino acids by RNA molecules seems to be banished from the modern translation system, it survives in other functional spheres of modern cells, in particular, as amino acid-binding riboswitches [96,97]. The structures of such riboswitches complexed with the cognate amino acids might shed light on the structures of primordial proto-tRNAs.

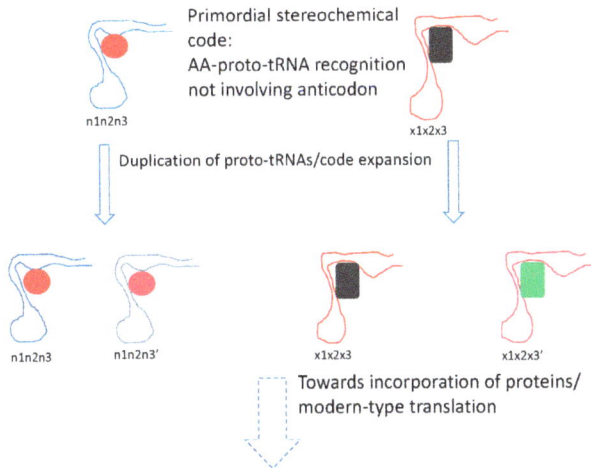

Figure 2. Putative primordial stereochemical code, evolution by code expansion and the standard code as frozen accident. Related amino acid (AA) are shown by similar colors. In the code expansion phase, the anticodons are shown changed in the third position (corresponding to the first position of the codon), resulting in recruitment of related amino acids.

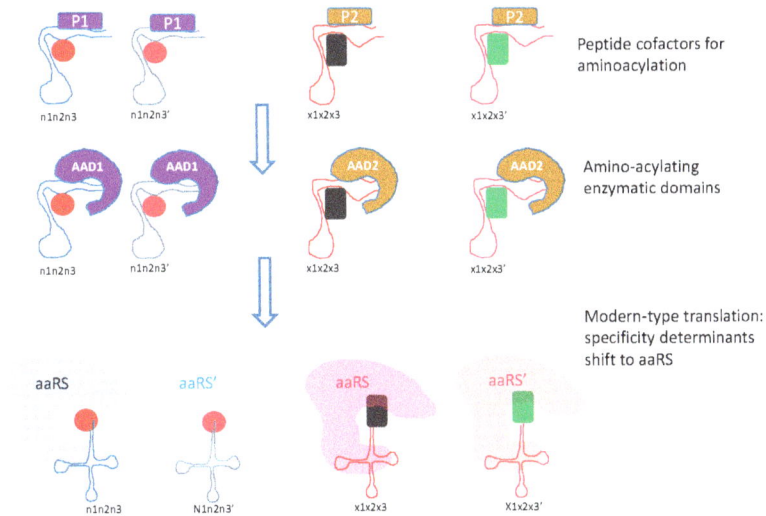

Figure 3. Transition from the RNA only to modern-type RNA-protein translation system. Two phases of evolution between the primordial, RNA only translation and the modern-type system are envisaged. At the first intermediate stage, peptides (denoted P) synthesized by the primitive translation system would serve as cofactors for amino acid binding and aminoacylation, and at the second stage, catalysis of aminoacylation would be relegated to catalytic aminoacylating domains (denoted AAD). Two forms of both the cofactor peptides and the AAD are shown, to provide continuity towards the two classes of aaRS.

6. Concluding Remarks

Five decades after the publication of Crick's seminal paper [1], frozen accident remains alive and well, and a major tenet of our thinking on the evolution of the code. It could be argued that frozen accident is not a theory and not even a hypothetical scenario but rather a meaningless non-explanation of the code origin. Although any detailed discussion on epistemology is beyond the scope of this article, I counter that random and (extremely) rare events are part and parcel of the evolution of life, even if this is unpalatable to some on philosophical grounds [88,98,99]. Thus, frozen accident actually is a falsifiable null hypothesis. The falsification of the frozen accident comes in the form of positive evidence of specific evolutionary processes that account for certain aspects of the code evolution. However, none of the three major theories of the code evolution has been fully successful in providing a definitive explanation although each has highlighted important features of the code. By combining findings from evolutionary analysis of translation system components, a distinct flavor of the stereochemical theory—the concept of the code expansion that is well grounded in data and theory—and the frozen accident idea, a simple, experimentally testable model of code evolution is proposed here. It owes most of the underlying ideas to Crick's 1968 paper and the concurrent work of Carl Woese.

Acknowledgments: E.V.K. is supported by intramural funds of the US Department for Health and Human Services (to the National Library of Medicine).

Conflicts of Interest: The author declares that he has no conflict of interest.

References

1. Crick, F.H. The origin of the genetic code. *J. Mol. Biol.* **1968**, *38*, 367–379. [CrossRef]
2. Tamura, K. The Genetic Code: Francis Crick's Legacy and Beyond. *Life* **2016**, *6*, 36. [CrossRef] [PubMed]
3. Watson, J.D.; Crick, F.H. Molecular structure of nucleic acids; a structure for deoxyribose nucleic acid. *Nature* **1953**, *171*, 737–738. [CrossRef] [PubMed]
4. Watson, J.D.; Crick, F.H. Genetical implications of the structure of deoxyribonucleic acid. *Nature* **1953**, *171*, 964–967. [CrossRef] [PubMed]
5. Crick, F.H. On protein synthesis. *Symp. Soc. Exp. Biol.* **1958**, *12*, 138–163. [PubMed]
6. Crick, F.H.; Barnett, L.; Brenner, S.; Watts-Tobin, R.J. General nature of the genetic code for proteins. *Nature* **1961**, *192*, 1227–1232. [CrossRef] [PubMed]
7. Nirenberg, M.W.; Matthaei, J.H. The dependence of cell-free protein synthesis in *E. coli* upon naturally occurring or synthetic polyribonucleotides. *Proc. Natl. Acad. Sci. USA* **1961**, *47*, 1588–1602. [CrossRef] [PubMed]
8. Nirenberg, M. Historical review: Deciphering the genetic code—A personal account. *Trends Biochem. Sci.* **2004**, *29*, 46–54. [CrossRef] [PubMed]
9. Woese, C. *The Genetic Code*; Harper & Row: New York, NY, USA, 1967.
10. Koonin, E.V.; Novozhilov, A.S. Origin and evolution of the genetic code: The universal enigma. *IUBMB Life* **2009**, *61*, 99–111. [CrossRef] [PubMed]
11. Sengupta, S.; Higgs, P.G. Pathways of Genetic Code Evolution in Ancient and Modern Organisms. *J. Mol. Evol.* **2015**, *80*, 229–243. [CrossRef] [PubMed]
12. Crick, F.H. The genetic code—Yesterday, today, and tomorrow. *Cold Spring Harb. Symp. Quant. Biol.* **1966**, *31*, 1–9. [CrossRef] [PubMed]
13. Haig, D.; Hurst, L.D. A quantitative measure of error minimization in the genetic code. *J. Mol. Evol.* **1991**, *33*, 412–417. [CrossRef] [PubMed]
14. Freeland, S.J.; Hurst, L.D. The genetic code is one in a million. *J. Mol. Evol.* **1998**, *47*, 238–248. [CrossRef] [PubMed]
15. Novozhilov, A.S.; Wolf, Y.I.; Koonin, E.V. Evolution of the genetic code: Partial optimization of a random code for robustness to translation error in a rugged fitness landscape. *Biol. Direct* **2007**, *2*, 24. [CrossRef] [PubMed]
16. Santos, M.A.; Moura, G.; Massey, S.E.; Tuite, M.F. Driving change: The evolution of alternative genetic codes. *Trends Genet.* **2004**, *20*, 95–102. [CrossRef] [PubMed]

17. Ling, J.; O'Donoghue, P.; Soll, D. Genetic code flexibility in microorganisms: Novel mechanisms and impact on physiology. *Nat. Rev. Microbiol.* **2015**, *13*, 707–721. [CrossRef] [PubMed]

18. Ambrogelly, A.; Palioura, S.; Soll, D. Natural expansion of the genetic code. *Nat. Chem. Biol.* **2007**, *3*, 29–35. [CrossRef] [PubMed]

19. Mukai, T.; Englert, M.; Tripp, H.J.; Miller, C.; Ivanova, N.N.; Rubin, E.M.; Kyrpides, N.C.; Soll, D. Facile Recoding of Selenocysteine in Nature. *Angew. Chem. Int. Ed. Engl.* **2016**, *55*, 5337–5341. [CrossRef] [PubMed]

20. Zhang, Y.; Baranov, P.V.; Atkins, J.F.; Gladyshev, V.N. Pyrrolysine and selenocysteine use dissimilar decoding strategies. *J. Biol. Chem.* **2005**, *280*, 20740–20751. [CrossRef] [PubMed]

21. Lobanov, A.V.; Turanov, A.A.; Hatfield, D.L.; Gladyshev, V.N. Dual functions of codons in the genetic code. *Crit. Rev. Biochem. Mol. Biol.* **2010**, *45*, 257–265. [CrossRef] [PubMed]

22. Yuan, J.; O'Donoghue, P.; Ambrogelly, A.; Gundllapalli, S.; Sherrer, R.L.; Palioura, S.; Simonovic, M.; Soll, D. Distinct genetic code expansion strategies for selenocysteine and pyrrolysine are reflected in different aminoacyl-tRNA formation systems. *FEBS Lett.* **2010**, *584*, 342–349. [CrossRef] [PubMed]

23. Sengupta, S.; Higgs, P.G. A unified model of codon reassignment in alternative genetic codes. *Genetics* **2005**, *170*, 831–840. [CrossRef] [PubMed]

24. Sengupta, S.; Yang, X.; Higgs, P.G. The mechanisms of codon reassignments in mitochondrial genetic codes. *J. Mol. Evol.* **2007**, *64*, 662–688. [CrossRef] [PubMed]

25. Xie, J.; Schultz, P.G. A chemical toolkit for proteins—An expanded genetic code. *Nat. Rev. Mol. Cell Biol.* **2006**, *7*, 775–782. [CrossRef] [PubMed]

26. Neumann, H.; Wang, K.; Davis, L.; Garcia-Alai, M.; Chin, J.W. Encoding multiple unnatural amino acids via evolution of a quadruplet-decoding ribosome. *Nature* **2010**, *464*, 441–444. [CrossRef] [PubMed]

27. Liu, C.C.; Schultz, P.G. Adding new chemistries to the genetic code. *Annu. Rev. Biochem.* **2010**, *79*, 413–444. [CrossRef] [PubMed]

28. Chin, J.W. Expanding and reprogramming the genetic code of cells and animals. *Annu. Rev. Biochem.* **2014**, *83*, 379–408. [CrossRef] [PubMed]

29. Koonin, E.V. Comparative genomics, minimal gene-sets and the last universal common ancestor. *Nat. Rev. Microbiol.* **2003**, *1*, 127–136. [CrossRef] [PubMed]

30. Szathmary, E.; Demeter, L. Group selection of early replicators and the origin of life. *J. Theor. Biol.* **1987**, *128*, 463–486. [CrossRef]

31. Szathmary, E.; Maynard Smith, J. From replicators to reproducers: The first major transitions leading to life. *J. Theor. Biol.* **1997**, *187*, 555–571. [CrossRef] [PubMed]

32. Koonin, E.V.; Martin, W. On the origin of genomes and cells within inorganic compartments. *Trends Genet.* **2005**, *21*, 647–654. [CrossRef] [PubMed]

33. Vetsigian, K.; Woese, C.; Goldenfeld, N. Collective evolution and the genetic code. *Proc. Natl. Acad. Sci. USA* **2006**, *103*, 10696–10701. [CrossRef] [PubMed]

34. Sengupta, S.; Aggarwal, N.; Bandhu, A.V. Two perspectives on the origin of the standard genetic code. *Orig. Life Evol. Biosph.* **2014**, *44*, 287–291. [CrossRef] [PubMed]

35. Aggarwal, N.; Bandhu, A.V.; Sengupta, S. Finite population analysis of the effect of horizontal gene transfer on the origin of an universal and optimal genetic code. *Phys. Biol.* **2016**, *13*, 036007. [CrossRef] [PubMed]

36. Doolittle, W.F. Phylogenetic classification and the universal tree. *Science* **1999**, *284*, 2124–2129. [CrossRef] [PubMed]

37. Koonin, E.V.; Makarova, K.S.; Aravind, L. Horizontal gene transfer in prokaryotes: Quantification and classification. *Annu. Rev. Microbiol.* **2001**, *55*, 709–742. [CrossRef] [PubMed]

38. Doolittle, W.F.; Bapteste, E. Pattern pluralism and the Tree of Life hypothesis. *Proc. Natl. Acad. Sci. USA* **2007**, *104*, 2043–2049. [CrossRef] [PubMed]

39. Puigbo, P.; Lobkovsky, A.E.; Kristensen, D.M.; Wolf, Y.I.; Koonin, E.V. Genomes in turmoil: Quantification of genome dynamics in prokaryote supergenomes. *BMC Biol.* **2014**, *12*, 66. [CrossRef] [PubMed]

40. Treangen, T.J.; Rocha, E.P. Horizontal transfer, not duplication, drives the expansion of protein families in prokaryotes. *PLoS Genet.* **2011**, *7*, e1001284. [CrossRef] [PubMed]

41. Iranzo, J.; Puigbo, P.; Lobkovsky, A.E.; Wolf, Y.I.; Koonin, E.V. Inevitability of genetic parasites. *Genome Biol. Evol.* **2016**, *8*, 2856–2869. [CrossRef] [PubMed]

42. Takeuchi, N.; Kaneko, K.; Koonin, E.V. Horizontal gene transfer can rescue prokaryotes from Muller's ratchet: Benefit of DNA from dead cells and population subdivision. *G3 (Bethesda)* **2014**, *4*, 325–339. [CrossRef] [PubMed]

43. Knight, R.D.; Freeland, S.J.; Landweber, L.F. Selection, history and chemistry: The three faces of the genetic code. *Trends Biochem. Sci.* **1999**, *24*, 241–247. [CrossRef]

44. Di Giulio, M. The origin of the genetic code: Theories and their relationships, a review. *Biosystems* **2005**, *80*, 175–184. [CrossRef] [PubMed]

45. Koonin, E.V.; Novozhilov, A.S. Origin and evolution of the universal genetic code. *Annu. Rev. Genet.* **2017**, *51*, in press.

46. Woese, C.R. The fundamental nature of the genetic code: Prebiotic interactions between polynucleotides and polyamino acids or their derivatives. *Proc. Natl. Acad. Sci. USA* **1968**, *59*, 110–117. [CrossRef] [PubMed]

47. Pelc, S.R.; Welton, M.G. Stereochemical relationship between coding triplets and amino-acids. *Nature* **1966**, *209*, 868–870. [CrossRef] [PubMed]

48. Root-Bernstein, R.S. On the origin of the genetic code. *J. Theor. Biol.* **1982**, *94*, 895–904. [CrossRef]

49. Dunnill, P. Triplet nucleotide-amino-acid pairing; a stereochemical basis for the division between protein and non-protein amino-acids. *Nature* **1966**, *210*, 1265–1267. [CrossRef] [PubMed]

50. Hendry, L.B.; Witham, F.H. Stereochemical recognition in nucleic acid-amino acid interactions and its implications in biological coding: A model approach. *Perspect. Biol. Med.* **1979**, *22*, 333–345. [CrossRef] [PubMed]

51. Shimizu, M. Molecular basis for the genetic code. *J. Mol. Evol.* **1982**, *18*, 297–303. [CrossRef] [PubMed]

52. Yarus, M. Amino acids as RNA ligands: A direct-RNA-template theory for the code's origin. *J. Mol. Evol.* **1998**, *47*, 109–117. [CrossRef] [PubMed]

53. Yarus, M. RNA-ligand chemistry: A testable source for the genetic code. *RNA* **2000**, *6*, 475–484. [CrossRef] [PubMed]

54. Yarus, M.; Caporaso, J.G.; Knight, R. Origins of the genetic code: The escaped triplet theory. *Annu. Rev. Biochem.* **2005**, *74*, 179–198. [CrossRef] [PubMed]

55. Yarus, M.; Widmann, J.J.; Knight, R. RNA-amino acid binding: A stereochemical era for the genetic code. *J. Mol. Evol.* **2009**, *69*, 406–429. [CrossRef] [PubMed]

56. Yarus, M. *Life from RNA World: The Ancestor Within*; Harvard University Press: Cambridge, MA, USA, 2010.

57. Wong, J.T. A co-evolution theory of the genetic code. *Proc. Natl. Acad. Sci. USA* **1975**, *72*, 1909–1912. [CrossRef] [PubMed]

58. Di Giulio, M. An extension of the coevolution theory of the origin of the genetic code. *Biol. Direct* **2008**, *3*, 37. [CrossRef] [PubMed]

59. Wong, J.T.; Ng, S.K.; Mat, W.K.; Hu, T.; Xue, H. Coevolution Theory of the Genetic Code at Age Forty: Pathway to Translation and Synthetic Life. *Life* **2016**, *6*, 12. [CrossRef] [PubMed]

60. Di Giulio, M. The lack of foundation in the mechanism on which are based the physico-chemical theories for the origin of the genetic code is counterposed to the credible and natural mechanism suggested by the coevolution theory. *J. Theor. Biol.* **2016**, *399*, 134–140. [CrossRef] [PubMed]

61. Goodarzi, H.; Nejad, H.A.; Torabi, N. On the optimality of the genetic code, with the consideration of termination codons. *Biosystems* **2004**, *77*, 163–173. [CrossRef] [PubMed]

62. Zhu, W.; Freeland, S. The standard genetic code enhances adaptive evolution of proteins. *J. Theor. Biol.* **2006**, *239*, 63–70. [CrossRef] [PubMed]

63. Torabi, N.; Goodarzi, H.; Shateri Najafabadi, H. The case for an error minimizing set of coding amino acids. *J. Theor. Biol.* **2007**, *244*, 737–744. [CrossRef] [PubMed]

64. Massey, S.E. A neutral origin for error minimization in the genetic code. *J. Mol. Evol.* **2008**, *67*, 510–516. [CrossRef] [PubMed]

65. Novozhilov, A.S.; Koonin, E.V. Exceptional error minimization in putative primordial genetic codes. *Biol. Direct* **2009**, *4*, 44. [CrossRef] [PubMed]

66. Salinas, D.G.; Gallardo, M.O.; Osorio, M.I. Local conditions for global stability in the space of codons of the genetic code. *Biosystems* **2016**, *150*, 73–77. [CrossRef] [PubMed]

67. Massey, S.E. Genetic code evolution reveals the neutral emergence of mutational robustness, and information as an evolutionary constraint. *Life* **2015**, *5*, 1301–1332. [CrossRef] [PubMed]

68. Massey, S.E. The neutral emergence of error minimized genetic codes superior to the standard genetic code. *J. Theor. Biol.* **2016**, *408*, 237–242. [CrossRef] [PubMed]

69. Pizzarello, S. The chemistry of life's origin: A carbonaceous meteorite perspective. *Acc. Chem. Res.* **2006**, *39*, 231–237. [CrossRef] [PubMed]

70. Zaia, D.A.; Zaia, C.T.; De Santana, H. Which amino acids should be used in prebiotic chemistry studies? *Orig. Life Evol. Biosph.* **2008**, *38*, 469–488. [CrossRef] [PubMed]

71. Cleaves, H.J., 2nd. The origin of the biologically coded amino acids. *J. Theor. Biol.* **2010**, *263*, 490–498. [CrossRef] [PubMed]

72. Burton, A.S.; Stern, J.C.; Elsila, J.E.; Glavin, D.P.; Dworkin, J.P. Understanding prebiotic chemistry through the analysis of extraterrestrial amino acids and nucleobases in meteorites. *Chem. Soc. Rev.* **2012**, *41*, 5459–5472. [CrossRef] [PubMed]

73. Ritson, D.J.; Sutherland, J.D. Synthesis of aldehydic ribonucleotide and amino acid precursors by photoredox chemistry. *Angew. Chem. Int. Ed. Engl.* **2013**, *52*, 5845–5847. [CrossRef] [PubMed]

74. Patel, B.H.; Percivalle, C.; Ritson, D.J.; Duffy, C.D.; Sutherland, J.D. Common origins of RNA, protein and lipid precursors in a cyanosulfidic protometabolism. *Nat. Chem.* **2015**, *7*, 301–307. [CrossRef] [PubMed]

75. Higgs, P.G.; Pudritz, R.E. A thermodynamic basis for prebiotic amino acid synthesis and the nature of the first genetic code. *Astrobiology* **2009**, *9*, 483–490. [CrossRef] [PubMed]

76. Jordan, I.K.; Kondrashov, F.A.; Adzhubei, I.A.; Wolf, Y.I.; Koonin, E.V.; Kondrashov, A.S.; Sunyaev, S. A universal trend of amino acid gain and loss in protein evolution. *Nature* **2005**, *433*, 633–638. [CrossRef] [PubMed]

77. Trifonov, E.N. The triplet code from first principles. *J. Biomol. Struct. Dyn.* **2004**, *22*, 1–11. [CrossRef] [PubMed]

78. Massey, S.E. A sequential "2-1-3" model of genetic code evolution that explains codon constraints. *J. Mol. Evol.* **2006**, *62*, 809–810. [CrossRef] [PubMed]

79. Higgs, P.G. A four-column theory for the origin of the genetic code: Tracing the evolutionary pathways that gave rise to an optimized code. *Biol. Direct* **2009**, *4*, 16. [CrossRef] [PubMed]

80. Francis, B.R. Evolution of the genetic code by incorporation of amino acids that improved or changed protein function. *J. Mol. Evol.* **2013**, *77*, 134–158. [CrossRef] [PubMed]

81. Fitch, W.M.; Upper, K. The phylogeny of tRNA sequences provides evidence for ambiguity reduction in the origin of the genetic code. *Cold Spring Harb. Symp. Quant. Biol.* **1987**, *52*, 759–767. [CrossRef] [PubMed]

82. Wolf, Y.I.; Aravind, L.; Grishin, N.V.; Koonin, E.V. Evolution of aminoacyl-tRNA synthetases—Analysis of unique domain architectures and phylogenetic trees reveals a complex history of horizontal gene transfer events. *Genome Res.* **1999**, *9*, 689–710. [PubMed]

83. Woese, C.R.; Olsen, G.J.; Ibba, M.; Soll, D. Aminoacyl-tRNA synthetases, the genetic code, and the evolutionary process. *Microbiol. Mol. Biol. Rev.* **2000**, *64*, 202–236. [CrossRef] [PubMed]

84. Aravind, L.; Anantharaman, V.; Koonin, E.V. Monophyly of class I aminoacyl tRNA synthetase, USPA, ETFP, photolyase, and PP-ATPase nucleotide-binding domains: Implications for protein evolution in the RNA. *Proteins* **2002**, *48*, 1–14. [CrossRef] [PubMed]

85. Aravind, L.; Mazumder, R.; Vasudevan, S.; Koonin, E.V. Trends in protein evolution inferred from sequence and structure analysis. *Curr. Opin. Struct. Biol.* **2002**, *12*, 392–399. [CrossRef]

86. Artymiuk, P.J.; Rice, D.W.; Poirrette, A.R.; Willet, P. A tale of two synthetases. *Nat. Struct. Biol.* **1994**, *1*, 758–760. [CrossRef] [PubMed]

87. Anantharaman, V.; Koonin, E.V.; Aravind, L. Comparative genomics and evolution of proteins involved in RNA metabolism. *Nucleic Acids Res.* **2002**, *30*, 1427–1464. [CrossRef] [PubMed]

88. Koonin, E.V. *The Logic of Chance: The Nature and Origin of Biological Evolution*; FT Press: Upper Saddle River, NJ, USA, 2011.

89. Tamura, K. Origins and Early Evolution of the tRNA Molecule. *Life* **2015**, *5*, 1687–1699. [CrossRef] [PubMed]

90. Wolf, Y.I.; Koonin, E.V. On the origin of the translation system and the genetic code in the RNA world by means of natural selection, exaptation, and subfunctionalization. *Biol. Direct* **2007**, *2*, 14. [CrossRef] [PubMed]

91. Woese, C.R. On the evolution of the genetic code. *Proc. Natl. Acad. Sci. USA* **1965**, *54*, 1546–1552. [CrossRef] [PubMed]

92. Szathmary, E. Coding coenzyme handles: A hypothesis for the origin of the genetic code. *Proc. Natl. Acad. Sci. USA* **1993**, *90*, 9916–9920. [CrossRef] [PubMed]

93. Yarus, M. The meaning of a minuscule ribozyme. *Philos. Trans. R. Soc. Lond. B Biol. Sci.* **2011**, *366*, 2902–2909. [CrossRef] [PubMed]
94. Chumachenko, N.V.; Novikov, Y.; Yarus, M. Rapid and simple ribozymic aminoacylation using three conserved nucleotides. *J. Am. Chem. Soc.* **2009**, *131*, 5257–5263. [CrossRef] [PubMed]
95. Kumar, R.K.; Yarus, M. RNA-catalyzed amino acid activation. *Biochemistry* **2001**, *40*, 6998–7004. [CrossRef] [PubMed]
96. Serganov, A.; Patel, D.J. Amino acid recognition and gene regulation by riboswitches. *Biochim. Biophys. Acta* **2009**, *1789*, 592–611. [CrossRef] [PubMed]
97. Serganov, A.; Patel, D.J. Metabolite recognition principles and molecular mechanisms underlying riboswitch function. *Annu. Rev. Biophys.* **2012**, *41*, 343–370. [CrossRef] [PubMed]
98. Monod, J. *Chance and Necessity: An Essay on the Natural Philosophy of Modern Biology*; Vintage: New York, NY, USA, 1972.
99. Lynch, M. *The origins of Genome Archiecture*; Sinauer Associates: Sunderland, MA, USA, 2007.

life

MDPI

Article

Intrinsic Properties of tRNA Molecules as Deciphered via Bayesian Network and Distribution Divergence Analysis

Sergio Branciamore [1],*, Grigoriy Gogoshin [1], Massimo Di Giulio [2] and Andrei S. Rodin [1],*

[1] Department of Diabetes Complications and Metabolism, Diabetes and Metabolism Research Institute, City of Hope, Duarte, 91010 CA, USA; ggogoshin@coh.org

[2] Early Evolution of Life Laboratory, Institute of Biosciences and Bioresources, CNR, 80131 Naples, Italy; massimo.digiulio@ibbr.cnr.it

* Correspondence: sbranciamore@coh.org (S.B.); arodin@coh.org (A.S.R.); Tel.: +1-626-218-3809 (S.B.); +1-626-218-3807 (A.S.R.)

Received: 9 December 2017; Accepted: 23 January 2018; Published: 8 February 2018

Abstract: The identity/recognition of tRNAs, in the context of aminoacyl tRNA synthetases (and other molecules), is a complex phenomenon that has major implications ranging from the origins and evolution of translation machinery and genetic code to the evolution and speciation of tRNAs themselves to human mitochondrial diseases to artificial genetic code engineering. Deciphering it via laboratory experiments, however, is difficult and necessarily time- and resource-consuming. In this study, we propose a mathematically rigorous two-pronged in silico approach to identifying and classifying tRNA positions important for tRNA identity/recognition, rooted in machine learning and information-theoretic methodology. We apply Bayesian Network modeling to elucidate the structure of intra-tRNA-molecule relationships, and distribution divergence analysis to identify meaningful inter-molecule differences between various tRNA subclasses. We illustrate the complementary application of these two approaches using tRNA examples across the three domains of life, and identify and discuss important (informative) positions therein. In summary, we deliver to the tRNA research community a novel, comprehensive methodology for identifying the specific elements of interest in various tRNA molecules, which can be followed up by the corresponding experimental work and/or high-resolution position-specific statistical analyses.

Keywords: tRNA identity; tRNA recognition; operational code; bayesian networks; information theory; distribution divergence

1. Introduction

A prototypical tRNA is a compact and ubiquitous molecule. Its primary function is that of an adapter mediating mRNA and growing protein sequence; as such, together with aminoacyl tRNA synthetases (aaRSes), tRNAs form the backbone of translation apparatus and embody the genetic code. In order to be effective, translation machinery requires unambiguous charging of tRNA molecules by the appropriate amino acids ("aa"); this, somewhat whimsically, is known as "second genetic code" (see [1] and references therein). It is the aaRSes that are physically responsible for charging tRNAs with amino acids.

What are the factors that determine the identity of tRNAs with respect to these pairings? How, exactly, is the "correct" tRNA recognized and aminoacetylated by cognate aaRS? It is reasonable to assume that this specificity is encoded in the tRNA molecule proper (perhaps localized around certain tRNA regions) and is then reflected in its 3D structure and chemical properties. Identifying such determinants and antideterminants (that, conversely, prevent incorrect recognition) within tRNA molecules is far from trivial, yet is of significant theoretical and practical importance (see [2] for an overview).

The straightforward approach to elucidating tRNA determinants and antideterminants would be via experimental work. This constitutes primarily combinatorial screening, where many tRNA mutants are checked for interaction with aaRSes. Specific combinations of nucleotide polymorphisms that correlate with changes in interaction status of tRNAs with cognate aaRSes are then noted. Intense experimental effort along the above lines led to identifying some such tRNA sequence features that are therefore assumed to be key elements of the "second genetic code" ([2]). These experimental efforts are ongoing ([3,4] and references therein). It should also be noted that, gratifyingly, our understanding of the underlying chemistry of amino acid activation is improving ([5,6]); recent progress in molecular dynamic simulations had also added to our insight into the tRNA – aaRS fundamental biology (see [7] for an overview). On a side note, much research had been done on the variability in the aaRSes themselves ([8,9]). This has substantial practical implications (see Datt and Sharma [10] for an overview of disease-associated aaRS mutation spectra).

However, our general grasp of the tRNA "determinant array" is very fragmented, at best. We can only speculate as to the completeness, universality and robustness of discovered determinants (single and multiple tRNA nucleotide polymorphisms) within the "second genetic code" context. To put it bluntly, we lack systemic approach to the problem.

Experimental approaches are very time- and resource-consuming. While "targeted evolution" of both tRNAs and aaRSes is technologically feasible ([11,12]) and happens to be of much practical importance (see [13] for a mt-tRNA mitochondrial multi-system disorder example), the direction in which such experimental work is pursued is often arbitrarily, if not anecdotally, mapped out.

Here, we propose a novel in silico approach for inferring important tRNA positions and features. Our methodology of choice is Bayesian (or Belief) Network (BN) modeling, which has been used by us ([14,15]) and others before in the domains with similar data structure. This methodology is augmented by estimation of the distribution divergence metrics for the tRNA positions, across different tRNA subclasses (subgroups), an information-theoretic approach. The primary data is a compendium of tRNA nucleotide sequences (of which there are many thousands presently available), coupled with both biological/chemical information (crystallographic data, biochemical behavior, etc.) and ontological data (existing large body of work reflected in recent literature on various aspects of tRNA chemistry, biology, evolution, and significance in general).

Our two main goals throughout this study were to (i) further improve the BN methodology (including both algorithms and software), predominantly in testing its generalization to novel data sets and types; and to (ii) gain better tools for identifying tRNA determinants (and studying tRNA evolution in general). The former is largely outside of the scope of this communication; we will concentrate on the latter. As our approach was purely data-driven, the rationale was to infer "interesting" (the most structurally important) positions/features of tRNAs regardless of the biological context. However, we have also followed up with the discussion of the discovered positions of interest using ontological (literature) knowledge. It should be noted that, as always in biology (and especially in the case of tRNAs), any "feature of interest" exists largely in the evolutionary sense, which was a constant consideration throughout the analyses and discussion of the results presented here.

In this study, we used a systems biology/probabilistic inference methodology (BNs) that in our opinion is well suited for the task; it is detailed elsewhere ([14,15]), but the rationale is briefly covered below in Materials and Methods. By applying BN analysis to the tRNA sequence data, we were able to recognize the interplay of structurally and biologically important features of the tRNA molecule. Furthermore, by extending the analysis to distribution divergence estimates across all tRNA positions and different tRNA subclasses, we were able to single out the important positions specific to each type of tRNA, which led us to identifying tRNA determinants and antideterminants.

2. Materials and Methods

Building tRNA probabilistic networks, as well as evolutionary studies of tRNAs in general, require assembling a reliable tRNA sequence alignment as the initial step. Resulting BNs are dependent on both the alignment and the number of sequences involved; more accurate alignment and higher number of sequences being of course desirable. In this study, we went to the tRNAdb database as a primary source ([16]); 9758 unambiguously aligned tRNA sequences were imported.

Each vertical position in the alignment is interpreted as a random variable with values ranging over all possible nucleotide states at this position in the tRNAs in the alignment. In other words, each vertical position is a discrete random variable with a number (label) assigned. These labels, or variable names, are thus invariant for all tRNA molecules. To avoid confusion with position numbering, Figure 1 depicts the layout of consensus tRNA sequence. The first row corresponds to the numbering scheme adopted throughout most of this study; position numbers in the first row are also the variable (node) names in the BNs. This is a universally accepted tRNA numbering standard ([17]). The master, or majority consensus, tRNA sequence (second row in Figure 1) contains notable fragments of tRNA molecule colored in red (acceptor stem), green (D-arm), blue (anticodon 131 arm) and yellow (T-arm). The same color scheme is used in the resulting BNs. Therefore, by consulting Figure 1 when evaluating these study results, it is easy to visually link to the specific positions and sections of the tRNA molecule.

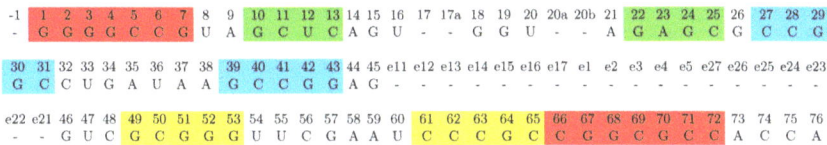

-1	1	2	3	4	5	6	7	8	9	10	11	12	13	14	15	16	17	17a	18	19	20	20a	20b	21	22	23	24	25	26	27	28	29
-	G	G	G	C	C	G	U	A	G	C	U	C	A	G	U	-	-	G	G	U	-	-	A	G	A	G	C	G	C	G	C	C	G

30	31	32	33	34	35	36	37	38	39	40	41	42	43	44	45	e11	e12	e13	e14	e15	e16	e17	e1	e2	e3	e4	e5	e27	e26	e25	e24	e23
G	C	C	U	G	A	U	A	A	G	C	C	G	G	A	G	-	-	-	-	-	-	-	-	-	-	-	-	-	-	-	-	-

e22	e21	46	47	48	49	50	51	52	53	54	55	56	57	58	59	60	61	62	63	64	65	66	67	68	69	70	71	72	73	74	75	76
-	-	G	U	C	G	C	G	G	G	U	U	C	G	A	A	U	C	C	C	G	C	C	G	G	C	G	C	C	A	C	C	A

Figure 1. tRNA sequence alignment. The first row corresponds to the "standard" numbering scheme. The second row is the consensus sequence. Structural parts of tRNA molecules are highlighted in color (see text for details).

Given the tRNA alignment, a BN, also formally known as PDAG (probabilistic directed acyclic graph), can be learned from it. Dependencies (intra-tRNA position interactions represented by the BN edges; they can also be colloquially thought of as "correlations" or "linkage" or "associations" between positions) suggested by the BN can subsequently be evaluated against a set of well-known structural features of tRNA. The purpose of doing so (building the BN and observing significant dependencies, as reflected by the network edges) is two-fold: to identify yet unknown but potentially interesting interactions, and to observe deviations from the established models. In a tRNA BN, network nodes correspond to the variables representing vertical positions in tRNA alignment, edges—to the dependencies between these variables. For a fuller understanding of BN representation, notions of conditional independence, edge directionality and strength, BN equivalence classes, and Markov neighborhood (or "blanket") of a node are useful; however, detailed theoretical treatment is outside the scope of this communication (see [15,18,19]). Briefly, BN is a graphical model of a joint multivariate probability distribution of random (both continuous and discrete) variables that reflects relationships of dependence and conditional independence among them. Absence of an edge between two nodes (variables) indicates conditional independence. Directionality of the edge (arrow) is, in this (tRNA) case, for mathematical convenience only and does not imply causation flow. Dependency (edge) strength can also be estimated; it should be interpreted as a relative support for the strength of a relationships between the variables. No absolute interpretation, such as *p*-value, is possible before the follow-up parametric statistical analysis; however, edge strengths can be directly compared to one another within a BN. Edge strengths are shown as numbers (next to the corresponding edges in the BN figures) that can be thought of as the likelihood ratios (marginal likelihood given the data, or model fit, of the BN with the edge present divided by the marginal likelihood of the BN with the edge absent). Local probability tables (conditional probabilities) can be estimated for pairwise frequency calculations between the two nodes; by observing these frequencies, we can, for example, speculate on the nature of different nucleotide pairings at different tRNA positions (e.g., whether they are canonical Watson–Crick pairings or not).

Parsing local probability tables can become cumbersome for the nodes with more than one or two edges; the problem is exacerbated when conditioning by the amino acid (or other tRNA subgroup stratification) is introduced. This is why we have also decided to complement the BN modeling with the simpler information theoretic-based pairwise analyses of the tRNA sequence data conditioned on the tRNA subclass variables, notably the amino acid variable. BN modeling remains the method of choice for higher-order abstraction; for scrutinizing tRNA position relationships at a finer level (and in smaller, for example stratified by amino acid, tRNA datasets), such pairwise comparisons (utilizing distribution divergence) might be more practical and statistically robust.

It remains to note that the actual BN search algorithm used by us was an iterative gradient method with restarts, and MDL (Minimum Description Length) was chosen as a model scoring criterion. Multinomial local probability model with four nucleotide states was used. Further technical details of our current BN modeling implementation can be found in [15,20]. Various BN reconstruction algorithm parameter adjustments were tried based on our previous experience with BNs (results not shown); the resulting BNs proved to be sufficiently robust. Our BN modeling software is directly available from the authors.

3. Results

3.1. Structural Information

The complete tRNA BN (using all tRNAdb sequences pooled together) is shown in Figure 2. Figure 2a is the "flattened" visualization that makes it easier to trace specific position dependencies. Figure 2b shows the same graph superimposed on the cloverleaf tRNA structure, for easier "biological" visualization. The interpretation of the dependencies (links, edges) between the tRNA positions in the network is necessarily equivocal—a dependency might have different (one or more) biological interpretations. Some of these dependencies, such as the ones directly explained by chemical bonds and physical interactions, are more straightforward and obvious. However, even these can be partially obscured by the probabilistic nature of BN reconstruction.

We will first consider the dependencies related to the secondary structure of tRNA. For example, in the acceptor stem, positions 1, 7 are linked with positions 72, 66, respectively; positions 10, 13 are linked with 25, 22 in D-arm; positions 27, 31 are linked with 43, 39 in anticodon arm; positions 49, 53 are linked with positions 66, 61 in the T-arm. These connections are expected and provide a solid point of reference (and a positive control, especially if one takes into account relative connection strengths) for further BN interpretation. Clearly, the BN modeling is able to identify structural relationships inherent to the secondary cloverleaf structure. At this point, we were curious if the procedure was also capable of suggesting principal tertiary structure interactions (known from the literature). The following relationships proved to be of particular interest:

First, note the interaction between position 26 and 44. These are often associated with a non-Watson–Crick pairing (usually GA, with the last base change from 33 degree to 45 degree angle).

Then, a parallel base pair interaction is indicated between position 15 in the d-loop and position 48 in the variable loop (bringing the D-loop and the variable loop together in the 3D configuration). A reverse Hoogsteen pairing is indicated with position 8 and position 14; it is known that this interaction, together with positions 15 and 48, is responsible for the linkage of the two main domains of tRNA molecule in the l-shaped configuration via precise docking of the D- and T psi C-arms). It should be noted that, although all four of these positions are largely invariant (or semi-invariant), the BN methodology is able to elucidate their relationships and suggest how all of them are linked together in maintaining the l-shaped structures of tRNA molecules. This is not surprising, considering that BN analysis is very sensitive to rare events ([15,20]. The latter are often ignored altogether in traditional statistical analyses (as well as in common classification algorithms, such as single decision trees in machine learning). Indeed, once a *p*-value (or, correspondingly, a decision tree classification decision) is generated, the rare event is simply no longer visible in the analysis schema; however, BN reconstruction process often creates an additional (comparatively weak) edge instead of just disregarding the "minority report" of a rare event. In such cases, low confidence in the edge (reflected as relatively low edge strength, as measured by the model marginal likelihood ratio) points to a rarity of the event (a particular tRNA mutation, in this context) rather than the statistical insignificance of the relationship.

Additionally, a triple base interaction is suggested between positions 9, 23 and 13. Another triple base interaction between 13, 22 and 46 is related to docking the variable loop onto the D-arm. However, another important interaction occurs between the D- and C-loops (link between G18G19 of D-loop with position 55c56). The BN shows the interactions between positions 18, 19, and 56, but not position 55. All of these relationships, highlighted by the BN, are essential to the determination of the core tertiary structure of the tRNA molecule. This is gratifying in itself; however, even a superficial glance at the Figure 2 reveals many more potential relationships in addition to what has been so far reported in the literature (which largely revolves around the intricacies of the tertiary structure determination and interactions).

Before following up on these potential relationships, however, one question that must be asked is whether the BN "overfits" the data (i.e., suggests spurious relationships reflecting random noise and data analysis artifacts). We have studied the question extensively in the broader BN modeling context ([14,15]). By using entropy-based model selection (and BN scoring) criteria, we can explicitly adjust the overfitting/underfitting balance (i.e., penalize less/more for the network complexity). The BN reconstruction engine used throughout this study was tuned more towards "underfitting", meaning that if the edge appears in the BN visualization, our confidence in it is sufficiently high (in other words, "if the edge is present, there is definitely something"). What the BN reconstruction doe, is simply underscores, agnostically, all information patterns contained in the data. Therefore, in our analyses of the results shown in Figure 2, it would behoove us to go beyond secondary and tertiary structure interpretations.

For example, consider the interplay between the positions 9, 12, 13, 22, 23 and 46. The "triple" interaction between 9, 27, 23 and 13, 22, 46 has been reported before. However, in addition to that, we observe a more complex pattern of interactions that hinges on position 9. We can say that position 9 is dependent on the state of position 13 (or 22, which is tightly linked to 13) and position 23 (or 12, which is tightly linked to 23). Which of these alternatives reflects the actual biological (mechanistic) relationship? We cannot tell from the BN alone; all we know is that the likelihood of 9-13 is slightly higher than that of 9-22 (and the likelihood of 9-23 is slightly higher than 9-12). However, this might be a sampling artifact. However, in any case, the BN suggests that there is a "global" pattern of interaction between all five nucleotides. Further untangling of the underlying biology would be an example of promising direction of laboratory research. One possible interpretation is that the crystal structures, in general, point to one specific configuration (the crystallized form) whereas in solution more than one tRNA structure can dynamically exist. tRNAs are flexible polymers, and what we see depicted in the BN are the multiple interactions that occur during small structural rearrangements of the molecule in vivo (and recall that BN modeling is particularly revealing of rare events/variants). These can reflect the fluctuations around the minimum free energy structure (often but not always the crystallized form)—or the phenomenon could be due to the alternative structures related to specific biological functions (e.g., conformational changes due to interaction with other molecules, such as ribosomes or aaRSes). Either way, more laboratory research is definitely indicated for this particular position cluster.

In general, it is our opinion that one could look at the tRNA molecule as a system (of interactions)—if so, the BN (such as in Figure 2) is a synthetic representation of this system, where a change in one base pair would affect numerous other positions either due to direct chemical interactions, or because it will alter the global stability of the molecule.

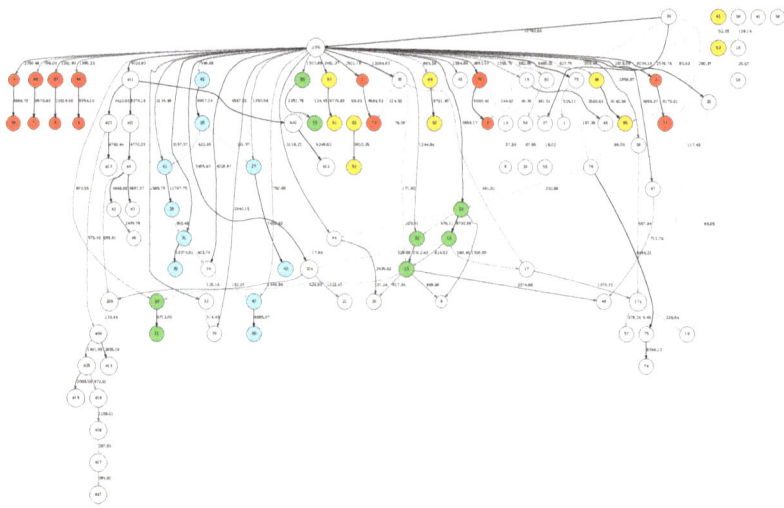

(a)

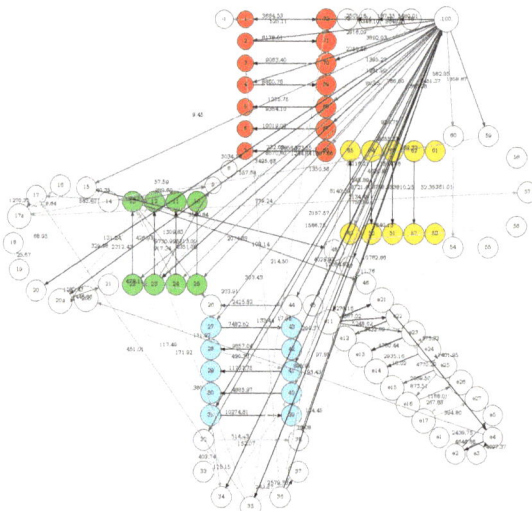

(b)

Figure 2. Bayesian network built from the full set of tRNA sequences. (**a**) direct visualization of the PDAG (Probabilistic Directed Acyclic Graph); (**b**) same, superimposed on the secondary tRNA structure. Nodes in the network correspond to the variables (specifically, tRNA positions, as enumerated in Figure 1, first row), and edges to the dependencies between the variables. "Boldness" of the edge is proportional to the dependency strength, also indicated by the number shown next to the edge. See text for BN construction details. *(Label "100" does not refer to a tRNA position but rather is a placeholder for the cognate aa variable, appearing here for the technical convenience reasons only; directionality of the edge (arrow) is for mathematical convenience reasons only as well, and does not imply causation.)*

To illustrate this outlook, let us revisit the interplay of the positions 12, 13, 22 and 23 at the far end of the d-arm close to the d-loop. The links between positions 12, 23 and 13, 22 are what we expect to see with the d-arm formation in the canonical cloverleaf tRNA structure. Interpretation of the connections 12, 13 and 22, 23, however, is not immediately obvious. A closer inspection of the BN, local probability tables, and tRNA alignments reveals that positions 13, 22 do not always form a canonical Watson–Crick (GU) pairing; sometimes, it is instead the two adenine residues that cannot pair. This means that the d-domain, in a subset of the tRNA molecules, presents a shorter arm and a larger loop. Analysis of the BN local probability tables reveals that in order to accommodate this structural variation (with a shorter and therefore less stable arm) positions 12, 23 have to simultaneously change from AT to the higher energy GC, to maintain the sufficient structure stability.

In other words, what the BN analysis is suggesting here (with the connections 12, 13 and 22, 23) are the constraints in the evolutionary dynamics of the d-domain of tRNA molecule. A local change in the sequence (and therefore in the structural configuration) must be physically and thermodynamically compatible with the biological structure and cannot take place unless the molecular "neighbors" are ready to accommodate the change event. Thus, the BN points us in the direction of these "neighbors", revealing the pattern of nucleotides changes in the linked positions. The non-canonical structure (with a larger loop and a shorter arm in the D-domain) is the results of one of the possible evolutionary paths of the tRNA molecule: these paths are reflected in the concerted changes in positions 12, 13, 22 and 23. One such evolutionary scenario, compatible with the observed BN, is as follows: initial mutations occur at positions 13, 22 (AU to GC). This event does not alter the secondary structure (it could be considered a neutral event), but represents a "pre-adaptation" to the next events that lead to the breakage of the stem at positions 12, 23 bringing about the altered secondary structure with a larger loop and a shorter stem.

While Figure 2 shows the BN built from all the available tRNA sequence data combined, obviously similar BNs can be constructed from the different subgroups (or (sub)classes; we will use these terms interchangeably throughout the manuscript) of tRNAs. Such subclasses can be stratified by the species groups, domains of life, amino acids, etc. *(Series of aa-stratified BNs are available directly from the authors, in pdf or dot formats.)* Then, one can compare resulting BNs by observing topological differences (presence or absence of particular edges between the nodes). Qualitative and quantitative analysis of these topological differences can prove very useful in elucidating various tRNA determinants. Such detailed analysis is outside of the scope of the present study (however, see the following two sections for the baseline analyses of amino acid- and domain of life-based stratifications), but it is our intention to carry out comprehensive tRNA subgroup analyses in the future, predominantly to clarify certain aspects of tRNA gene histories. One caveat here, though, is the necessarily smaller samples, leading to the less robust BNs. For such smaller samples, "downgrading" from high-level-abstraction of graphical multivariate analysis (BNs, in their essence) to series of statistically rigorous pairwise tests might be advantageous. In the following section, we describe precisely such an approach.

3.2. Amino Acids Specificity

As depicted in Figure 2, the full-set tRNA BN is reflective of the overall organization of tRNA molecules, comprising all position associations that can be extracted from the data (inclusive of all types of tRNAs). Therefore, some of the observed relationships are common to all (or most of) the tRNAs present in the database, whereas others apply only to selected subsets. For example, in a group of tRNAs specific for a given amino acid, a singular base (or a specific combination of bases) could be primarily responsible for amino acid determination by aaRSes.

Here, we set out to investigate the positional relationships specific to particular classes of amino acids. Simply put, this analysis utilizes the notion that if the position i is, in some sense, "important" for a specific class of tRNAs, the conditional distribution of nucleotides at ith position within the class will be different from the unconditional distribution of nucleotides at the same position over the complete set of tRNAs. It is possible to estimate corresponding distribution dissimilarity (AKA distribution divergence, distance); furthermore, we assume that the higher the dissimilarity, the more likely is the associated position to play a significant biological role for the specific subgroup of tRNAs considered. Otherwise, we assume the position to be relatively "unimportant" for this tRNA subgroup—although it might still have an overall functional significance, for all tRNA molecules in general.

Formally, given the probability space $(\omega, \dashv, \mu)$ and a set of categorical random variables with range defined by the set of nucleotides and a gap symbol, $s = (a, c, g, u, \_)$

$$x_i : \omega \to s,$$

we denote the distribution at ith position of the alignment by

$$q_i = \mu(x_i). \tag{1}$$

Note that the gap state here is used for convenience in interpreting the results in Figures 3 and 4 (and Supplemental Figures) only. It was not used in the BN construction/interpretation. Instead, during the BN reconstruction, dependencies between positions 17a, 20ab, e1-27 were treated as (comparatively) rare events, under multinomial local probability models in correspondingly reduced datasets.

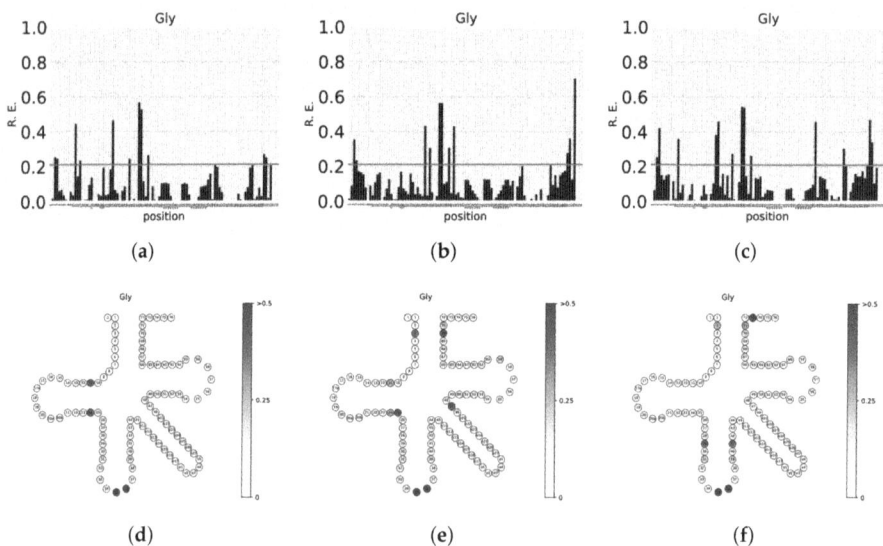

(a)   (b)   (c)

(d)   (e)   (f)

Figure 3. Position "importance" profile for Gly tRNAs, shown for three life domains: Archaea (**a,d**), Bacteria (**b,e**) and Eukarya (**c,f**). Relative Entropy is shown as function of tRNA position (**a–c**) (enumerated as in Figure 1, first row), or visualized as color intensity superimposed over the secondary tRNA structure (**d–f**). Significance cutoff limit is shown as a red line in (**a–c**)—see text for discussion.

Given the categorical random variable y with the same sample space ω and range a defined by the set of amino acid classes $a = \{a_j\}$,

$$y : \omega \to a,$$

we define the conditional distribution of nucleotides at each position i for a given class of amino acids a_j as

$$p_{ij} = \mu(x_i \cap [y = a_j]). \tag{2}$$

Our goal is then to estimate the dissimilarity between p_{ij} and q_i using some appropriate dissimilarity or distance measure. For this purpose, we select the (symmetrised and smoothed) measure of relative entropy l; its square root is a proper metric (distance function), which is a desirable property. We define

$$l(p_{ij}, q_i) = \frac{1}{2}d(p_{ij}, m_{ij}) + \frac{1}{2}d(q_i, m_{ij}), \tag{3}$$

where the operator $d(x, y)$ is

$$d(x, y) = \sum_k x(k) \log\left(\frac{x(k)}{y(k)}\right) = h(x, y) - h(x), \tag{4}$$

and

$$m_{ij} = \frac{1}{2}\left(p_{ij} + q_i\right). \tag{5}$$

(Below, we will occasionally refer to $l(p_{ij}, q_i)$ as simply l_{ij}, where appropriate.) The quantity denoted by l_{ij} is the symmetrised measure of relative entropy (RE), also known as Jensen–Shannon divergence, or JSD ([21,22]; conveniently, its square root defines a mathematically correct distance function. Because the image of l is always a finite interval $im(l) = [0, \max(l)]$, the results (inter-distributional differences) are easy to interpret—when l is close to 0, the distributions that are being compared are very similar, whereas, when l is close to its maximum value, the distributions have little to nothing in common. Conveniently, this allows us to superimpose an absolute scale on the degree of "importance" that a particular position has within the context of a given class of tRNAs—not only can we contrast the relative importances between different positions, but we can also estimate how important any given position is in absolute terms.

A series of plots (Figure 3 for Gly, the simplest, most basic, amino acid, example; Supplemental Figures S1–S22 for all amino acids) shows the "importance" profile for each class of tRNA in y, partitioned into the three life domains, Archaea (Arc), Bacteria (Bac), and Eukarya (Euk). A brief glance at the charts reveals that each tRNA class possesses a unique "fingerprint" pattern of RE values assigned to the tRNA positions. Throughout all the classes, however, certain general features are easily identifiable: for instance, codon position (34, 35, 36) shows high "importance" in all tRNA groups; on the other hand, invariant and semi-invariant positions are associated with low "importance" (see Figure 4 for a heat plot summary visualization of tRNA position importance, by life domain (Figure 4a–c), tRNA class, and nucleotide position). The next obvious question is: just how high is sufficiently (significantly) high? Obviously, positions 35, 36 are significant, and, for example, position 33 is not. What about position 34? Throughout most of this study (see also Table 1 below), we have set a somewhat arbitrary significance cutoff value of RE, at 0.2 (marked by the red horizontal lines in Figure 3 and Supplemental Figures S1–S22), largely following current literature and visual analysis of the distributions. In order to impose a rigorous quantitative cutoff measure, however, we can generate a set of random samples y_j^* from y to obtain

$$p_{ij}^* = \mu(x_i \cap y_j^*).$$

This operation is expected to have the effect of scrambling any probabilistic relationship between y and x_i (unless there is a systematic bias in the data). We then define

$$l_{ij}^* = l(p_{ij}^*, q_i)$$

and obtain a sequence $\{l_{ij}^*\}_{j=1}^m$ induced by the random sampling of y, which should correspond to the distribution h_0 of random noise in the data. In our experiments, setting m at 10^5 was sufficient for convergence. Whenever the observed value of l_{ij} exceeds this "random noise" threshold, it is reasonable to suspect that there is a non-trivial signal. It appears that, for all amino acid classes, in almost all positions, this is indeed the case. Therefore, it is highly improbable that most observed

l_{ij} values could have resulted from the random processes. To put it bluntly, almost nothing about most tRNA positions is even close to random.

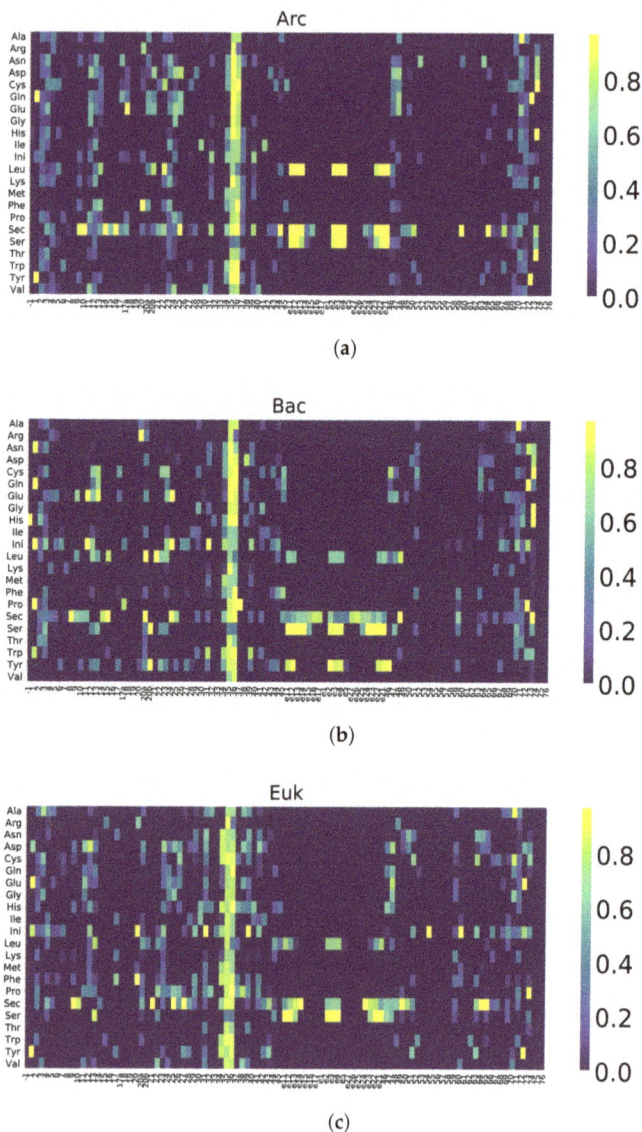

Figure 4. Summary visualization of the tRNA position "importance", for all aa tRNA subclasses, shown for three life domains: Archaea (**a**); Bacteria (**b**) and Eukarya (**c**). Higher values correspond to "hotter" colors. tRNA position numbering is as in Figure 3. This is a summary visualization of the detailed plots presented in Supplemental Figures S1–S22.

Of course, this is not at all unexpected from the purely biological viewpoint—this observation is simply a confirmation of the fact that the tRNAs for a certain amino acid class are not independent and that they share specific structural and sequence information.

What was not expected, however, is that, at almost every position, the distribution between the tRNA classes is highly non-random. Two factors likely attribute to this effect: first, the vestiges of the phylogenetic relationship between the tRNAs; and, second, structural and functional aspects associated with each tRNA class. After having observed the non-random signal associated with a particular tRNA position (as detailed in Supplemental Figures S1–S22 and summarized in Figure 4), we now need to establish whether it is not only "high", but also "specific" (to the tRNA class). The non-specific signal can be attributed to some other kind of underlying relationship (e.g., a structural one); in general, such signal is a second- (or higher-) order relationship compared to the tRNA class-driven one. It can also be interpreted as a "background" signal, present irrespective of the tRNA class stratification. It is intrinsic to the tRNA structure in general but has little to no relevance regarding amino acid specificity. These higher-order signals are unlikely to be directly responsible for any specific tRNA function but will be strong enough to show as non-random effects. In this context, we posit that the positions with "high" and "specific" signals are the ones with the highest likelihood of being "important" (for the biological function of a particular tRNA class). They are preserved within, and discriminate between, specific tRNA groups. In contrast, in "non-specific" positions, non-random signal simply either reflects the echoes of the phylogenetic relationships, not yet completely dissolved, or points to the general tRNA structural relationship patterns.

In summary, by selecting the positions with "high" and "specific" RE signal, we zero in on the positions crucial for the biological functions and features of specific tRNA groups (as opposed to the "background", or higher-order, signal in non-specific positions associated with the phylogenetic, structural, etc., aspects of all the tRNAs).

One way to separate the latter (not directly related to the amino acid specificity) positions would be to compute RE after conditioning the data for the states in position i, and repeating for all n positions, thus obtaining a new collection of distributions:

$$z_{iqr} = \mu \left(x_q \cap \{ x_i = r \} \right), \tag{6}$$

where r is one of the possible states (nucleotides/gap) at the position i with position $q \neq i$. By using this procedure, one can establish the baseline RE distribution associated with each position.

3.3. Identification, in the Three Domains of Life, of the tRNA Amino Acid Specificity Identity Determinants

Typically, in the literature, the notion "tRNA identity determinants" refers to the nucleotide positions that are recognized by aaRSes, and are involved in the loading of tRNAs. Here, we expand on that definition. By "identity determinants" we mean that all of the nucleotide positions that are potentially predictive with respect to identifying a particular tRNA subclass. Table 1 summarizes the quantitative results of our analyses (by a given amino acid, and the three domains of life).

Table 1. Determinant tRNA positions classified by the amino acid and the three domains of life (Archaea, Arc; Bacteria, Bac; Eukarya, Euk). Only positions with the Relative Entropy (RE) value higher than the preset cutoff value (set at RE approx. 0.2, as illustrated in Figure 3 and Supplemental Figures S1–S22) are shown. Note that the preset cutoff value is higher than that suggested by the "random noise" $\{l_{ij}^*\}_{j=1}^m$ sequence (see text). For compatibility with the literature, the position numbering system follows that of [23,24]. It corresponds to the first row in Figure 1. Paired bases in the secondary structure of tRNA are shown as connected by a dash; for example, the 3-70 pair indicates that in the secondary structure of the tRNA molecule, the bases at position 3 and 70 should be paired [23,24]. Similarly, "=" indicates tertiary interactions among tRNA nucleotides. Question marks indicates the uncertain positions, that is to say positions that might not be important for the tRNA molecule. "Confirmed" indicates that the nucleotides identified as "important", or identity determinants, in the experimental analyses [2,25] were also significant (high RE values) in our analysis. "Low significance" means that the "experimentally proven to be important/identity determinants" nucleotides were not associated with high RE values. On the contrary, "Suggested" positions exhibit high RE values but so far have not been registered in the experimental analysis literature.

Amino Acid	Confirmed	Low Significance	Suggested	Domain
Ala	2-71; 3-70; 4-69; 20; 64; 73 3-70		2-71; 3-70; 4-69; 9; 12-23; 13-22 20a; 30-40; 35; 36; 44; 47 17; 20a; 29-41; 35; 36; 44; 51-63 2-71; 4-69; 5-68; 9; 12-23; 13-22; 20a; 27-43 29-41; 31-39; 32; 35; 36; 38; 59	Arc Bac Euk
Arg	20; 20A;	38; 73	4-69; 20; 35; 36; 73 4-69; 5-68; 20a; 35; 36 4-69; 15; 20; 35; 36; 48; 71	Arc Bac Euk
Asn	73		2-71; 3-70; 9; 11-24; 12-23; 17; 17a 20a; 22; 31-39; 34; 35; 36; 37 46; 47; 51-63; 59; 73 1-72; 2-71; 3-70; 12-23; 31-39 32; 34; 35; 36; 51-63 2-71; 4-69; 13-22; 17; 27-43; 29-41 31-39; 34; 35; 36; 38; 49-65 50-64; 51-63; 59; 73	Arc Bac Euk
Asp	25; 38; 73 9-12-23; 25; 38; 73	2-71; 10 10	2-71; 3-70; 6-67; 11-24; 12-23; 13-22; 17a 20b; 20a; 20; 25; 28-42; 34; 35 36; 44; 46; 47; 49-65; 64; 73 11-24; 20a; 31-39; 34; 35; 36 43; 44; 50-64; 51-63; 65 1-72; 11-24; 13-22; 20a; 26; 28-42 29-41; 31-39; 34; 35; 36; 46 47; 49-65; 50-64; 59; 63; 71	Arc Bac Euk
Cys	2-71; 3-70; 13-22; 46; 73 12-23; 73	15; 48 20a	3-70; 4-69; 5-68; 12-23; 13-22; 17 17a; 20; 21; 24; 27-43; 34 35; 36; 45; 46; 47; 73 12-23; 17; 29-41; 34; 35; 36 43; 45; 47; 51-63; 71 2-71; 3-70; 6-67; 7-66; 9; 13-22; 26; 29-41; 31-39; 34; 35; 36; 37 38; 51-63; 59; 68; 69	Arc Bac Euk
Gln	1-72; 38; 73	2-71; 3-70; 10-25; 37	1-72; 2-71; 3-70; 4-69; 11-24; 12-23; 13-22 17; 17a; 20b; 20a; 25; 34; 35 36; 37; 44; 46; 47; 73 12-23; 13-22; 20a; 34; 35; 36 44; 45; 46; 51-63; 65 2-71; 3-70; 6-67; 7-66; 11-24; 12-23 13-22; 26; 29-41; 31-39; 34; 35 36; 44; 46; 47; 52-62; 73	Arc Bac Euk
Glu	11-24; 13; 46; 47; 71	1-72; 22; 33; 37	2-71; 3-70; 11-24; 12-23; 13-22; 17a; 20a; 20b 25; 34; 35; 36; 46; 47; 49-65 3-70; 4-69; 5-68; 7-66; 9; 12-23 17; 20a; 30-40; 34; 35; 36 38; 45; 49-65; 51-63 1-72; 2-71; 3-70; 5-68; 11-24; 12-23 13-22; 25; 26; 31-39; 34; 35 36; 38; 47; 59	Arc Bac Euk
Gly	2-71; 3-70; 73 2-71; 3-70	1-72; 10-25 73	2-71; 3-70; 11-24; 13-22 31-39; 35; 36; 49-65 29-41; 31-39; 35; 36; 63 11-24; 25; 31-39; 35; 36; 47; 59	Arc Bac Euk

Table 1. *Cont.*

Amino Acid	Confirmed	Low Significance	Suggested	Domain
His	-1; 73	-1; 73	-1; 2-71; 3-70; 5-68; 11-24; 12-23 34; 35; 36; 37; 50-64; 73	Arc
			2-71; 3-70; 4-69; 6-67; 31-39 32; 34; 35; 36; 38; 63	Bac
			2-71; 9; 11-24; 12-23; 13-22; 26 30-40; 31-39; 32; 34; 35; 36 37; 38; 44; 45; 46; 47	Euk
Ile	12-23; 29-41	4-69; 24; 37; 38; 73	2-71; 3-70; 11-24; 12-23; 29-41; 31-39 34; 35; 36; 37; 46; 73	Arc
			3-70; 6-67; 13-22; 20a; 27-43; 28-42 34; 35; 36; 44; 51-63	Bac
			4-69; 17; 20a; 28-42; 29-41 30-40; 34; 35; 36; 60	Euk
Ini	2-71; 3-70; 32	33; 37	1-72; 2-71; 9; 11-24; 17; 17a; 20 20b; 27-43; 31-39; 34; 35; 36; 37 47; 51-63; 57; 64; 73	Arc
			1-72; 5-68; 6-67; 11-24; 12-23; 17a 26; 27-43; 29-41; 31-39; 34; 35 36; 44; 57; 59; 73	Bac
			1-72; 2-71; 3-70; 4-69; 5-68; 6-67; 7-66 12-23; 20; 20a; 22; 27-43; 29-41; 31-39 33; 34; 35; 36; 38; 46; 51-63 54; 59; 60; 73	Euk
Leu	20A; 73	20; 38;	2-71; 3-70; 4-69; 5-68; 9; e11 12-23; 13-22; 20a; 20b; e21; 31-39 35; 36; 37; 44; ; 46	Arc
			2-71; e5; 9; e11; 12-23; 13-22; 15; 20a; 21 e21; 35; 36; 44; ; 46; 47; 48; 73	Bac
			4-69; e5; e11; 12-23; 13-22; 20a 20b; e21; 29-41; 35; 36; 37 44; ; 45; 47; 49-65; 68	Euk
Lys	73?		2-71; 3-70; 4-69; 9; 11-24; 12-23; 22 31-39; 34; 35; 36; 37; 46; 73	Arc
			4-69; 5-68; 7-66; 12-23; 20a; 26 31-39; 34; 35; 36; 73	Bac
			2-71; 7-66; 9; 12-23; 13-22; 17; 20a; 29-41 31-39; 34; 35; 36; 44; 59; 70	Euk
Met	73 20	4-69; 5-68; 38 73	31-39; 34; 35; 36; 37; 73 31-39; 34; 35; 36; 71 1-72; 12-23; 31-39; 34; 35; 36; 60; 64	Arc Bac Euk
Phe	27-43; 31-39; 44; 45; 59 20; 31-39; 37	20?; 28-42; 30-40; 37; 39?; 43?; 60; 73? 73?	9; 12-23; 13-22; 20; 20a; 34; 35 36; 37; 45; 46; 47; 73	Arc
			3-70; 12-23; 17; 20a; 34; 35 36; 39; 43; 51-63; 73	Bac
			2-71; 4-69; 5-68; 6-67; 9; 12-23 13-22; 17; 20a; 29-41; 34; 35 36; 51-63; 59; 60; 73	Euk
Pro	72; 73	15; 48	2-71; 3-70; 6-67; 11-24; 12-23; 13-22 17a; 25; 35; 36; 37; 46	Arc
			1-72; 2-71; 3-70; 17a; 35 36; 37; 44; 59	Bac
			2-71; 11-24; 12-23; 13-22; 20a; 25 26; 27-43; 29-41; 31-39; 32; 35 36; 37; 38; 49-65; 73	Euk
Sec	2-71; 3-70; 4-69; 5-68; e5; 7-66 8; 9; 10-25; 11-24; e11; e17 e21; e27; 45; 48; 73; e24; e25 e26; 50-64; 64?; 66?; 68?	1-72; e2; e3; e4; 6-67; 12-23; e12; e13 13-22; 20; e22; e23	3-70; 4-69; 7-66; 9; 10-25; e11; 11-24; 12-23 13-22; 14; 15; 16; 17a; 17; 20b; 20 20a; 21; e21; 27-43; 28-42; 31-39; 34; 35 36; 37; 44; ; 48; 49-65; 50-64; 59 67; 68; 72; 73	Arc
			14; 15; 16; 20a; 29-41; 31-39; 34; 35 36; ; 59; 63; 64; 66; 68	Bac
			2-71; 4-69; 5-68; 9; 10-25; e11; 14; 20a e21; 21; 23; 26; 27-43; 28-42; 29-41; 31-39 34; 35; 36; 38; 44; ; 46; 47 48; 49-65; 50-64; 51-63; 59; 66; 67; 73	Euk
Ser	2-71; 3-70; e4-69;11; e21; 44; 73 e4; e5?;e12; e13; e14; e15; e16; e22 e23; e24; e25; e26; 69? e2; e3; e11; e21; e2; e3; e4; e12; e13; e22; e23	11-24	2-71; 4-69; 5-68; e11; 12-23; e21; 22; 24 35; 36; 44; ; 46; 47; 73	Arc
			5-68; 12-23; 13-22; 20a; 20b; 35 36; ; 46; 47; 51-63; 59	Bac
			4-69; 13-22; 20a; 23; 27-43; 35; 36; 44 ; 46; 47; 49-65; 51-63; 59; 73	Euk
Thr	2-71; 3-70	1-72; 73 1-72	2-71; 3-70; 11-24; 12-23; 35 36; 37; 46; 73	Arc
			20a; 35; 36	Bac
			31-39; 35; 36; 73	Euk

Table 1. *Cont.*

Amino Acid	Confirmed	Low Significance	Suggested	Domain
			2-71; 3-70; 6-67; 22; 27-43	Arc
			31-39; 34; 35; 36; 50-64	
Trp	1-72; 3-70; 73	2-71; 5-68; 9	15; 20a; 29-41; 31-39; 34; 35; 36; 48	Bac
			2-71; 15; 20a; 31-39; 34; 35	Euk
			36; 43; 48; 52-62; 65	
			1-72; 4-69; 9; 12-23; 13-22; 31-39; 34	Arc
			35; 36; 37; 46; 47; 51-63; 73	
Tyr	73		e5; 6-67; 10-25; e11; 12-23; 13-22; 17; 20	Bac
			20a; 20b; e21; 27-43; 28-42; 31-39; 34; 35	
			36; 44; ; 46; 59; 71	
	1-72	73	12-23; 17; 27-43; 28-42; 31-39	Euk
			34; 35; 36; 51-63; 70	
			2-71; 3-70; 4-69; 5-68; 6-67; 11-24; 12-23	Arc
			20a; 30-40; 31-39; 35; 36; 47	
Val	73	3-70; 4-69	13-22; 35; 36	Bac
			3-70; 11-24; 12-23; 13-22; 27-43	Euk
			31-39; 35; 36; 38; 60	

Here, we will discuss Gly tRNA as an example (see also Figure 3 a–c for the exact RE values). The rationale between this choice is two-fold: first, Gly is obviously the simplest possible, most basic, amino acid. Second, it is well-represented in the tRNA database, thus making this example particularly robust. Consider bacterial Gly tRNAs (Figure 3b). Immediately, positions 2-71 and 3-70 are confirmed as important ones for the tRNA identity. However, positions 10-25 and 1-72, previously recognized by means of experimental analyses [2,25], do not appear to be identity determinants in our analysis. Interestingly, 1-72, 2-71 and 3-70 are all present in the same ancestor stem; while the experimental analysis supports all three, our information theory-based analysis implies that the 1-72 pair is not really involved in determining the identity of Gly tRNAs in the Bacteria domain. Is the experimental analysis pinpointing 1-72 simply incorrect, and thus should be further validated? A similar scenario unfolds in the case of positions 10-25, supported by the experimental analysis; instead, our analysis suggests 29-41 and 31-39. These latter pairs apparently constitute the kind of identity that might not involve the aaRSes at all. Moving on to Eukarya (Figure 3c), 2-71 and 3-70 are confirmed by our analysis as well; 1-72 and 10-25 are not supported by either experimental or information-theoretic analyses. Neither are they supported by our analysis in Archaea Gly tRNAs (Figure 3a). This seems to strengthen the suggestion that 1-72 and 10-25 pairs are not in fact involved in determining the identity of Gly tRNAs in all domains, in spite of the experimental (aaRSes) evidence. On the contrary, positions 31-39 (in addition to 2-71 and 3-70) are supported by our analysis for all three domains. This hints at a strong signal because it is persistently present in smaller, relatively independent (or at least sufficiently heterogeneous) tRNA subsets.

Another representative example is the Ala tRNAs, where our analysis confirmed all known critical position (e.g., position 3-70 for Bacteria and Eukarya). For Ala tRNA, our analysis further suggests potentially important positions, namely 31-39 in the anticodon stem and 32,38 in the anticodon loop for Eukarya, and 20a for Archaea, Bacteria and Eukarya. Such analyses can of course be extended to all amino acid identity determinants, dovetailing with the existing and ever-growing literature. This largely statistical approach should ideally precede (or augment) the experimental work, in suggesting the possible (and likely) identity determinants to be pursued and validated experimentally. For instance, just by looking at Figure 3, positions 31-39 would be a strong candidate in the case of Gly tRNAs. Detailing chemistry and biochemistry of particular tRNAs and their positions is outside the scope of this manuscript—rather, it is a general future research direction for both ourselves and investigators elsewhere, armed with the methodology presented in this communication. (Specifically, we intend to follow up with a comprehensive, more detailed, analysis of the tRNA families for which both phylogenetic/alignment data and atomic structures of their complexes with aaRSes are available.)

4. Discussion and Future Directions

While probabilistic directed acyclic graph (PDAG) representation (as realized in the BN) is a conveniently compressed, informative and intuitively appealing visualization of the probabilistic relationships observable in the data, when it comes to fine detail scrutiny, high order abstraction can be a shortcoming. The information contained in conditional node distribution is difficult to summarize and interpret in its totality because of the multitude of all possible states. For instance, for a node with the three edges, one would have to scrutinize a 27x3 conditional probability table only to get a partial picture. In addition, it would be a partial picture because formation of the comprehensive conditional probability table would require inference over the complete Markov blanket of the node (which, in most cases, would include more than the immediate neighbor nodes) (see [18,19] for a comprehensive discussion of conditional independence and "Markov blankets" in directed graphs).

Recall that one of the motivations of our research was to study amino acid group-specific changes in tRNAs, or, to paraphrase, to develop a procedure for (1) conditioning on the aa variable; and (2) somehow comparing, in a quantifiable way, the resulting group-specific dependence structures. Building BNs for each and every aa class is one such procedure; however, quantifying topological differences between resulting networks is a difficult problem. Indeed, it is not enough to just "count" the presence or absence of a specific edge in the network—one would have to establish which particular set of differences pushes the model outside of the BN equivalence class [18,19]. While doable, this necessitates further BN methodological development, which is outside the scope of this paper (but is a research direction we are pursuing presently). Of course, there is also an issue of decreased sample sizes (when stratifying by aa variable). *(We remind here that all of the BN built for the different tRNA subclasses (stratified by aa and three life domains) are available directly from the authors.)*

An alternative way to deal with the complexity of the situation (i.e., "top-down" analysis from the high order abstraction of the BN topology to the individual tRNA determinants), described in detail in the previous section, required making two crucial simplifications. First, we restrict further ("downstream") analysis to pairwise relationships (conditioned on the aa class variable) only. This step compresses much of the dynamics into one low order approximation for every group that can be easily assessed and compared by means of simple charts (as in Figure 3 and Supplemental Figures S1–S22) and 2D color arrays (as in Figure 4). Second, we switch from the more comprehensive model scoring function (the MDL principle that underlies our BN structure learning algorithm) to a Mutual Information (MI)-based criterion that retains the desired feature of information theoretic entropy in order to condense the values of a given multidimensional probability distribution to a single distance-like value (namely, RE).

In summary, our procedural approach is a three-pronged one:

(1) Build the BNs for the tRNA sub- and super-classes of interest. Visually compare and contrast topologies. Interpret the intra-tRNA features (dependencies between the positions, pairwise and more complex) and inter-tRNA class differences.
(2) In parallel, use measures of distribution divergence to identify aa class-specific tRNA identity positions.
(3) For such positions, go back to the results of (1) to identify and interpret dependencies between these positions and other tRNA sites.

Ideally, this process can be further automated by developing routines for directly querying and performing statistical inference within Markov blankets (conditional probability tables) of nodes of interest, and for direct comparison of BN topologies by differentiating the equivalence classes associated with them. This is an ongoing research direction.

Our approach, as implemented in this study, is conceptually similar to the recent data-driven analyses of tRNA databases [26,27]. Galili and colleagues used a decision-tree based classifier to perform a variable selection (embedded in classification) to derive the *minimal* set of tRNA identity/informative positions for each amino acid in Arc domain. This is also a machine learning-based approach; decision tree-based classifiers are in many ways a complementary technique to the BN modeling. However, single decision trees (such as Classification and Regression Trees (CART) used in the analysis) are notorious for getting stuck into the local extremums, especially at the earlier stages of decision process (corresponding to the more important tRNA positions, in our case). Therefore, generated tRNA identity/informative position sets are, while undeniably compact, potentially biased, and might not be particularly robust. This is especially true for smaller datasets with more "difficult" dimensionality (such as aa-specific tRNA subclasses). Ideally, such sets should be at least complemented by the variable sets generated via methods espousing more exhaustive search (BNs, or ensemble decision tree-based classifiers). Depending on the extent of decision tree pruning, single decision tree-based classifiers can also be relatively blind to the rare variants (which is not a drawback that BNs share). In our opinion, it would be a very fruitful research direction to adapt the analyses in [26] to more robust ensemble classifiers, such as Random Forests with double-loop cross-validation [28]. At this time, however, we believe that the robustness and exhaustiveness of the variable sets generated by our approach is preferable to the compactness of the sets generated by a single decision tree—especially in the context of the exploratory, hypothesis-generating analysis activity. In statistical terms, the latter is simply too false negative-prone. Galili et al. [26] clearly recognize the shortcomings of such "greedy" algorithms as CART; however, they conclude that combinatorial efficiency and compactness of the results outweigh such drawbacks. We will re-emphasize here that CART and BNs, in this context (of predominantly exploratory analytic activity), complement (rather than compete with) each other.

Zamudio and José ([27]) used mutual entropy to discover isoacceptor tRNA positions that form "clusters" with respective anticodons. (This is mathematically similar to SNP linkage disequilibrium—except, of course, that the underlying biology, whatever it might be, is likely different in most cases.) In statistical terms, this presents a univariate testing alternative to our analysis (which is multivariate by design), with all the intrinsic advantages and disadvantages thereof. The primary advantage is that it allows to "manually" zero in on the primary intra-tRNA relationships of interests (such as identifying tRNA isoacceptor sites that are highly related to anticodon). Consequently, the primary disadvantage is the inability to capture more complex (than pairwise), non-additive, relationships. Similarly, although pairwise testing is obviously very computationally efficient, as a strategy it is essentially a series of independent tests that has to be carried out "manually", which is not conducive to the automated hypothesis generation (a hallmark of machine learning methods such as BNs and, to a somewhat lesser degree, decision tree-based classifiers). In our view, BN analysis is the method of choice for automatically identifying (and visualizing) potentially interesting intra-tRNA relationships. These can be further scrutinized using univariate testing such as developed by Zamudio and José [27], or conditional probability tables associated with the specific nodes/edges in the BN context.

Our approach can also be highly useful in addressing the specific problem of predicting tertiary (3D) tRNA interactions. Briefly, the problem can be defined as follows: thorough analysis of tRNA sequences (such as carried out in this study, and by us and others before) can be used to derive the basic secondary (2D) structure of tRNA. Standard 2D structure prediction algorithms aim to find thermodynamically stable 2D tRNA structures that minimize free energy given the primary tRNA sequence. This is a straightforward search/optimization task, and it is largely tractable within the tRNA sequences domain, assuming standard complementary base pairings hold (see [29] for a recent example). The resulting 2D structure is represented by a familiar tRNA "cloverleaf" consisting of the helix (stem) and the loop regions—the former subject to canonical Watson–Crick and GU pairing rules—the latter, less constrained. We now move on to the prediction of 3D tRNA structures (also known as "l-shape"), and this is a much more daunting task. Part of the intrinsic difficulty is that 3D structure prediction usually relies on the 2D structure (as a starting point), to which thermodynamics- and kinetics-driven operators (reflecting relatively minor free energy changes) are subsequently applied. This, of course, suffers greatly from the cumulative "dead reckoning" effect.

Another complication is that 3D tRNA structures do not exist in a vacuum, but are shaped in the process of interacting with other molecules within different cellular environments (mitochondria, ribosomes, etc.). In addition, yet another (related) complication has to do with the formal definition of operators (reflecting energy changes in 2D/3D structure search/optimization process). They are based on tabulated thermodynamic parameters estimated under "standard" fixed conditions (temperature, pH, ionic strength), which might or might not hold under different circumstances (varying cellular environments, interacting proteins and other RNAs, etc.).

The work of Mustoe et al. [30] is a representative recent example of in silico 3D tRNA analysis that had to be complemented by experimental work in order to reach unequivocal conclusions. Here, just as with the more specific tRNA – aaRS coupling problem, experimental methods become essential. Expensive and time-consuming procedures (crystallographic and spectroscopic techniques) can obtain, or confirm, structural information, but even then there is no guarantee that the results generalize to in vivo. In particular, there is usually a number of alternative, however similar, optimal structures, and the one that reaches the absolute minimum (in free energy) in silico or in vitro is not necessarily the one actually functioning in vivo.

Thus, we return to the notion of the automated data-driven discovery of "interesting" polymorphisms/structural features/interactions within the tRNA molecules that would otherwise be difficult to capture because of an altogether unknown biological context, or prohibitively high predictive algorithm complexity, or modeling assumption violations, or simply sheer time and expense of experimental work.

However, another benefit of the tRNA BN analysis lies in its potential utility for tRNA phylogenetic reconstruction. Applying standard nucleotide distance metrics, or nucleotide transition probabilities (usually a cornerstone of phylogenetic reconstruction) to the tRNA molecules is questionable for a number of reasons ([31] for recent discussion). However, information contained in structural (topological) differences between BNs estimated for specific subclasses of tRNAs can be used to measure phylogenetic distances between these subclasses, which can be further translated into the phylogenetic hierarchy. In fact, unlike with some ad hoc metrics, differences in BNs by their very nature capture the "holistic" (more complete, comprehensive) picture of the differences between a given pair of subclasses—indeed, the sole purpose of BNs is to reflect the significant probabilistic relationships contained in the "flat" data, regardless of the underlying nature of these relationships. Assuming that these observed probabilistic relationships reflect true biological interactions (and not random noise)—a reasonable assumption for the sufficiently large datasets, such as the currently available tRNA data—they should be a viable foundation for the analysis of tRNA gene histories.

Such BN-based approach to tRNA phylogeny is conceptually similar to, and would complement, the approach espoused in [31], which prioritizes "anchor elements" based on syntenic tRNA properties as the most useful for deriving phylogenetic information. The rationale behind the latter strategy is that relatively stable synteny anchors are less susceptible to the randomizing and obfuscating effects of the rapid-churn concerted evolution of gene families (such as tRNAs). Similarly, topological differences between BNs built from the different tRNA subclasses are less sensitive to the evolutionary (background) noise, and reflect robust features that have been/are being fixated during the correspondent tRNA subclasses' formation. One of the intriguing future directions of our work is to trace tRNA gene histories using BN topological distances and see how the results compare to the synteny-based phylogenies of tRNAs in, for example, fruit flies and primates [31].

5. Conclusions

If there is one thing that can be said with certainty about the tRNA world, both extant and extinct, is that it is/was arguably much more complex (quantitatively and qualitatively) than what we used to think as recently as two to three decades ago ([32] and references therein). Amusingly, it has been suggested [32] that the ongoing evolution of the multitude of tRNA species can best be described by analogy with an artificial intelligence system (specifically, an interplay of minimax algorithms).

This might be a stretch; if anything, tRNA populations can be better approximated by the "genetic algorithms" (which, in turn, were developed to mimic natural genetic/evolution processes in a first place). However, whimsicality aside, we do believe that artificial intelligence methodology (specifically, machine learning techniques) is the best way to approach secondary data analysis of growing tRNA sequence databases, be it for the purposes of tRNA structural research, identity elements cataloging, unraveling tRNA-disease associations, or tRNA evolutionary studies.

This brings us to the problem of genetic code and translation machinery origins, a subject unequivocally linked with the tRNAs and their (early) evolution. A recent review by Koonin and Novozhilov ([33]) suggests, among other things, that "theoretical study of the genetic code as a cryptographic problem has largely run its course". In our opinion, this sentiment, while having a defeatist flair about it, is fundamentally correct. The consensus understanding, at this time, revolves around the "synthesis" of stereochemical affinity hypothesis, and frozen accidents alongside universal genetic code code expansion. The former largely hinges on the "important" tRNA sites other than the anticodon, automated discovery of which is of course the primary goal of this study. This should be followed by the experimental work as well as dynamical modeling, along the lines suggested by Koonin and Novozhilov [33], Carter and Wills [34], Martinez-Rodriguez et al. ([6]), to name just a few.

Such experimental work is time-consuming and costly; not coincidentally, one of the primary deliverables of this study is the list of "suggested" tRNA positions of interest (third column in Table 1). Moreover, it is possible to "rank" the suggested positions using the RE metric. For any specific subgroup of tRNAs, investigators can identify the suggested (and so far experimentally unverified) tRNA positions and simultaneously rank/select them using RE and/or various RE cutoff criteria. The resulting selected positions can then be scrutinized by identifying other tRNA positions associated with them in the appropriate tRNA subgroup BNs. After subsequently evaluating (and further prioritizing) these selected positions from the evolutionary, biochemical, biological and structural (2D, 3D) viewpoints, one can proceed to the thus more precisely guided experimental work.

On a philosophical note, even unequivocal experimental validation of the processes suspected to have taken place during the genetic code/translation apparatus origin does not yet prove that these events actually took place. There are many possible, and perfectly realistic, alternative pathways leading to the extant universal genetic code; the sheer volume of related literature (and hypotheses and scenarios suggested therein) attests to the fact. Experimentally reproducing one of them does not preclude the possibility of another, and does not necessarily bring us that much closer to the definitive proof of "what has really happened". Similarly, for in silico work, the Occam's razor approach (which underlies much of the machine learning research by directing us to seek the most compact system/model fitting the observed data the best) is subject to a parallel logical fallacy, that of assuming that nature (and evolution) deals with globally optimized, at any given moment, biological systems ("greedy search" of locally optimized states being significantly closer to the reality of evolutionary processes). At any rate, while the purely "cryptographic" approach to the elucidation of universal genetic code origins might feel unsatisfying [33], we strongly believe that non-trivial relationships and associations observed in the extant molecules (principally tRNAs and aaRSes) can and should point to the more likely (statistically and experimentally) scenarios of code origin, or at least to certain likely evolutionary steps along the way. Much of the recent work in the field has been pursued following this paradigm [27,35–37].

Detailing plausible scenarios of genetic code origin is outside the scope of this study; rather, our goal here was to present a useful instrument and methodological framework for identifying statistically significant relationships between the positions within the tRNA molecule and between different subclasses of tRNAs. It is our intention for this novel methodology to assist the investigators in pinpointing non-random patterns of change in tRNAs, at both micro- and macro-evolutionary levels.

Supplementary Materials: The following are available online at http://www.mdpi.com/2075-1729/8/1/5/s1, Figures S1–S22: Position "importance" profile for all aa tRNA subclasses, shown for three life domains, Archaea (**a**,**b**), Bacteria (**c**,**d**) and Eukarya (**e**,**f**). Relative Entropy is shown as function of tRNA position (**a**,**c**,**e**) (enumerated as in Figure 1, first row), or visualized as color intensity superimposed over the secondary tRNA structure (**b**,**d**,**f**). Significance cutoff limit is shown as a red line in (**a**,**c**,**e**)—see text for discussion.

Acknowledgments: This work was supported by the Susumu Ohno Chair in Theoretical and Computational Biology (held by A.S.R.), a Susumo Ohno Distinguished Investigator fellowship (to G.G.), and City of Hope funds (to S.B., G.G. and A.S.R.). We are thankful to Arthur D. Riggs and Peter P. Lee for many useful discussions on the applicability of BN and information-theoretic data analysis methodology to the different types of biological data. We are enormously grateful to the anonymous reviewers, for numerous suggestions that led to the improvement of the manuscript, and for positing promising venues for future research.

Author Contributions: S.B. conceived the study, carried out the analyses, interpreted the results and drafted the manuscript. G.G. conceived the study, carried out the analysis, and contributed to interpreting the results and drafting the manuscript. M.D.G. contributed to carrying out the analyses, interpreted the results and contributed to drafting the manuscript. A.S.R. conceived the study, contributed to carrying out the analyses, interpreted the results and drafted the manuscript.

Conflicts of Interest: The authors declare no conflict of interest.The founding sponsors had no role in the design of the study; in the collection, analyses, or interpretation of data; in the writing of the manuscript, and in the decision to publish the results.

References

1. Schimmel, P. Development of tRNA synthetases and connection to genetic code and disease. *Protein Sci.* **2008**, *17*, 1643–1652.

2. Giegé, R.; Eriani, G. Transfer RNA Recognition and Aminoacylation by Synthetases. In *eLS*; John Wiley & Sons, Ltd.: Hoboken, NJ, USA, 2014.

3. Eriani, G.; Karam, J.; Jacinto, J.; Richard, E.M.; Geslain, R. MIST, a Novel Approach to Reveal Hidden Substrate Specificity in Aminoacyl-tRNA Synthetases. *PLoS ONE* **2015**, *10*, e0130042.

4. Cvetesic, N.; Gruic-Sovulj, I. Synthetic and editing reactions of aminoacyl-tRNA synthetases using cognate and non-cognate amino acid substrates. *Methods* **2017**, *113*, 13–26.

5. Sapienza, P.; Li, L.; Williams, T.; Lee, A.; Carter, C.W., Jr. An Ancestral Tryptophanyl-tRNA Synthetase Precursor Achieves High Catalytic Rate Enhancement without Ordered Ground-State Tertiary Structures. *ACS Chem. Biol.* **2016**, *11*, 1661–1668.

6. Martinez-Rodriguez, L.; Erdogan, O.; Jimenez-Rodriguez, M.; Gonzalez-Rivera, K.; Williams, T.; Li, L.; Weinreb, V.; Collier, M.; Chandrasekaran, S.; Ambroggio, X.; et al. Functional Class I and II Amino Acid-activating Enzymes Can Be Coded by Opposite Strands of the Same Gene. *J. Biol. Chem.* **2015**, *290*, 19710–19725.

7. Li, R.; Macnamara, L.; Leuchter, J.; Alexander, R.; Cho, S. MD Simulations of tRNA and Aminoacyl-tRNA Synthetases: Dynamics, Folding, Binding, and Allostery. *Int. J. Mol. Sci.* **2015**, *16*, 15872–15902.

8. Adrion, J.; White, P.; Montooth, K. The Roles of Compensatory Evolution and Constraint in Aminoacyl tRNA Synthetase Evolution. *Mol. Biol. Evol.* **2016**, *33*, 152–161.

9. Fang, P.; Guo, M. Structural characterization of human aminoacyl-tRNA synthetases for translational and nontranslational functions. *Methods* **2017**, *113*, 83–90.

10. Datt, M.; Sharma, A. Novel and unique domains in aminoacyl-tRNA synthetases from human fungal pathogens Aspergillus niger, Candida albicans and Cryptococcus neoformans. *BMC Genom.* **2014**, *15*, 1609.

11. Amiram, M.; Haimovich, A.; Fan, C.; Wang, Y.; Aerni, H.; Ntai, I.; Moonan, D.; Ma, N.; Rovner, A.; Hong, S.; et al. Evolution of translation machinery in recoded bacteria enables multi-site incorporation of nonstandard amino acids. *Nat. Biotechnol.* **2015**, *33*, 1272–1279.

12. Terasaka, N.; Iwane, Y.; Geiermann, A.; Goto, Y.; Suga, H. Recent developments of engineered translational machineries for the incorporation of non-canonical amino acids into polypeptides. *Int. J. Mol. Sci.* **2015**, *16*, 6513–6531.

13. Perli, E.; Fiorillo, A.; Giordano, C.; Pisano, A.; Montanari, A.; Grazioli, P.; Campese, A.; Di Micco, P.; Tuppen, H.; Genovese, I.; et al. Short peptides from leucyl-tRNA synthetase rescue disease-causing mitochondrial tRNA point mutations. *Hum. Mol. Genet.* **2016**, *25*, 903–915.

14. Rodin, A.; Gogoshin, G.; Litvinenko, A.; Boerwinkle, E. Exploring genetic epidemiology data with Bayesian networks. In *Handbook of Statistics*; Rao, C., Chakraborty, R., Eds.; Elsevier: St. Louis, MI, USA, 2012; Volume 28, pp. 53–76.

15. Gogoshin, G.; Boerwinkle, E.; Rodin, A. New Algorithm and Software (BNOmics) for Inferring and Visualizing Bayesian Networks from Heterogeneous Big Biological and Genetic Data. *J. Comput. Biol.* **2017**, *24*, 340–356.
16. Juhling, F.; Morl, M.; Hartmann, R.; Sprinzl, M.; Stadler, P.; Putz, J. tRNAdb 2009: Compilation of tRNA sequences and tRNA genes. *Nucleic Acids Res.* **2009**, *37*, D159–D162.
17. Quigley, G.; Rich, A. Structural domains of transfer RNA molecules. *Science* **1976**, *194*, 796–806.
18. Heckerman, D. *A Tutorial on Learning with Bayesian Networks*; MSRFF-TR-95-06; Microsoft Research: Redmond, WA, USA, 1995.
19. Pearl, J. *Causality: Models, Reasoning, and Inference*; Cambridge University Press: Cambridge, UK; New York, NY, USA, 2009.
20. Zhang, X.; Branciamore, S.; Gogoshin, G.; Rodin, A.S.; Riggs, A.D. Analysis of high-resolution 3D intrachromosomal interactions aided by Bayesian network modeling. *Proc. Natl. Acad. Sci. USA* **2017**, *114*, E10359–E10368.
21. Kullback, S.; Leibler, R.A. On information and sufficiency. *Ann. Math. Stat.* **1951**, *22*, 79–86.
22. Lin, J. Divergence measures based on the Shannon entropy. *IEEE Trans. Inf. Theory* **2002**, *37*, 145–151.
23. Sprinzl, M.; Horn, C.; Brown, M.; Ioudovitch, A.; Steinberg, S. Compilation of tRNA sequences and sequences of tRNA genes. *Nucleic Acids Res.* **1998**, *26*, 148–153.
24. Laslett, D.; Canback, B. ARAGORN, a program to detect tRNA genes and tmRNA genes in nucleotide sequences. *Nucleic Acids Res.* **2004**, *32*, 11–16.
25. Giegé, R.; Sissler, M.; Florentz, C. Universal rules and idiosyncratic features in tRNA identity. *Nucleic Acids Res.* **1998**, *26*, 5017–5035.
26. Galili, T.; Gingold, H.; Shaul, S.; Benjamini, Y. Identifying the ligated amino acid of archaeal tRNAs based on positions outside the anticodon. *RNA* **2016**, *22*, 1477–1491.
27. Zamudio, G.; Jose, M. Identity Elements of tRNA as Derived from Information Analysis. *Orig. Life Evol. Biosph.* **2017**, doi:10.1007/s11084-017-9541-6.
28. Rodin, A.; Litvinenko, A.; Klos, K.; Morrison, A.; Woodage, T.; Coresh, J.; Boerwinkle, E. Use of wrapper algorithms coupled with a random forests classifier for variable selection in large-scale genomic association studies. *J. Comput. Biol.* **2009**, *16*, 1705–1718.
29. Saffarian, A.; Giraud, M.; Touzet, H. Modeling alternate RNA structures in genomic sequences. *J. Comput. Biol.* **2015**, *22*, 190–204.
30. Mustoe, A.; Liu, X.; Lin, P.; Al-Hashimi, H.; Fierke, C.; Brooks, C.L., Jr. Noncanonical secondary structure stabilizes mitochondrial tRNA(Ser(UCN)) by reducing the entropic cost of tertiary folding. *J. Am. Chem. Soc.* **2015**, *137*, 3592–3599.
31. Velandia-Huerto, C.; Berkemer, S.; Hoffmann, A.; Retzlaff, N.; Marroquin, L.C.R.; Hernandez-Rosales, M.; Stadler, P.; Bermudez-Santana, C. Orthologs, turn-over, and remolding of tRNAs in primates and fruit flies. *BMC Genom.* **2016**, *17*, 617.
32. Schimmel, P. The emerging complexity of the tRNA world: Mammalian tRNAs beyond protein synthesis. *Nat. Rev. Mol. Cell Biol.* **2018**, *19*, 45–58.
33. Koonin, E.; Novozhilov, A. Origin and Evolution of the Universal Genetic Code. *Annu. Rev. Genet.* **2017**, *51*, 45–52.
34. Carter, C.W., Jr.; Wills, P. Interdependence, Reflexivity, Fidelity, Impedance Matching, and the Evolution of Genetic Coding. *Mol. Biol. Evol.* **2017**, *35*, 269–286.
35. Wills, P.; Carter, C.W., Jr. Insuperable problems of the genetic code initially emerging in an RNA world. *Biosystems* **2017**, doi:10.1016/j.biosystems.2017.09.006.
36. Di Giulio, M. The aminoacyl-tRNA synthetases had only a marginal role in the origin of the organization of the genetic code: Evidence in favor of the coevolution theory. *J. Theor. Biol.* **2017**, *432*, 14–24.
37. Di Giulio, M. Some pungent arguments against the physico-chemical theories of the origin of the genetic code and corroborating the coevolution theory. *J. Theor. Biol.* **2017**, *414*, 1–4.

MDPI AG

St. Alban-Anlage 66

4052 Basel, Switzerland

Tel. +41 61 683 77 34

Fax +41 61 302 89 18

http://www.mdpi.com

Life Editorial Office

E-mail: life@mdpi.com

http://www.mdpi.com/journal/life

www.ingramcontent.com/pod-product-compliance
Lightning Source LLC
Chambersburg PA
CBHW051851210326
41597CB00033B/5853